STUDY GUIDE

CLAUDIA DOUGLASS
Central Michigan University

BIOLOGY OF HUMANS

CONCEPTS, APPLICATIONS, AND ISSUES

GOODENOUGH McGUIRE WALLACE

Upper Saddle River, NJ 07458

Editor-in-Chief, Science: John Challice
Executive Editor: Gary Carlson
Project Manager: Crissy Dudonis
Vice President of Production & Manufacturing: David W. Riccardi
Executive Managing Editor: Kathleen Schiaparelli
Assistant Managing Editor: Becca Richter
Production Editor: Rhonda Aversa
Supplement Cover Manager: Paul Gourhan
Supplement Cover Designer: Joanne Alexandris
Manufacturing Buyer: Ilene Kahn
Cover Image Credits: Father Teaching Son Basketball - Jim Erickson/Corbis; Spiraling Roller Coaster - Robert Landau/Corbis; Beach Volleyball - Michael Kevin Daly/Corbis; American Gymnast - Wilson Competes Corbis/Bettman

Pearson Prentice Hall
Pearson Education, Inc.
Upper Saddle River, NJ 07458

Printed in the United States of America

10 9 8 7 6 5 4 3 2

ISBN 0-13-143784-4

Pearson Education Ltd., *London*
Pearson Education Australia Pty. Ltd., *Sydney*
Pearson Education Singapore, Pte. Ltd.
Pearson Education North Asia Ltd., *Hong Kong*
Pearson Education Canada, Inc., *Toronto*
Pearson Educación de Mexico, S.A. de C.V.
Pearson Education—Japan, *Tokyo*
Pearson Education Malaysia, Pte. Ltd.

Contents

Chapter 1

Science and Society

Objectives

After reading the text and studying the material in this chapter, you should be able to:

- List the characteristics of life.
- Identify characteristics that are uniquely human.
- Describe how evolutionary relationships are reflected in modern systems of classification.
- List the classification of humans from the kingdom to species level.
- Define and give examples of the following groups of organisms: population, community, ecosystem and biosphere.
- List the steps in the scientific method.
- Design an experiment with a control and experimental group that follows the scientific method.
- Differentiate between deductive and inductive reasoning.
- Explain the importance of clinical trials on humans and the necessary informed consent of subjects.
- Explain the importance of critical thinking.
- Describe the questions one might ask while using the critical thinking approach.

Chapter Summary

It would be nice to provide a single definition for life, but always there are exceptions to the definitions. Nonetheless, all living things can be characterized as containing the molecules of life (nucleic acids, proteins, carbohydrates and lipids), being composed of cells, capable of reproduction, using energy and raw materials, responding to their environment, and maintaining homeostasis. As a population, living things have **adaptive traits** that allow them to survive, reproduce and ultimately evolve.

Scientists assume that shared characteristics among organisms indicate a common ancestry. Humans share many characteristics with other animals showing greatest similarity to other vertebrates and mammals. However, humans are unique in that they have a relatively large brain, stand upright on two legs, have an opposable thumb, and possess a culture.

Life has many levels of organization beyond the individual. Many individuals of the same species living together in a common geographical area and interbreeding are called a **population**. When several populations interact they form a **community**, for example the community of organisms in a rotting log. The sum total of all the living things in a given area and their interaction with the physical environment is called an **ecosystem**. Fields, forests and ponds are ecosystems teaming with life that depends upon the non-living portion of the environment. All organisms exist within the much larger **biosphere,** that part of Earth where life is found. Just as there is the hydrosphere, atmosphere and lithosphere; there is also the biosphere of which we are a part.

Humans are curious and ask many questions. A logical approach to gathering information and reaching conclusions includes these steps of the scientific method: 1) make careful observations and ask questions about those observations; 2) develop a testable **hypothesis** that provides an explanation about the observations and answers to the questions; 3) make a prediction and design an experiment with a **control group** to determine if the hypothesis is correct; 4) draw a conclusion based on the results of the experiment; and 5) ask more questions and design additional experiments to further refine your conclusions. On rare occasions, after many, many experiments that support related hypotheses, a **theory** may be formed that offers a broad-ranging explanation for a set of phenomena.

Two types of reasoning, inductive and deductive, complement the scientific method as a means to inquiry. **Inductive reasoning** involves the accumulation of facts through observation until finally there are enough facts to draw a conclusion or develop a testable hypothesis. **Deductive reasoning** involves making a general statement, often in the format of an "if-then" statement, then drawing more specific conclusions from it. With deductive reasoning, a series of observations leads to a general statement that often sets the stage for further experimentation.

Scientists use all of these methods when they test medicines or other materials such as cosmetics that will be used by humans. The first tests are on laboratory animals and must follow strict guidelines. If these tests go well, then the first clinical trials begin using humans, all of whom must provide **informed consent** to participating in the studies. Clinical trials consist of several experimental groups which receive different dosages of the drug and a control group that receives a **placebo**. It is absolutely important that both groups be treated identically except for one **variable**. When it might be risky to use humans as research subjects, information can be gathered *about* humans by making other observations of their behavior. One potential problem with these studies is the lack of control. Sometimes more than one variable may influence the results and hamper our ability to draw conclusions. The second variable is called a **confounding variable** because it confounds, or confuses, the conclusions.

Everyone must evaluate scientific claims and make the best possible decision. Critical-thinking skills help us analyze information and make informed decisions. These skills apply to all areas of our life. A critical thinker asks the following questions about claims: 1) Is the information consistent with information gathered from other sources? 2) How reliable is the source of the information? 3) Was the information obtained through proper scientific procedures? 4) Were the results of the experiment interpreted correctly? and 5) Are there other possible explanations for the results? It is very important that you critically evaluate the claims made by others that influence your life, society in general, and the biosphere of which you are a major player. It is everyone's responsibility to ask questions, evaluate the data and make informed decisions.

Key Concepts

- All living things share basic characteristics.
  - ✓ Living things contain nucleic acids, proteins, carbohydrates, and lipids.
  - ✓ Living things are composed of cells.
  - ✓ Living things reproduce.
  - ✓ Living things use energy and raw materials.
  - ✓ Living things respond.
  - ✓ Living things maintain homeostasis.
  - ✓ Populations of living things evolve and have adaptive traits.
- Living organisms share an evolutionary history evidenced by shared characteristics.
- Life has many levels of organization beyond the individual including the population, community, ecosystem and biosphere.
- The scientific method is circular in that all new discoveries lead to more questions.
- Most often the scientific method begins with observations which lead to testable hypotheses, the design and conduct of an experiment, and finally conclusions.
- Scientists identify experimental variables that change and are compared to a control group.
- Inductive (specific to general) and deductive reasoning (general to specific) reasoning help solve problems.
- Critical thinking helps us analyze information and make informed decisions based upon the consistency of the information, the reliability of the source, the methods used to collect data and the interpretation of results.

Study Tips

It is important that you develop good study skills right now at the beginning of this course. Many of the study tips in this study guide will help you in other courses as well. First, make sure that you understand the requirements of the course and know when major assignments are due and when tests are scheduled. Mark them in your calendar now. It is important to study science on a regular basis. A good rule of thumb is to study two hours for each hour you are in class and to do it as soon after the class as possible.

Get organized. Have a notebook or folder for each class. Decide if you will integrate notes over the text material with your lecture notes. If so, leave room on the page for both sets of notes. As you wait for class to begin, review the notes from the previous lecture. Following the lecture, review your notes, flush out ideas that may be too brief and identify questions you may have for the professor. Some students find it beneficial to rewrite their notes after the lecture to add detail and enhance their understanding of the concept.

Read the assigned material before coming to class. Outline the material or highlight it in the book. Read the text with your lecture notes by your side so that you can identify important concepts and integrate the material. Be sure to pay particular attention to the figures and photos. Explain to yourself or a friend the meaning of each chart or graph. Read the summary material at the end of the chapter and answer the chapter questions. This study guide is organized in the same manner as the book. Use it to help you review the material as it is presented in class.

As you study the material in this chapter list the characteristics of life and note those that are unique to humans. List the various levels of organization in one column on your paper and give examples common to your area in a second column. Next draw a flow chart of the scientific method and design a simple experiment with a control and experimental group as an example. Finally, rephrase the questions that are the focus of critical thinking and put them in the format of a checklist. Take one controversial topic or claim made in an advertisement and analyze it using the keys to critical thinking.

Review Questions

A. Complete this list of the characteristics of life by filling in the blanks.

1. Living things contain the following biological molecules: nucleic acids, proteins, lipids, and carbohydrates
2. Living things are composed of cells
3. Living things reproduce.
4. Living things use energy and raw materials
5. Living things respond.
6. Living things maintain homeostasis
7. Populations of living things evolve and have adaptive traits

B. Review the examples in Figure 1-2 and the text discussion. Each statement below demonstrates a characteristic of life. Identify that characteristic of life and write it in the space provided.

1. In the park, you see a family with three young children watching the tadpoles in a shallow pond. ______________________________
2. A squirrel is seen eating a nut. ______________________________
3. You burn yourself and some skin peels off. Looking at it under the microscope, you see that it is a very thin sheet of epithelial cells. ______________________________
4. At the football game, you are sitting in the stands shivering while the players are wiping the sweat from their faces. ______________________________

5. Several organisms were collected from a tropical area. A chemical analysis showed that all of the organisms contained a protein and the nucleic acid, RNA. ____________________

6. Walking through the woods early in the morning, you hear animals running, but don't see them. When you pick up a rotting log, you see many invertebrates scurrying away from the light. ____________________

7. People who fish know that different species live at different depths in the same lake and that some river species will be found along the shore while others may be in the middle in the current. ____________________

C. Match the level of classification (Column A) with the specific classification for humans (Column B).

Column A	Column B
1. kingdom	a. primates
2. phylum	b. vertebrates
3. subphylum	c. Homo
4. class	d. animals
5. order	e. chordates
6. genus	f. *Homo sapiens*
7. species	g. mammals

D. Complete the following table by defining the word in the first column and giving an example. Refer to Figure 1-6 for ideas.

Term	**Definition**	**Example**
Population	All the organisms of a same species living in the same area.	
Community		
Ecosystem		Pond, forest, bog
Biosphere		All living things on Earth.

A. Hypothesis
B. prediction
C. Results
D. Conclusion
E. Revise hypothesis
Make new predictions + test.

*E. Label the diagram below with the steps in the scientific method.

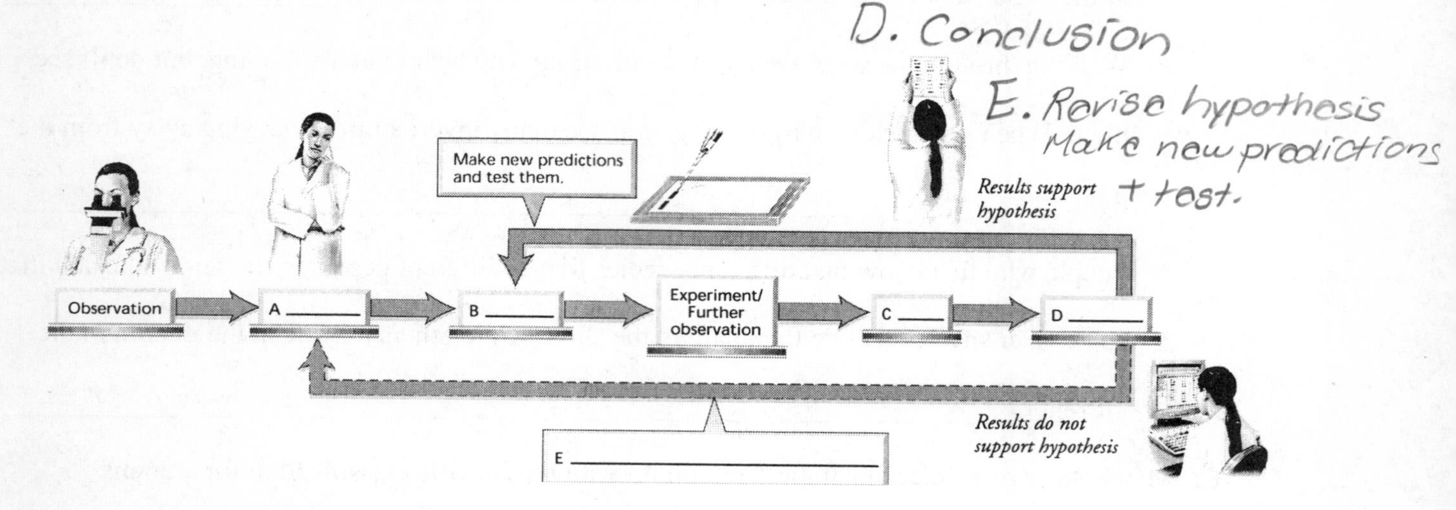

*F. Carefully read the following description of the testing of a new product, SMARTPIES. Fill in the blanks using the following terms: conclusions, control, experiment, experimental, hypothesis, informed consent, observations, variable, and theory.

Several instructors mentioned that as they studied the results of a recent exam, they noticed that all of the students from the Far Quad had much higher scores than the students living in the Near Quad. These observations led one professor to ask the question, "What is making the students in the Far Quad smarter?"

Discussing the observations with colleagues, the professor thought of many possible reasons – maybe all the students with high scores on the entrance exams were placed into the Far Quad, maybe the students in the Near Quad were ill more often and missed more classes, or maybe there was something in the diet of those in the Far Quad that was enhancing their intelligence. Since the professor wanted to narrow the considerations to only one hypothesis she began to do some inquiring. Housing said that students were randomly assigned and Health Services said that there was no correlation between housing unit and illness. The dietician reported that students in Far Quad were getting SMARTPIES, a new product that was being served in only that dorm.

The professor made the prediction that SMARTPIES was causing the difference in test scores and designed an experiment to test the hypothesis. The experiment was explained to all the students, and they were asked to sign an informed consent document agreeing to participate in the study. Since students gave their student number as they entered the line to pick up their food, it was easy to be sure that those with even numbers received SMARTPIES for dessert and those with odd numbers received another dessert that looked just like SMARTPIES but contained different ingredients. The students who received the authentic SMARTPIES were part of the experimental group and those who received the look-alike PIES were part of the Control group. The experiment went on for a month and grades were collected and averaged for each group.

Data analysis showed that the students who ate the authentic SMARTPIES did in fact have higher grades while those who ate the look-alike PIES did not show the same achievement. Since the only Variable that changed between the two groups was the dessert and since both desserts were the same except for a few secret ingredients, the professor was able to draw Conclusions from this study. In an interview, the professor cautioned that based on only one experiment, she could not develop a theory regarding SMARTPIES and achievement.

*G. Put an I next to the statements that require inductive reasoning and a D next to those that are examples of deductive reasoning.

D If fatty foods lead to high cholesterol, then people who eat a lot of bacon, hamburgers and French fries should have higher cholesterol than those who do not eat those foods.

I Exercise reduces the resting heart rate.

I Some students appear more nervous during testing situations.

D If antacids reduce the pH level of a solution, then they should relieve acid stomach.

H. Critical thinking leads to informed decisions. Underline the key words in each of the guiding questions listed below. Then describe in a few sentences what you might use for resources or how you might go about seeking clarification.

1. Is the information consistent with information gathered from other sources?

____

____

2. How reliable is the source of information?

____

____

3. Was the information obtained through proper scientific procedures?

____

____

4. Were the results of the experiment interpreted correctly?

____

____

5. Are there other possible explanations for the results?

____

____

Critical Thinking Questions

Read each of the following questions carefully. If it helps you, underline or list each component to the question. Use complete sentences in your explanation.

1. With your classmates, brainstorm some possible questions based on the world around you that might lead to an experiment. Write them down and circle the two or three that most clearly lend themselves to experimentation. Discuss one of the questions and develop a hypothesis, then design an experiment that would test the hypothesis.

2. Suppose that you are working for a pharmaceutical company and that your new drug to relieve migraines has been approved for clinical trials. Design an experiment with several experimental groups to test different dosages of the drug and at least one control group. Provide detailed instructions to the testers including what to include in the informed consent

document, how many subjects to include in each group and how long the subjects should take each treatment. What errors might remain in your experiment?

3. Choose a common drug and locate the manufacturer's fact sheet on the Internet or in an advertisement. Read the information surrounding the clinical trials. Identify the experimental and control groups, the number of subjects who completed the study, the range of dosage, and the age range of the subjects. What surprised you the most about the information presented in the fact sheet?

4. As you walk around campus, try to identify as many of the characteristics of life as you can in the organisms that you observe. Which characteristics are easiest to observe? Which are most difficult?

5. As a living organism, you demonstrate the characteristics of life. List those characteristics on a note card. It is not as easy to define death. This definition impacts the lives of others in cases of organ transplants. If someone can be declared "dead" while the organs are still functioning they are much more useful. Once all body functions have ceased, then some organs are no longer useful for transplant. Turn the card over and list criteria that you might use to define death. Compare your thoughts with those of others in your class.

Practice Test

Choose the one best answer to each question that follows. As you work through these items, explain to yourself why each answer you discard is incorrect and why the answer you choose is correct.

1. Which of the following is not a characteristic of life?
   a. adaptation to environmental changes
   b. complex chemical organization
   c. differentiation from cells to tissues
   d. reproduction

2. Metabolism is the ____________.
   a. output of energy usually as heat
   b. rate at which food is consumed
   c. sum total of all chemical reactions
   d. way that an organism uses energy

3. When an organism loses homeostasis it is ____________.
   a. dead
   b. dormant
   c. hibernating
   d. shivering

4. Humans are the only organism to demonstrate ____________.
   a. communication
   b. courtship behavior
   c. culture
   d. signaling behavior

*5. Through the evolutionary processes, organisms can ____________.
(a.) adapt to their environment
b. eliminate their cell structure
c. eliminate their need for an energy source
d. stop the need to reproduce

*6. Humans belong to which kingdom?
(a.) animals
b. humans
c. primates
d. vertebrates

*7. Humans share the characteristics of an opposable thumb, forward-looking eyes, and a well-developed brain with other ____________.
a. chordates
b. mammals
(c.) primates
d. vertebrates

*8. Which of the following include organisms all of only one species?
a. community
b. ecosystem
c. kingdom
(d.) population

*9. Given the following list of common and scientific names, which two birds are most closely related?

Black duck	*Anas rubripes*
Clay-colored sparrow	*Spizella pallida*
Green-winged teal	*Anas carolinensis*
White-crowned sparrow	*Zonotrichia leucophrys*
Wood duck	*Aix sponsa*

a. Black duck and wood duck
(b.) Black duck and green-winged teal
c. Clay-colored sparrow and white-crowned sparrow
d. Wood duck and green-winged teal

*10. Which of the following groups is the largest and most diverse?
a. Community
(b.) Ecosystem
c. Population
d. Species

*11. Which of the following statements is not true regarding science?
(a.) Because science is based on facts it cannot change.
b. Data can be descriptive and does not always need to include numbers.
c. Scientists obtain information in many different ways.
d. The scientific method can vary slightly from one investigation to another.

12. Which of the following statements could be a hypothesis?
   a. All subjects over age 50 showed no growth using hormone supplements.
   b. Girls will be more responsive to the new instructional methods.
   c. Which instructional method will be most effective?
   d. Why is the sky blue?

13. A pharmaceutical company tested a new drug to relieve arthritis. Some of the subjects received a red and pink capsule with the drug while others received a red and pink capsule with filler and no drug. Those receiving the drug are called the ____________ group while the subjects receiving the placebo are called the ____________ group.
   a. affected, unaffected
   b. control, experimental
   c. experimental, control
   d. hypothetical, theoretical

14. A city wanted to determine whether fluoride was an effective additive against tooth decay. The city was large enough that there were several sources of water that served different areas. What is the best control for their experiment?
   a. babies with no tooth decay
   b. fluoride in the toothpaste of some residents
   c. water with different strengths of fluoride added
   d. water with no fluoride added

15. Most often the results of an experiment lead to ____________.
   a. more questions
   b. publications in scientific journals
   c. theories
   d. well-developed controlled experiments

16. Which of the following is a commonly accepted theory?
   a. Abstractions of mitotic theories
   b. Cell theory of life
   c. Theory of vitamin C
   d. Unifying theory of energy

17. Which of the following is an example of deductive reasoning?
   a. Bees pollinate apple orchards.
   b. Hormones affect growth.
   c. If pollen causes allergies, then when the pollen levels rise so will the symptoms of allergies.
   d. If tires can be recycled to save trees, then we should do it.

18. Which of the following statements are characteristic of clinical trials?
   a. Subjects must give informed consent.
   b. They only include volunteers.
   c. They are completed after successful animal testing.
   d. All of the above are true.

19. A critical thinker ____________.
   a. asks questions.
   b. designs experiments.
   c. presents conclusions.
   d. provides original data.

20. As a critical thinker it is important to realize that all advertisements contain ____________.
   a. advertising slogans
   b. bias
   c. false data
   d. imperfect conclusions

Short Answer Questions

Read each of the following questions carefully. If it helps you, underline or list each component to the question. Formulate your answer and jot down the main points. Then use complete sentences in your explanation.

1. A fire exhibits many characteristics of life. It can reproduce as it moves across acres of land; it can respond to the environment as it "jumps" creeks and becomes fierce in the wind, it transforms energy as it makes heat, etc. Intuitively you know that a fire is not alive. List the characteristics of life and use that list to explain why a fire is not alive.

2. List the steps of the scientific method. Put them in order and connect them with arrows as appropriate. Explain why you arranged the arrows as you did.

3. While taking care of mice in the animal room, one of the workers noticed that a new vitamin-enhanced food resulted in more growth. The worker wanted to investigate her findings further. How would you advise her to design the experiments? What if she wanted to continue the experimentation with humans? What would she need to do then?

4. While researching the keys to a long life, you came across a lot of resources and advice. What steps would you use to determine the best path to take? How would you identify the most valuable advice? What kinds of statements would alert you to potential problems?

Answer Key

Review Questions

A. 1. Living things contain the following biological molecules: nucleic acids, proteins, carbohydrates, and lipids.
2. Living things are composed of cells.
3. Living things reproduce.
4. Living things use energy and raw materials.
5. Living things respond.
6. Living things maintain homeostasis.
7. Populations of living things evolve and have adaptive traits.

B. 1. Living things reproduce.
2. Living things use energy and raw materials.
3. Living things are composed of cells.
4. Living things maintain homeostasis.
5. Living things contain the following biological molecules: nucleic acids, proteins, carbohydrates and lipids.
6. Living things respond.
7. Populations of living things evolve and have adaptive traits.

C. 1. d
2. e
3. b
4. g
5. a
6. c
7. f

D.

Term	Definition	Example
Population	All the organisms of a same species living in the same area.	Starlings at the county square, field of sunflowers, tadpoles in a vernal pond
Community	All of the species that can interact in a particular geographic area.	All of the organisms living in an alpine meadow, all species found in a hollow log
Ecosystem	The area where living organisms interact with one another and their physical environment.	Pond, forest, bog
Biosphere	That part of the Earth where life is found.	All living things on Earth.

E. A. Hypothesis
B. Prediction
C. Results
D. Conclusion
E. Revise hypothesis, make new predictions, and test them with new experiments

F. Several instructors mentioned that as they studied the results of a recent exam, they noticed that all of the students from the Far Quad had much higher scores than the students living in the Near Quad. These observations led one professor to ask the question, "What is making the students in the Far Quad smarter?"

Discussing the observations with colleagues, the professor thought of many possible reasons – maybe all the students with high scores on the entrance exams were placed into the Far Quad, maybe the students in the Near Quad were ill more often and missed more classes, or maybe there was something in the diet of those in the Far Quad that was enhancing their intelligence. Since the professor wanted to narrow the considerations to only one hypothesis she began to do some inquiring. Housing said that students were randomly assigned and Health Services said that there was no correlation between housing unit and illness. The dietician reported that students in Far Quad were getting SMARTPIES, a new product that was being served in only that dorm.

The professor made the prediction that SMARTPIES was causing the difference in test scores and designed an experiment to test the hypothesis. The experiment was explained to all the students, and they were asked to sign an informed consent document agreeing to participate in the study. Since students gave their student number as they entered the line to pick up their food, it was easy to be sure that those with even numbers received SMARTPIES for dessert and those with odd numbers received another dessert that looked just like SMARTPIES but contained different ingredients. The students who received the authentic SMARTPIES were part of the experimental group and those who received the look-alike PIES were part of the control group. The experiment went on for a month and grades were collected and averaged for each group.

Data analysis showed that the students who ate the authentic SMARTPIES did in fact have higher grades while those who ate the look-alike PIES did not show the same achievement. Since the only variable that changed between the two groups was the dessert and since both desserts were the same except for a few secret ingredients, the professor was able to draw conclusions from this study. In an interview, the professor cautioned that based on only one experiment, she could not develop a theory regarding SMARTPIES and achievement.

G. D
I
I
D

H.
1. Gather as much information as possible. Do library research, read the published results of other people's experiments. Interview experts.
2. Think about who might gain from the information. Is there bias in the presentation? Is the information/study supported by a well-respected organization?
3. Review the experimental design. Was there a control group? Were there enough subjects? Did the collection of data follow a set of standard procedures?
4. Reread the study/experiment and determine if the results are based on the data. Are the results exaggerated? Are there results that are ignored?
5. Think about the study/experiment/claim and try to determine if there might be other reasons for the conclusions that have been drawn. Were there factors that could have influenced the results that were not reported? Were the results absolutely the result of the variables in the experiment or could they have been the result of something else?

Critical Thinking Questions

1. Your questions might be generated by something you noticed in your environment (Is there a higher concentration of pollutants from car exhaust at an intersection?) or yourself (Does running lower blood pressure?). After choosing your question, discuss it with your friends and determine a hypothesis – a statement that can be tested. (Cardiovascular exercise, specifically running, will lower blood pressure.) Identify what the variables (independent variable is the duration and speed of the running and the dependent variable is the blood pressure) you will test and then design an experiment with appropriate numbers of subjects and a control or comparison group. (The experiment might include two groups of similar age and gender. One group will continue their usual

activities but will not engage in cardiovascular training while the other group will engage in a planned running program. Blood pressures will be taken before the program begins and after one month. The change in blood pressure between the two groups will be compared.)

2. The informed consent document should present the purpose of the experiment, the possible benefits, and the possible harmful effects of the treatment plan. It must include a place for a patient signature. An appropriate experimental design would include a placebo group and several other groups receiving different dosages of the product.
3. Identify an appropriate pharmaceutical from television or print advertisements. Then use the Internet to search for the fact sheet that shows in detail the clinical trials. It should include the number of subjects and the negative side effects experienced by subjects. Record the range of dosages and the characteristics of each test group.
4. Intuitively, it is easy to distinguish between living and nonliving things. It is obvious that living things respond as birds fly away when you approach. It is also sometimes easy to note that living things reproduce if you see bunnies or ducklings on the pond. However, without testing, it is not obvious that living things are composed of complex biochemicals or even that they are made of cells. If organisms are seen eating, you know that they are taking in energy and if they are moving, you know that the energy is converted from potential energy to kinetic energy, but the cellular use of energy is not seen. For the most part, homeostasis is taken for granted, but an athlete covered in sweat or a spectator shivering on a cold day are reminders that the body must maintain a constant environment. The processes of evolution and adaptation are much too slow to notice in our immediate environment.
5. The characteristics of life can be copied from your text. Use the Internet to determine the different perspectives on the definition of death. Death is hard to define because it is a processes and not a single event. Death is usually defined as occurring when an organ system that is essential for life fails. Therefore someone is pronounced dead when the respiratory or circulatory system cease or when all functions of the brain and brain stem stop.

Practice Test

1. c
2. c
3. a
4. c
5. a
6. a
7. c
8. d
9. b
10. b
11. a
12. b
13. c
14. d
15. a
16. b
17. c
18. d
19. a
20. b

Short Answer Questions

1. In spite of a convincing explanation, fire is not alive. It does not possess all of the characteristics of life including reproduction or adaptation to the environment. It is not composed of biological macromolecules or cells. However, it does, like many other things, transform energy from one form to another.
2. The correct order of the steps in the scientific method is shown in Figure 1-9 as observation, hypothesis development, experimentation and conclusion. It is important to know that the conclusions from one experiment lead to another and therefore an arrow should go from conclusion to the development of new hypotheses and more experimentation.
3. It would be important to design the experiment with both a control group that received a similar food with no vitamin enrichment and an experimental group that was fed the enriched food. It is

important to have enough subjects in each group, to design the experiment to reduce the impact of extraneous variables, to feed the mice for a sufficient period of time and to carefully collect data. One must consider all the implications of clinical trials and informed consent in order to conduct the experiment on humans.

4. As a critical thinker it is important for you to ask questions, verify information, identify bias, and make informed decisions.

Chapter 2

The Chemistry of Life

Objectives

After reading the text and studying the material in this chapter, you should be able to:

- Describe the parts of an atom.
- Define an element and determine its atomic weight and atomic number.
- Explain how an isotope differs from an atom and give examples of the harmful and beneficial aspects of radiation.
- Define a compound.
- Differentiate between covalent, ionic and hydrogen bonds in terms of strength and the sharing of electrons.
- Describe the unique characteristics of water that make it valuable to biological systems.
- Describe what happens when an acid or a base is added to water.
- Define pH, explain the range of the pH scale and tell which values indicate acid and which values indicate base.
- Describe the structure of a polymer including its formation through dehydration synthesis and its breakdown through hydrolysis.
- Give the structure, biological purpose, and examples of a carbohydrate, lipid and protein.
- Explain the purpose of enzymes and how they function.
- Describe ATP as the energy currency of the cell.

Chapter Summary

Everything that takes up space and has mass is called **matter**. All matter is made of **atoms**, each containing a nucleus with protons and neutrons surrounded by a cloud of electrons. An atom is the smallest unit of an element that shows the same characteristics as the element. An element is made of many atoms that are all the same.

An **element** is a form of matter that cannot be broken down into simpler substances. It is made of many atoms that are all the same. Carbon, oxygen, hydrogen and nitrogen are the most common of the 20 elements found in the human body. Each element has an atomic number and atomic weight. The atomic number is the number of protons in the nucleus and the atomic weight is the number of protons plus the number of neutrons.

Elements always must contain the same number of protons, but the number of neutrons may vary. Elements with the same number of protons but different number of neutrons are called **isotopes**. Common isotopes of carbon include ^{12}C, ^{13}C and ^{14}C. If the isotope is unstable and emits radiation it is called a **radioisotope**. Radiation can be very harmful causing the death of cells or very beneficial as evidenced by its use in medicine.

Two or more elements may combine to form a compound. The atoms in a compound are held together by covalent, ionic or hydrogen bonds. **Covalent bonds** are the strongest bonds and form when two or more atoms share the electrons in their outer shells. A **molecule** is a chemical structure held together by covalent bonds. The chemical structure shows the number of each element forming the molecule.

An **ion** is an atom or group of atoms with a positive or negative electrical charge. Ionic bonds, weaker than covalent bonds, result from the attraction of oppositely charged ions, rather than shared electrons.

When the electrons of a covalent bond are shared unequally, the bond is called polar and the resulting molecules are called polar molecules. Water is a polar molecule. **Hydrogen bonds** are the attraction formed between a slightly positively charged hydrogen atom and another slightly negatively charged atom. Hydrogen bonds account for the unique properties of water and the geometric shape of many biological molecules.

Acids and **bases** react differently to water. Acids release hydrogen ions (H^+) when placed in water and bases produce hydroxide ions (OH^-) when added to water. pH is the negative logarithm of the concentration of the H^+ ion in solution. The lower the pH on the **pH scale**, the greater the acidity and the higher the pH, the more basic a solution. A **buffer** prevents dramatic changes in pH. Many body fluids have the buffering capacity to maintain a constant internal environment.

The giant molecules of life are called biological **macromolecules**. They are long chains called **polymers** made of repeating units called **monomers**. When polymers are made, water is removed and the reaction is called a **dehydration synthesis**. Conversely, when the same molecules are broken apart, water is added and the reaction is a **hydrolysis**.

Carbohydrates are a polymer made of **monosaccharides** composed only of C, H, and O. They are sugars and starches and provide fuel for the body. Sucrose and lactose are examples of **disaccharides** or double sugars. Long chains of monosaccharides are called **polysaccharides**. **Cellulose** is a common example of an indigestible polysaccharide made of repeating units of glucose.

Lipids are water-insoluble molecules made of C, H, and O that store long-term energy, protect vital organs, and form cell membranes. Fats and oils are examples of triglycerides, a polymer made of one molecule of glycerol and three fatty acids. The fact that phospholipid molecules have one end that is water soluble (hydrophilic) and another end that is water insoluble (hydrophobic) is critical to their function as a membrane. Steroids are a unique group of lipids that consist of four ring compounds.

Proteins are chains of amino acids that provide structure, transport and movement for the body. Amino acids consist of a central carbon atom bound to a hydrogen (H) atom, an amino group (NH_2) and a carboxyl group (COOH) in addition to a unique side chain called a radical (R). Proteins have four distinct levels of structure—primary, secondary, tertiary and quaternary—that affect their function in the body. Changes in the chemical environment of a protein can cause it to lose its structure, or denature, resulting in a loss of function. A special group of proteins are called enzymes and they serve as catalysts for chemical reactions. They speed up a reaction while not being consumed. Enzymes bind to substrates at a specific active site forming an enzyme-substrate complex. Sometimes **cofactors**, often called **coenzymes**, bind at the active site to facilitate the reaction.

Nucleotides are the building blocks of another polymer called nucleic acids. The sequencing of the nucleotides in DNA and RNA determine the sequence of amino acids in a protein. A special nucleotide is adenosine triphosphate (ATP), a molecule capable of storing energy in its phosphate-to-phosphate bonds. All energy from the breakdown of molecules such as glucose must be channeled through ATP before the body can use it, thus it is often described as the energy currency of cells.

Key Concepts

- Matter is anything that takes up space and has mass.
- Mass is the measure of how much matter is contained within an object.
- Atoms contain protons, neutrons and electrons.
- The atomic number of an atom is the number of protons found in the nucleus.
- Isotopes are two or more atoms with the same number of protons but different number of neutrons. If they are unstable and emit particles or energy in the form of radiation, they are called radioisotopes.
- Radiation can cause great damage to cells, or when used in small amounts, can greatly benefit humans.
- An element is a form of matter, such as iron or oxygen, that cannot be broken down into simpler substances.
- Chemical bonds connect two or more elements to form a compound.
- Covalent bonds are the strongest bonds and occur when electrons are shared between atoms.
- Ionic bonds form when electrons are transferred between atoms creating oppositely charged ions.
- Hydrogen bonds are weak attractions between a hydrogen atom with a slight positive charge and another atom with a slight negative charge.
- A molecule is a chemical structure held together by covalent bonds.
- Water is essential to life. It has the unique characteristics of cohesiveness, high heat of vaporization and the ability to act as a solvent.
- Acids and bases are defined by what happens when they are added to water. An acid releases hydrogen ions (H^+) in water and a base produces hydroxide ions (OH^-) in water.
- The pH scale ranges from 0 to 14 with a pH of 7 being neutral, a pH of less than 7 being acidic, and a pH of greater than 7 basic.
- Since biological systems operate in a very narrow range of pH, they need buffers to prevent dramatic changes in pH.
- Large biological macromolecules are polymers made of many monomers linked together. They are formed through dehydration synthesis and broken down by hydrolysis.
- Carbohydrates, including monosaccharides, disaccharides and polysaccharides provide quick energy for the body.
- Lipids are water insoluble and provide long-term energy, insulation, protection for internal organs, and form membranes.
- Amino acids are joined by peptide bonds to form proteins. Proteins are the materials for tissue repair and growth. Enzymes are a unique group of proteins that act as catalysts.
- Nucleotides are the monomers of nucleic acids (for example, DNA and RNA) and ATP, the energy currency of the cell.

Study Tips

The information in this chapter will be very challenging for students who have a weak background in chemistry. It will be very important for you to study this material on a daily basis as it is covered in class. First, get an overview of the material from the outline on the first page of the chapter. Note that the first section deals with basic chemistry and the second section applies those concepts to biological systems. Therefore it is clear that the content of the second section builds upon the first.

As you read the text, highlight important concepts and words. Be sure you understand what the words mean and how one definition differs from another. For example what is the difference between atom and atomic mass, between a proton and a neutron, between RNA and DNA? Make notes as to the questions you have or material that needs clarifying.

This chapter has many figures and tables. The figures are very important and add to your understanding of the text. Take time to relate the figures to the content described in the text. As you understand the material better, explain the details of the figure to yourself or a friend to test your understanding. The tables summarize the information and often bring clarity to it as different concepts are compared.

Finally, review the end of the chapter summary and complete this study guide. Again note any areas that are not clear and ask your instructor for more explanation. When you feel you are ready, attempt the practice test at the end of this study guide chapter. If you miss items, review that material to improve your understanding.

Review Questions

A. Use the following terms to label the figure below: electron, neutron, nucleus, and proton.

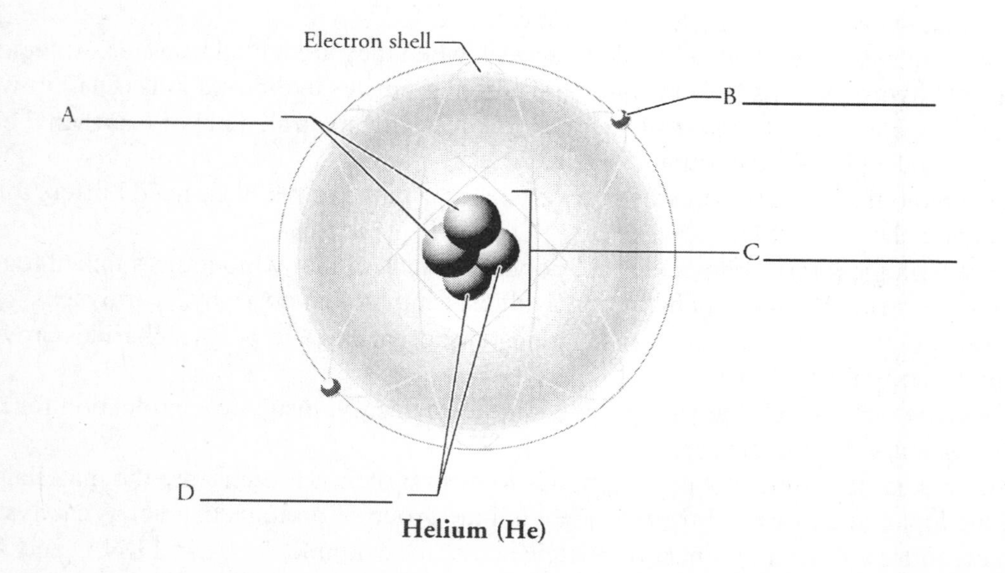

Helium (He)

B. Match the term in Column A with the definition or example in Column B.

Column A	Column B
1. atom	a. measure of how much matter is contained within an object
2. atomic number	b. two or more elements, for example HCl
3. compound	c. uncharged particle found in the nucleus of an atom
4. electron	d. unstable isotope emitting radiation
5. element	e. elements with the same number of protons but a different number of neutrons
6. isotopes	f. found in the nucleus of an atom and carries a positive charge
7. mass	g. fundamental unit of an element
8. neutron	h. number of protons in the nucleus of an atom
9. proton	i. found outside the nucleus of an atom and carrying a negative charge
10. radioisotope	j. form of matter that cannot be broken down by chemical means: for example, iron or oxygen

C. Complete the following table with the missing information about chemical bonds.

Type of Bond	**Basis for Attraction**	**Strength**	**Example**
	Sharing of electrons between atoms		
Ionic			NaCl
Hydrogen		Weak	

D. Fill in the blank.

1. Two or more elements combine to form a ___________.

2. A ___________ bond is formed when atoms share ___________ in their outermost shell.

3. $C_6H_{12}O_6$, glucose is an example of a ___________ that has ___________ atoms of oxygen.

4. Carbon dioxide is held together by two ___________ bonds, which means that carbon shares ___________ electrons with each oxygen.

5. The most common molecule found in the human body is ___________.

6. ___________ bonds are very weak attractions. They are responsible for the structure of ___________.

7. It takes a great deal of energy to raise the temperature of water. This is called a high ___________.

8. Because water has a very high ___________ of ___________, it takes a great deal of heat to make water evaporate and thus it can serve as a coolant as it evaporates from our skin.

9. An acid releases ___________ ___________ when added to water and a ___________ produces hydroxide (OH^-) when placed in water.

10. Carbonic acid is a common ___________ in the human system, which means that it prevents dramatic changes in pH.

E. Put these 5 substances in order from the most acidic to the least acidic.

_____ Soda crackers (pH = 8)

_____ Urine

_____ Blood

_____ Distilled water

_____ Stomach contents (pH = 2.7)

F. Complete the following reactions, then explain the importance of these reactions.

Dissociation of Water

H – O – H ⟷ ☐ + ☐

Water — Hydrogen ion — Hydroxide ion

Reaction of Carbon Dioxide and Water to form a Buffer

☐ + ☐ ⟷ H_2CO_3 ⟷ ☐ + ☐

Carbon dioxide — Water — Hydrogen ion — Bicarbonate ion

G. Complete the following table with information on biological polymers.

Polymer	Monomer	Function	Example
Carbohydrate	Glucose, fructose, galactose	Quick energy	
	Glycerol and fatty acids		
		Build and repair tissue	
Nucleic Acid		Carry genetic information	

H. Each statement below pertains to nucleic acids. After the statement write the nucleic acid (DNA, RNA, ATP) that may be an example of that characteristic. Some statements may apply to more than one nucleic acid.

1. Two twisted strands form the backbone. ____________

2. Contains the sugar ribose. ___________

3. Contains the base adenine. ___________

4. Exists as a single strand. ___________

5. Releases usable energy when the terminal phosphate is removed. ___________

I. Label the following diagram of DNA. What would be different if this was a diagram of RNA?

____

____

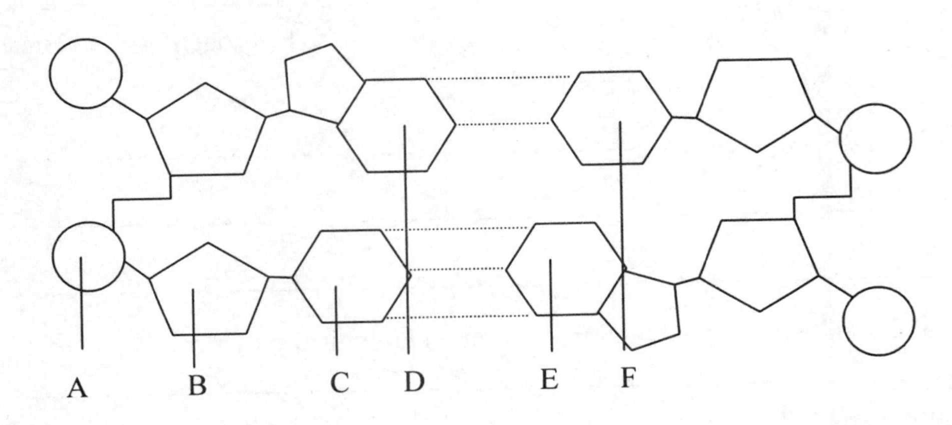

CRITICAL THINKING QUESTIONS

Read each of the following questions carefully. If it helps you, underline or list each component to the question. Use complete sentences in your explanation.

1. Compare an isotope with a radioisotope. Construct a T-graph listing the positive aspects of radioisotopes (and radiation) on one side and the negative aspects on the other side. Can you come to a conclusion as to the value of radiation and its uses?

2. What characteristics of cellulose make it the fiber in "high-fiber foods"? Why is a high fiber diet good and what foods contain high quantities of fiber?

3. Lipids are very important to our bodies. However, they often get a "bad rap" because they are high in calories. This is a double-edged sword. Explain why it is efficient for the body to store energy in fats and also why the calories increase so quickly in foods high in fat.

4. As you have read, proteins are three dimensional molecules. It is the shape of a protein that often determines its function and if the shape changes, the molecule no longer may be of value. All proteins including enzymes are sensitive to fluctuations in temperature. It is no surprise that most human proteins function best at body temperature. Using this knowledge, why are high fevers so very dangerous? What might happen to the protein structure and the effectiveness of enzymes under high temperatures?

PRACTICE TEST

Choose the one best answer to each question that follows. As you work through these items, explain to yourself why each answer you discard is incorrect and why the answer you choose is correct.

Use the diagram at the right to answer questions 1 – 2.

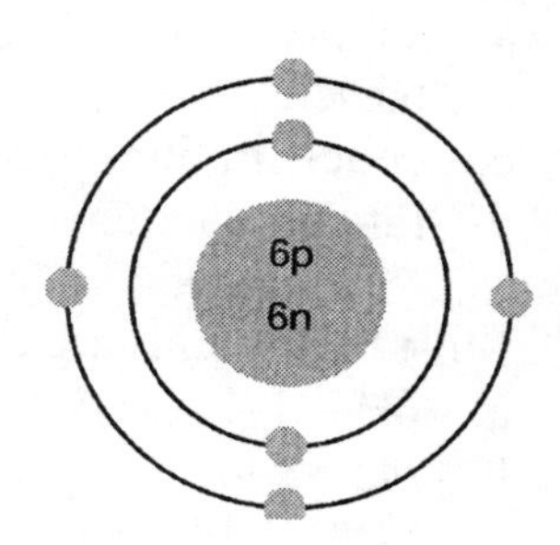

1. What is the charge of the carbon atom at the right?
   a. negative
   b. neutral
   c. positive
   d. not enough information is given

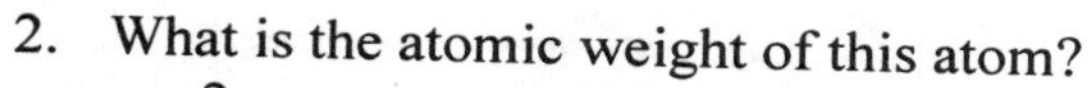

2. What is the atomic weight of this atom?
   a. 2
   b. 6
   c. 12
   d. 18

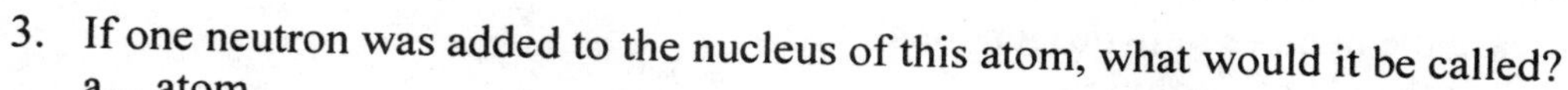

3. If one neutron was added to the nucleus of this atom, what would it be called?
   a. atom
   b. compound
   c. isotope
   d. molecule

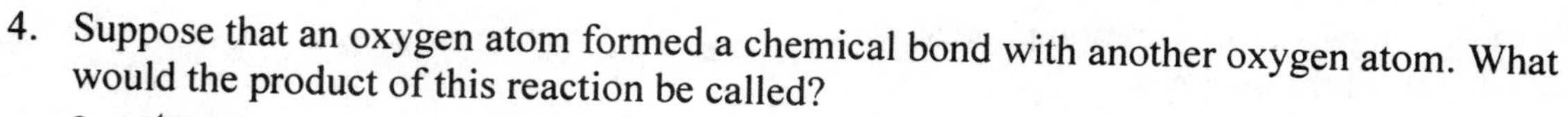

4. Suppose that an oxygen atom formed a chemical bond with another oxygen atom. What would the product of this reaction be called?
   a. atom
   b. compound
   c. element
   d. molecule

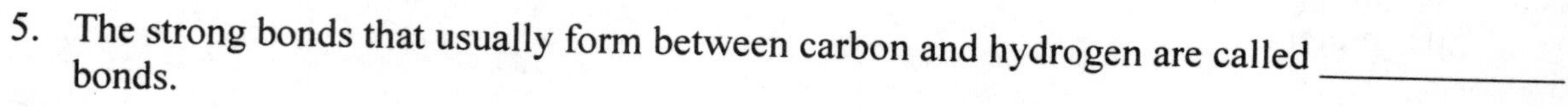

5. The strong bonds that usually form between carbon and hydrogen are called ____________ bonds.
   a. covalent
   b. elemental
   c. hydrogen
   d. ionic

6. Weak bonds that form between water molecules and between the nucleotide bases of DNA and RNA are called ____________ bonds.
   a. covalent
   b. fluid
   c. hydrogen
   d. ionic

7. Which property of water provides the cooling effect of sweating?
   a. cohesiveness
   b. high heat capacity
   c. high heat of vaporization
   d. surface tension

8. The pH of a solution is an indication of its ____________.
   a. acidity
   b. basicity
   c. concentration
   d. rate of reaction

9. Most body fluids are found in which range of the pH scale?
   a. pH = 1-4
   b. pH = 5-8
   c. pH = 9-14
   d. pH exactly neutral

10. Buffers are important to human systems because they ____________.
   a. assist in chemical reactions
   b. bring a solution to a pH of 9.8
   c. increase the number of hydrogen ions
   d. stabilize the pH of a solution

11. Polymers are ____________.
   a. broken down by hydrolysis
   b. macromolecules
   c. made by a dehydration process
   d. all of the above

12. Monosaccharides, disaccharides, and polysaccharides are all ____________.
   a. carbohydrates
   b. lipids
   c. monomers
   d. proteins

13. Carbohydrates are important because they ____________.
   a. are the building blocks of proteins
   b. insulate internal organs
   c. provide quick energy
   d. provide long-term energy

14. Which of the following elements occur in proteins but not in carbohydrates or lipids?
    a. carbon
    b. hydrogen
    c. oxygen
    d. nitrogen

15. A biologist studying proteins would identify the simplest unit as a(n) ____________.
    a. adenine
    b. amino acid
    c. fatty acid
    d. glucose

16. Triglycerides are so named because they are formed by a reaction between three fatty acid molecules and one ____________.
    a. amino acid
    b. glucose
    c. glycerol
    d. lipid

17. Which of the following foods contain the most energy per gram?
    a. beans
    b. butter
    c. cellulose
    d. sucrose

18. Enzymes are one type of protein. They function to ____________.
    a. build strength
    b. form hormones
    c. repair tissue
    d. speed reactions

19. An analysis of the nucleotides in a complex molecule yielded equal amounts of adenine and thymine. What was the molecule?
    a. adenosine triphosphate
    b. deoxyribonucleic acid
    c. ribonucleic acid
    d. not enough information to tell

20. All energy must be channeled through ____________ to be useful within the cells of the human body.
    a. amino acids and proteins
    b. ATP
    c. fatty acids and glycerol
    d. glucose and carbohydrates

Short Answer Questions

Read each of the following questions carefully. If it helps you, underline or list each component to the question. Formulate your answer and jot down the main points. Then use complete sentences in your explanation.

1. Compare protons, neutrons, and electrons with respect to their charge, mass and location within an atom.

2. Water is very important to the body. Describe the characteristics of water that make it so valuable and give an example of how those characteristics are demonstrated in humans.

3. Compare the structure, location and function of DNA and RNA. DNA is found only in the nucleus of the cell while RNA carries genetic messages out of the nucleus to the rest of the cell. How does the structure of the molecule relate to its function?

4. Explain why adenosine triphosphate (ATP) is considered to be the energy currency of the cell. Draw analogies to other areas where only one form of energy can be used (most gasoline-driven engines) or where only one form of currency is accepted (many snack machines).

Answer Key

Review Questions

A. A. proton
   B. electron
   C. nucleus
   D. neutron

B. 1. g
   2. h
   3. b
   4. i
   5. j
   6. e
   7. a
   8. c
   9. f
   10. d

C.

Type of Bond	Basis for Attraction	Strength	Example
Covalent	Sharing of electrons between atoms	Strongest	CH_4
Ionic	Transfer of electrons creates oppositely charged ions that are attracted to one another	Strong	NaCl
Hydrogen	Attraction between a partially positively charged hydrogen atom and a partially negatively charged atom in another molecule	Weak	Attraction between hydrogen atom of water and oxygen atom of another water molecule

D. Fill in the blank.

1. Two or more elements combine to form a compound.
2. A covalent bond is formed when atoms share electrons in their outermost shell.
3. $C_6H_{12}O_6$, glucose is an example of a molecule that has 6 atoms of oxygen.
4. Carbon dioxide is held together by two double bonds, which means that carbon shares four electrons with each oxygen.
5. The most common molecule found in the human body is water.
6. Hydrogen bonds are very weak attractions. They are responsible for the structure of water.
7. It takes a great deal of energy to raise the temperature of water. This is called a high heat capacity.
8. Because water has a very high heat of vaporization, it takes a great deal of heat to make water evaporate and thus it can serve as a coolant as it evaporates from our skin.
9. An acid releases hydrogen ions when added to water and a base produces hydroxide (OH^-) when placed in water.
10. Carbonic acid is a common buffer in the human system, which means that it prevents dramatic changes in pH.

E. Stomach contents (pH = 2.7)
Urine
Distilled water
Blood
Soda crackers (pH = 8)

F.

$H - O - H \longleftrightarrow$ H^{+} + OH

When water dissociates, it produces hydrogen ions that are positively charged and hydroxide ions that are negatively charged. This is what happens when acids or bases are added to water and what gives the solution a characteristic pH.

CO_2 + H_2O $\longleftrightarrow$ carbonic acid $\longleftrightarrow$ H^{+} + $HCO3^{-}$

This is the buffering system of blood that maintains a relatively constant pH. It occurs when CO_2 enters the blood that is mostly water.

G.

Polymer	Monomer	Function	Example
Carbohydrate	Glucose, fructose, galactose	Quick energy	Starches, pasta, bread, sugar
Lipid	Glycerol and fatty acids	Long-term energy, insulation, cushioning, membranes	Fats, oils, plasma membrane, steroids
Protein	Amino acids	Build and repair tissue	Fish, meat, beans
Nucleic Acid	Nucleotide bases	Carry genetic information	RNA, DNA, ATP

H. 1. DNA
2. RNA
3. DNA, RNA, ATP
4. RNA
5. ATP

I. It would be single stranded, contain the sugar ribose and the base uracil and contain no thymine.
A. phosphate
B. deoxyribose
C. cytosine
D. adenine
E. guanine
F. thymine

Critical Thinking Questions

1. Atoms that have the same number of protons but a different number of neutrons are called isotopes. Isotopes that emit radiation are called radioisotopes. Make a two-column table, or simply a T-graph, with one column being positive aspects of radioisotopes and the other being negative aspects of radioisotopes. Positive aspects of radiation include use in x-rays, diagnostic procedures, and cancer

therapy. Negative aspects include damage to reproductive cells and future offspring, lowering the number of white blood cells, and serious damage to the skin and other organs.

2. Cellulose is a long-chain polysaccharide found in the cell walls of plants. Humans cannot digest cellulose, and thus it passes unchanged through our digestive tract collecting free radicals on the way. It is an important form of dietary fiber that might reduce the incidence of colon cancer.
3. Energy is stored in the carbon to hydrogen bonds of a chemical. Fats store twice the calories per gram as carbohydrates or proteins, thus we can store much more energy in fat with much less bulk.
4. All proteins are sensitive to high temperatures and since enzymes are one group of proteins, they too are affected by high temperatures. High temperatures break chemical bonds and cause proteins to denature or change shape and break down. When complex proteins change shape, they are no longer able to function. Thus high temperatures can cause the protein structure of neurotransmitters, sperm, hormones and enzymes to break down.

Practice Test

1. b	11. d
2. c	12. a
3. c	13. c
4. d	14. d
5. a	15. b
6. c	16. c
7. c	17. b
8. a	18. d
9. b	19. b
10. d	20. b

Short Answer Questions

1. Protons carry a positive charge, electrons a negative charge and neutrons no charge. Both protons and neutrons have an atomic mass unit of one and are found in the nucleus. Electrons have negligible mass and are located in the orbital shells around the nucleus.
2. Humans are at least two-thirds water. The hydrogen bonds of water cause it to stick together or demonstrate cohesion. Its charged nature also allows it to dissolve and carry a great variety of molecules in our circulatory system. The high heat capacity helps to maintain a relatively constant temperature so that as the external temperature fluctuates, we maintain a nearly constant internal temperature. The high heat of vaporization provides cooling when we sweat. Lastly, water is a liquid at room temperature.
3. DNA is a twisted double helix while RNA is a single short strand. DNA is made of the paired bases adenine and thymine, and cytosine and guanine. Uracil replaces thymine in RNA. DNA remains in the nucleus while the shorter pieces of RNA allow it to move into the cytoplasm and function in protein synthesis.
4. ATP is the only source of energy that the cell can use. All other energy whether just consumed as food or stored as lipids must be converted to ATP to be used. It is the energy currency of the cell because it is the only form of energy the cell will accept, much like gasoline being the only form of energy used by most cars. Putting a candy bar into the tank will not power the automobile. Likewise, foreign currency is usually not accepted in soda and candy machines found in this country; it is the wrong currency.

Chapter 3

The Cell

Objectives

After reading the text and studying the material in this chapter, you should be able to:

- Compare prokaryotic and eukaryotic cells.
- Explain why the surface area-to-volume ratio dictates cell size.
- Draw and label a plasma membrane and describe the function of each component.
- List the functions of the plasma membrane including its role in exocytosis and endocytosis.
- Compare movement of materials across the plasma membrane including simple diffusion, facilitated diffusion, osmosis, and active transport.
- Explain the differences between hypertonic, hypotonic and isotonic solutions and predict the movement of water in or out of a cell in each case.
- Describe the function and structural features of each of the following organelles: nucleus, endoplasmic reticulum, Golgi complex, lysosomes and mitochondria.
- Compare the structure and function of the three fibers that make up the cytoskeleton.
- Explain in general terms how the food we eat is turned into cellular energy as ATP.
- Describe the three phases of cellular respiration indicating the products of each phase.
- Explain why the electron transport chain generates such a large quantity of ATP.
- Differentiate between fermentation in plants and in animals.
- Compare cellular respiration to fermentation as methods to harvest energy for cellular use.
- Describe ATP as the energy currency of the cell.

Chapter Summary

We are one huge conglomerate of cells. And although we begin life as only one cell, that cell has differentiated into many specialized cells since then. There are two basic types of cells: **eukaryotic cells** that have membrane-bound **organelles**, and **prokaryotic cells** that do not. Eukaryotic cells are far more complex because they contain compartments, which are also membrane-bound, called organelles. Humans are made entirely of eukaryotic cells. Cells vary in size but can never exceed the volume that can be nourished by materials passing through the surface membrane. Thus it is a cell's **surface area-to-volume ratio** that determines when it must divide.

The **plasma membrane** is the outer boundary of the cell and controls the movement of substances in and out of the cell. The primary component of all cell membranes is phospholipids. This lipid bilayer separates the **extracellular fluid** from the material inside the cell contained in the **cytoplasm**. Proteins, cholesterol, and carbohydrates are also part of the membrane giving it the qualities of a **fluid mosaic**.

The primary functions of the membrane are to separate the internal contents of the cell from everything else and to control what moves in and out of the cell. The structure of the membrane allows it to be selectively permeable and to function in cell recognition. **Receptors** facilitate communication between cells and **cell adhesion molecules** help cells to form tissues and organs.

Movement across the membrane is accomplished by **simple diffusion**, **facilitated diffusion** and **active transport** that require the input of energy. **Osmosis** is the movement of water across a membrane. To facilitate the movement of large particles into the cell, the membrane encloses the substance in a **vesicle**, a process called **endocytosis**. **Exocytosis** is the process whereby large molecules leave the cell.

The plasma membrane is used within the cell to delineate compartments called organelles, each with its own function. The **nucleus** contains the genetic information of the cell. It is surrounded by a **nuclear envelope** that allows communication through **nuclear pores**. **Chromosomes** contained in the nucleus are made of DNA. The **nucleolus** is involved in the production of ribosomal RNA, a component of **ribosomes** used during protein synthesis.

The **endoplasmic reticulum** is important for cell communication through a network of channels. The **rough endoplasmic reticulum** contains ribosomes that guide the production of cell products. They are packaged in vesicles and transferred to the **Golgi complex** for export. The **smooth endoplasmic reticulum** lacks ribosomes and is involved in the production of phospholipids and detoxification. The enzymes within **lysosomes** breakdown macromolecules, diseased cells, and invaders.

Mitochondria are the site of cellular respiration. This double-membrane organelle contains inner foldings, the **cristae** that provide increased membrane surface for cellular respiration.

Microtubules, a part of the **cytoskeleton**, provide shape and support for the cell. Microtubules are also responsible for the structure and movement of **cilia** and **flagella**. **Microfilaments** are made of actin and function in muscle contraction. They also are responsible for the movement of **pseudopodia**. **Intermediate filaments** are ropelike fibers that maintain cell shape and anchor organelles.

Cellular respiration and fermentation generate cellular energy. Cellular respiration requires oxygen to breakdown glucose in a three-step process. **Glycolysis** occurs in the cytoplasm splitting glucose into two pyruvate molecules generating a net gain of two ATP and two NADH molecules. Within the mitochondria CO_2 is removed from each pyruvate forming 2 acetyl CoA molecules during the **transition reaction**. Acetyl CoA enters the **citric acid cycle** releasing two ATP, two $FADH_2$ and six NADH molecules. The final step is the **electron transport chain** where the electrons of $FADH_2$ and NADH are transferred from one protein to another, until they reach oxygen, releasing energy that results in 32 ATP.

Fermentation is the breakdown of glucose without oxygen and takes place entirely in the cytoplasm. It begins with glycolysis and uses pyruvate or one of its derivatives as the final electron receptor. It is very inefficient resulting in only two ATP. **Lactic acid fermentation** takes place in the muscles while **alcohol fermentation** is used in bread making and the production of alcohol.

Key Concepts

- Prokaryotic and eukaryotic cells are the two main types of cells. Eukaryotic cells are more complex and have membrane-bound organelles. Humans are made entirely of eukaryotic cells.
- The surface-to-volume ratio determines when a cell needs to divide.
- The plasma membrane is made of a phospholipid bilayer with proteins and cholesterol molecules interspersed. Surface carbohydrates are used for cell identification.
- The functions of the plasma membrane are to maintain the cell's integrity, regulate movement of substances into and out of the cell, aid in cell-to-cell recognition, promote communication between cells, and stick cells together to form tissue and organs.
- The plasma membrane functions in endocytosis and exocytosis.
- Molecules also cross the plasma membrane along a concentration gradient by simple diffusion and facilitated diffusion. Molecules concentrate by active transport, which requires energy.
- Water moves across the plasma membrane by osmosis.
- Organelles are contained in the cytoplasm and are surrounded by a membrane.
- The nucleus contains the genetic information for the organism and the nucleolus produces ribosomal RNA used in the synthesis of proteins.
- Smooth endoplasmic reticulum functions in communication and in the production of plasma membrane. Rough endoplasmic reticulum contains ribosomes and functions to produce proteins.
- The Golgi complex packages cell products for export from the cell.
- Lysosomes contain substances that can digest diseased or dying cells as well as invaders.
- Mitochondria are the energy-processing center of the cell.
- Microtubules, microfilaments and intermediate filaments support the cell structurally. Microtubules and microfilaments also function in cell movement.
- Cells use two pathways, cellular respiration and fermentation, to break down glucose and turn it into usable energy in the form of ATP.
- Cellular respiration requires oxygen and produces 36 ATP through the three phases of glycolysis, the citric acid cycle and the electron transport chain.
- Fermentation is the breakdown of glucose without oxygen. In animals the end product is lactic acid, while in plants it is alcohol. In both cases only two ATP are generated.

Study Tips

The material in this chapter is highly visual – that is, you can gain much from studying the pictures and figures. Sometimes students skip over the illustrations, but remember that a picture is worth a thousand words and that the author carefully chose each figure, drawing, and photo to illustrate important concepts.

As you read each section, study the figures, tables and pictures that are referred to in the text. First read the caption and highlight important terms. Then look at the figure in its entirely to determine the main topic, the symbols and the flow of the figure. Next connect the figure to the caption identifying each concept. Refer again to the text to be sure you have related all of the text information to the figure. Finally, explain the process or structure contained in the figure to yourself. Recopy the figure or drawing onto a note card and use it for study at a later time.

As you study the information about the plasma membrane and cell organelles, be sure to relate structure to function. As you learn the function of each organelle, identify it in the figures and electron micrographs noting special structures that help it to do its job more efficiently. Study the diagrams relating to osmosis and then construct some of your own with different concentrations. Finally, it is very important that you understand the similarities and differences between cellular respiration and fermentation. Write both processes out on note cards, making the flowcharts similar for easy comparison.

Lastly, find a study partner. Identify each cell feature on a figure or photo and explain its function to one another. Compare the different ways materials move in and out of a cell. Point out the features of the plasma membrane that facilitate movement of materials. Use the micrographs in the text to explain the function of the cytoskeleton. Finally, construct a T-graph to compare and contrast the reactions dealing with cellular energy. Be sure you know the reactants, products and resulting energy output in each case. Explain the reactions to your study partner including where energy is generated and the advantages of aerobic cellular respiration. Take the practice test and review the items you missed with your study partner. Ask your instructor about difficult concepts.

Review Questions

A. Complete the following paragraph describing the components of the plasma membrane and their functions.

   The main component of the plasma membrane is ____________ molecules. They are amphipathic with a water-loving end that is ____________ in nature and a water-hating end that is ____________ in nature. The two hydrophobic ends are in the center of the plasma membrane while the hydrophilic ends contact the extracellular fluid and the ____________. The largest proteins found in the membrane are called ____________ proteins. Smaller proteins attached to the surface of the membrane are called ____________ proteins. Molecules of ____________ are found throughout the membrane. Glycoprotein and glycolipid molecules are protein and lipid molecules with extra bits of ____________ attached and are located on the membrane surface.

B. Complete the following table with the missing information about the movement of materials across the cell membrane.

Mechanism	Uses Cellular Energy	Description
Simple diffusion		Movement of small molecules from a higher concentration to a lower concentration.
	No	
	No	Movement of water from a higher concentration of water to a lower concentration of water.
Active Transport		
Endocytosis	Yes	
	Yes	Membrane-bound vesicle from inside the cell fuses with the plasma membrane releasing the contents of the vesicle.

C. Label the drawing below using the terms hypotonic, isotonic and hypertonic. Then draw an arrow showing the direction of movement of most of the water and describe what would happen to a cell in that situation. In each case, the contents of the membranous bag are similar to that of a cell, 98% water, 2% salt.

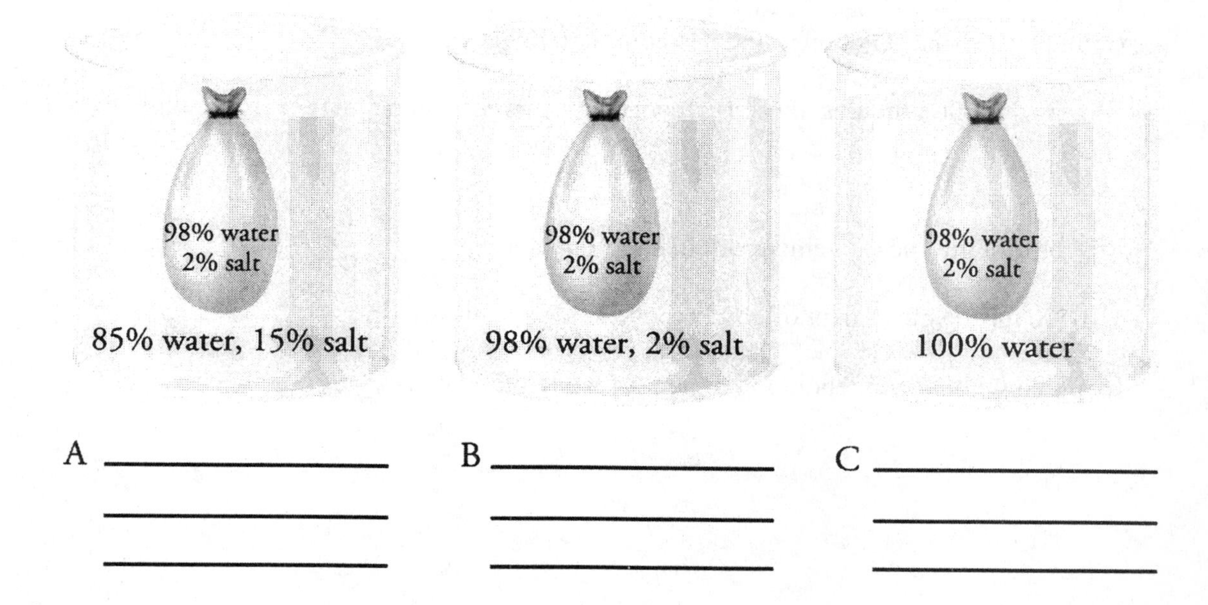

A ____________________ B ____________________ C ____________________

____________________ ____________________ ____________________

____________________ ____________________ ____________________

D. Match the term in Column A with the function or specialized structure in Column B.

Column A	Column B
1. chromosomes	a. inner folds of the mitochondrial membrane
2. cristae	b. site of protein synthesis
3. cytoskeleton	c. made of DNA, contains most of the genetic material of the cell
4. Golgi bodies	d. specialized region of the nucleus responsible for the production of ribosomal RNA
5. lysosomes	e. produces phospholipids and functions in detoxification of poisons
6. mitochondria	f. threadlike structures made of DNA and associated proteins
7. nucleolus	g. contains digestive enzymes capable of destroying a cell
8. nucleus	h. modifies, sorts and repackages proteins
9. rough ER	i. network of fibers providing support and movement
10. smooth ER	j. site of cellular respiration

E. Indicate which element of the cytoskeleton would function in each situation below. The terms, intermediate filaments, microfilaments and microtubules, may be used more than once.

1. Contain actin and function in muscle contraction. ____________
2. Along with microtubules, these filaments help support and maintain the shape of the cell.

   ____
3. Responsible for the movement of cilia and flagella. ____________
4. Provide for the movement of chromosomes during cell division. ____________
5. Involved with the amoeboid movement of white blood cells. ____________

F. Complete the following table with information summarizing the energetics of the cell. Classify the products as carbon products, ATP, NADH or $FADH_2$.

Process	Reactants	Net Products per Initial Glucose	
		carbon molecule	ATP Generated
	Glucose	2 pyruvate	
Transition reaction		2 acetyl CoA 2 CO_2	0
Citric Acid Cycle		4 CO_2	2
	2 acetyl CoA	No change	

G. Complete the following reaction of the breakdown of glucose during cellular respiration.

$$C_6H_{12}O_6 + \square \longrightarrow 6CO_2 + \square + ATP$$

__________ oxygen __________ water __________

H. Compare and contrast the two types of fermentation. Include the reactants, products and location of both processes.

____

____

Critical Thinking

Read each of the following questions carefully. If it helps you, underline or list each component to the question. Use complete sentences in your explanation.

1. List the functions of the plasma membrane. Then explain which components of the membrane assist with each function relating structure to function.

2. Review the function and structure of the organelles. In a small group, identify or create a model analogous to the cell that has functional subunits similar to organelles. For example, consider your campus – the administration would be analogous to the nucleus

with the chancellor or president the nucleolus, the university mail system might be similar in function to the smooth endoplasmic reticulum, and so on.

3. Athletes in all sports focus on aerobic conditioning. Use your knowledge of aerobic cellular respiration and lactic acid fermentation to explain why performance is improved if more oxygen can reach the cells for longer periods of time due to excellent conditioning.

4. Peel a raw potato and then slice it into ¼" slices. Next prepare a glass of salt solution by filling a drinking glass about ¾ full of water and stirring in as much salt as you can get to dissolve. Next find two shallow, flat-bottomed dishes and put about an inch of distilled or tap water into one of them. Pour the salt solution into the other. Put the same number of potato slices into each dish and observe them several times over the next hour. Record your results noticing whether the pieces are rigid or limp. Rinse the slices that were in the salt solution and pat them dry. Now put them into a dish containing fresh tap or distilled water. Record what happens in the next hour. How do the slices compare with those previously in the tap/distilled water? How to they compare to when they were freshly cut?

PRACTICE TEST

Choose the one best answer to each question that follows. As you work through these items, explain to yourself why each answer you discard is incorrect and why the answer you choose is correct.

1. Which of the following would be found in a eukaryotic cell but not a prokaryotic cell?
   a. chromosomal material
   b. cytosol
   c. nuclear membrane
   d. plasma membrane

2. As the surface area of a cell increases, the volume of the cell ____________.
   a. decreases proportionally
   b. increases proportionally
   c. increases at a faster rate
   d. remains the same

3. Which of the following most likely give the plasma membrane the characteristic of being fluid?
   a. carbohydrates
   b. cholesterol
   c. phopholipids
   d. proteins

4. When a membrane is said to be selectively permeable, this means that ____________.
   a. half of the membrane is permeable and the other half is not.
   b. only large molecules can pass through.
   c. the cell regulates what passes in and out.
   d. the membrane is permeable part of the time.

5. The process whereby small molecules pass through a membrane by moving from an area of high concentration to an area of low concentration is called ___________.
   a. active transport
   b. diffusion
   c. endocytosis
   d. exocytosis

6. Why will a carrot slice when placed in tap water for several hours become very stiff?
   a. cellular materials were made filling the cell
   b. salt has left the cell due to diffusion
   c. water has entered the cell due to osmosis
   d. water has left the cell due to osmosis

7. If drops of blood are added to a 10% salt solution, what will happen to the blood cells?
   a. They will burst.
   b. They will shrink.
   c. They will swell but not burst.
   d. Nothing, they will remain the same.

8. Exocytosis and endocytosis are accomplished by the __________.
   a. endoplasmic reticulum
   b. Golgi complex
   c. nucleus
   d. plasma membrane

9. The ____________ is surrounded by a double membrane with pores and contain genetic material.
   a. endoplasmic reticulum
   b. mitochondria
   c. nucleus
   d. plasma membrane

10. Which of the following performs packaging and secretion functions?
   a. endoplasmic reticulum
   b. Golgi complex
   c. lysosomes
   d. mitochondria

11. A cell with a great deal of granular ER would most likely be heavily involved in ________.
   a. active transport
   b. cell division
   c. lipid synthesis
   d. protein synthesis

12. Bursting of which of the following would most likely kill a cell?
   a. endoplasmic reticulum
   b. lysosome
   c. mitochondria
   d. plastic

13. The pancreas is an important organ that secretes both digestive enzymes and hormones. What organelle would you expect to find in a higher concentration in the cells of this organ than in other cells?
    a. Golgi complex
    b. mitochondria
    c. plasma membrane
    d. smooth endoplasmic reticulum

14. Which of the following would be active in muscle contractions?
    a. intermediate filaments
    b. long filaments
    c. microfilaments
    d. mircotubules

15. Which of the following is not a product of cellular respiration?
    a. carbon dioxide
    b. energy
    c. oxygen
    d. water

16. When is most of the ATP produced during aerobic respiration?
    a. citric acid cycle
    b. electron transport chain
    c. glycolysis
    d. transition reaction

17. The three processes when taken together that are called cellular respiration are ____________.
    a. citric acid cycle, cyclic phosphorylation and fermentation
    b. glycolysis, fermentation and the Kreb's cycle
    c. glycolysis, citric acid cycle and electron transport chain
    d. lactic acid cycle, citric acid cycle and fermentation

18. The big advantage to getting oxygen to the mitochondria for cellular respiration is the ____________.
    a. carbon dioxide given off
    b. energy produced
    c. heat generated
    d. water released to the cells

19. When your exercise exceeds your aerobic conditioning, your muscles may ache due to the presence of ____________.
    a. ATP
    b. alcohol
    c. glucose
    d. lactic acid

20. Anaerobic respiration requires ___________.
    a. carbon dioxide
    b. energy
    c. glucose
    d. oxygen

SHORT ANSWER QUESTIONS

Read each of the following questions carefully. Jot down the main points you want to include in your answer. Then write a well-organized explanation. Use a table if it will save words and clearly make comparisons.

1. Describe how materials move in and out of cells, then explain why most cells are very small.

2. Draw and label the plasma membrane. Then write a short explanation as to why it is often called a fluid mosaic model. Which components of the membrane make it "fluid"?

3. Choose any two organelles and describe their function and any special structures that help accomplish that function.

4. Compare cellular respiration and fermentation in terms of reactants, products and energy output. When does the human body use each form of energy conversion?

ANSWER KEY

REVIEW QUESTIONS

A. The main component of the plasma membrane is phospholipid molecules. They are amphipathic with a water-loving end that is hydrophilic in nature and a water-hating end that is hydrophobic in nature. The two hydrophobic ends are in the center of the plasma membrane while the hydrophilic ends contact the extracellular fluid and the cytosol. The largest proteins found in the membrane are called integral proteins. Smaller proteins attached to the surface of the membrane are called peripheral proteins. Molecules of cholesterol are found throughout the membrane. Glycoprotein and glycolipid molecules are protein and lipid molecules with extra bits of carbohydrates attached and are located on the membrane surface.

B.

Mechanism	Uses Cellular Energy	Description
Simple diffusion	No	Movement of small molecules from a higher concentration to a lower concentration.
Facilitated diffusion	No	Movement of molecules from a higher concentration to a lower concentration with the aid of a carrier or channel protein.
Osmosis	No	Movement of water from a higher concentration of water to a lower concentration of water.
Active Transport	Yes	Movement of molecules against the concentration gradient resulting in a higher concentration.
Endocytosis	Yes	Plasma membrane surrounds molecules or small organisms forming a vesicle that enters the cell.
Exocytosis	Yes	Membrane-bound vesicle from inside the cell fuses with the plasma membrane releasing the contents of the vesicle.

C. A. Hypertonic: the bag loses more water than it gains, and shrivels.
B. Isotonic: the bag gains and loses the same amount of water, and maintains its shape.
C. Hypotonic: the bag gains more water than it loses, and swells.

D. 1. f
2. a
3. i
4. h
5. g
6. j
7. d
8. c
9. b
10. e

E. 1. microfilaments
2. intermediate filaments

3. microtubules
4. microtubules
5. microfilaments

F.

Process	Reactants	Net Products per Initial Glucose	
		carbon molecule	ATP Generated
Glycolysis	Glucose	2 pyruvate	2
Transition reaction	2 pyruvate	2 acetyl CoA 2 CO_2	0
Citric Acid Cycle	2 acetyl CoA	4 CO_2	2
Electron Transport Chain	2 acetyl CoA	No change	32

G.

$$C_6H_{12}O_6 + \boxed{6O_2} \longrightarrow 6CO_2 + \boxed{H_2O} + ATP$$

glucose oxygen carbon dioxide water energy

H. Both types of fermentation occur in the absence of oxygen. Glucose enters glycolysis where it is broken down into pyruvate. Then in the case of lactic acid fermentation two lactate molecules form. In the case of alcohol fermentation, two ethanol molecules form plus two molecules of CO_2. The only energy generated is two ATP from glycolysis.

Critical Thinking

1. The plasma membrane is composed of lipids, proteins and carbohydrates. Phospholipids found in the membrane have hydrophilic heads and hydrophobic tails that form a double layer. The structure of the membrane as a double layer of phospholipids keeps the membrane from dissolving in water while regulating the materials that pass through it. It allows it to be semipermeable. Proteins are interspersed in the membrane as are cholesterol molecules giving the membrane flexibility. Some proteins assist with cell communication. Surface glycoproteins provide cell recognition.
2. Any variety of models is appropriate just so long as the function of the organelle is maintained in the analogous part. A computer, automobile, airport, city, country, university, classroom or business can be used as a model. Include a control center (nucleus), a manufacturing site (ribosomes), communication mechanism (smooth ER), power plant (mitochondria), some enclosure (membrane), and maybe a packaging facility (Golgi apparatus).
3. Cardiovascular conditioning improves the body's ability to get oxygen to the cells. Oxygen is required for aerobic respiration and aerobic respiration results in much more ATP than anaerobic respiration.
4. This experiment demonstrates that water can cross cell membranes. The potato is made of cells. The distilled water is a hypotonic solution and water will enter the cells causing them to swell. The salt solution is hypertonic and water will leave the cells, causing them to shrink. Once the limp potato slices have been rinsed and put into clean water, the cells will become rigid as they swell with water.

Practice Test

1. c	11. d
2. c	12. b
3. d	13. a
4. c	14. c
5. b	15. c
6. c	16. b
7. b	17. c
8. d	18. b
9. c	19. d
10. b	20. c

Short Answer Questions

1. Molecules move in and out of cells by 1) simple diffusion in which molecules randomly move from a region of high concentration to low concentration, 2) facilitated diffusion in which molecules move from a region of higher concentration to a region of lower concentration with the aid of membrane proteins, 3) osmosis, a special case of diffusion, in which water is the only substance moving across the membrane, and 4) active transport in which molecules move across the membrane with the aid of a carrier protein and energy. Cells engulf water by pinocytosis and food molecules by phagocytosis, two means of endocytosis. Large molecules encapsulated in a vesicle eave cells by exocytosis when the membrane of the vesicle fuses with the cell membrane and spills its contents out of the cell. Most cells are small because materials passing in and out of the cell must be able to reach the center using only the means described here.
2. Any drawing similar to that in Figure 3-3 is sufficient. The drawing should show the double phospholipid layer, imbedded protein and cholesterol molecules, and glycoprotein and carbohydrate molecules. The protein and phospholipid molecules are flexible or "fluid."
3. Any two organelles chosen from Table 3-4 are good. The explanation of function needs to be in more detail than that found in Table 3-4 using additional information from the text.
4. Cellular respiration is summarized in Table 3-6 and Figure 3-31. The reactants for cellular respiration are glucose and oxygen. The products are carbon dioxide and water. In addition about 36 ATP are produced. Fermentation is summarized in Figure 3-32. The reactant is glucose with no oxygen. In plants the products are a two-carbon ethanol and carbon dioxide, while in animals it is a three-carbon lactic acid. Either type of fermentation yields only two ATP per glucose molecule. Humans use aerobic respiration during sustained activity and anaerobic respiration resulting in lactic acid when oxygen is scarce or for short bursts of strenuous activity.

Chapter 4

Body Organization and Homeostasis

Objectives

After reading the text and studying the material in this chapter, you should be able to:

- Define a tissue, organ and organ system and give examples of each.
- Describe epithelial tissue and compare the different types in terms of structure and function.
- Explain how the structure of the different types of connective tissue relates to its function.
- Compare the different types of muscle tissue in terms of structure, location and control.
- Describe the function of nerve tissue, the structure of a neuron and role of neuroglia.
- Compare the structure of the three types of junctions between cells.
- Identify the major body cavities and the organs and systems they house.
- List the four types of membranes, their location and function.
- Relate the structure of the skin to its ability to carry out its various functions.
- Explain the factors that affect skin color and the causes of albinism and vitiligo.
- Describe the structure and function of nails and hair.
- Compare the types of glands found in the skin, their products and location.
- Give examples of how the integumentary system functions with other organ systems.
- Define homeostasis and explain its importance to life.
- Describe and give an example of a negative feedback system.

Chapter Summary

The organization of the human body increases in complexity from cells to organ systems. Multicelled organisms require specialized cells to perform specific tasks. These cells then organize into tissues, organs and systems.

A **tissue** is a group of cells that work together to accomplish a common function. There are four primary tissue types. **Epithelial tissue** covers the body surfaces, lines cavities and organs, and forms glands. **Connective tissue** serves as a storage site for fat, participates in our immunity, and provides support and protection for our organs. **Muscle tissue** is responsible for movement and **nervous tissue** coordinates body activities.

Epithelial tissue is composed of tightly packed cells that form a continuous layer. **Membranous epithelium** forms the outer layer of the skin, covers some internal organs, and lines the blood vessels and several systems. **Glandular epithelium** is specialized to secrete a product. If the glands secrete the product directly onto the body surfaces through ducts, they are called **exocrine glands**; if their products are secreted into spaces outside the cells, not passing through ducts, they are called **endocrine glands**.

Epithelium can also be classified according to the shape of the cells and the number of layers. **Squamous epithelium** is made of flattened, or scale-like, cells. This shape allows for diffusion of materials, a slick surface to reduce friction and protection for internal organs. **Cuboidal epithelium** not surprisingly is made of cube-shaped cells that are specialized for secretion and absorption. **Columnar epithelium** is tall and rectangular and also is specialized for secretion and absorption, lining the small intestine. In addition, each of these shapes may occur in a single layer called **simple epithelium** or stacked in several layers called **stratified epithelium**.

Connective tissue is the most abundant and widely distributed tissue in the body. Its cells are contained in an extracellular **matrix** of **collagen**, **elastic**, and **reticular fibers.** It is the characteristics of this matrix that determine the nature of connective tissue. **Loose connective** (**areolar** and **adipose tissue**) tissue contains many cells and fewer, loosely woven fibers and is therefore used to cushion organs and to provide insulation. **Dense connective tissue** is made of tightly woven fibers and is found in ligaments, tendons and the dermis. The protein fibers in loose and dense connective tissue are produced by fibroblasts. Bone, blood and cartilage are considered to be specialized connective tissue. **Cartilage** is tough but flexible serving as a cushion between bones. **Hyaline**, **elastic** and **fibrocartilage** are three types of cartilage differing in flexibility and location. **Bone** protects and supports internal structures, and along with muscles facilitates movement, stores lipids, calcium, and phosphorus, and produces blood cells. Bone tissue is classified as compact or spongy depending upon its underlying structure. The **formed elements** of **blood (white cells, red cells and platelets)** are suspended in a liquid matrix called **plasma**. **Skeletal**, **cardiac** and **smooth muscle** tissue vary in their structure, location and control mechanism. Nerve tissue makes up the brain, spinal cord, and nerves. **Neurons** are nerve cells that convert stimuli into nerve impulses that they conduct to other neurons, muscle cells or glands. Neurons are supported, insulated, and protected by **neuroglia**.

The cells that make up tissues are held together by three types of junctions: **tight junctions** forming a leak-proof seal, **adhesion junctions** resembling a riveted joint and **gap junctions** that leave small spaces between the cells.

An **organ** is a group of tissues that work together to perform a specific function. In turn organs work together to form an **organ system**. The body is divided into four cavities that contain various groups of organs: **thoracic**, **abdominal**, **cranial** and **spinal cavity**. Body cavities and the surfaces of organs are covered with membranes that protect the underlying tissues. **Mucous membranes** line tissues that open to the exterior of the body, **serous membranes** line the thoracic and abdominal cavities and the organs within them, **synovial membranes** line the cavities of joints, and the **cutaneous membrane** forms the skin covering the outside of the body.

The **integumentary system** is composed of the skin and its derivatives. It provides physical and chemical protection, prevents water loss, regulates body temperature, synthesizes vitamin D, receives stimuli and is an organ of excretion. The skin is made of two layers, the **epidermis**, a thin outer layer, and the **dermis**, a thicker inner layer containing nerves, blood vessels and glands. The **hypodermis** is a layer of loose connective tissue beneath the skin connecting it to other tissues.

Skin color is determined by blood flow and the distribution and quantity of **melanocytes**, the cells that produce the pigment called **melanin**. A person's genetic makeup determines the type of melanin that is produced or whether no melanin is produced at all as in the case of **albinism**. **Vitiligo** is a condition where patches of white skin appear where the melanocytes have disappeared from the skin. Skin color is also influenced by the amount of **carotene** in the skin and the **hemoglobin** that is carried in the blood vessels.

Many diverse structures are derived from skin. **Hair** is dead skin cells filled with keratin and formed into a column. Its primary function is protection. The hair shaft extends above the scalp and the root extends into the dermis or hypodermis where it is imbedded in a follicle. **Nails** are also dead cells designed to protect the tips of our toes and fingers. **Oil glands** are found all over the body except the palms of the hands and soles of the feet. **Sweat glands** produce sweat which helps in the regulation of body temperature. **Mammary glands** and **wax glands** are modified sweat glands.

Homeostasis is the relatively constant internal condition required for life. A **negative feedback mechanism** receives information from a **receptor** that detects a change in the internal or external environment. Then a **control center** such as the brain sends out a response and the **effector** carries out the response returning the system to homeostasis again. The thermostat for the body is located in the hypothalamus. **Hyperthermia**, abnormally elevated body temperature, and **hypothermia**, abnormally low body temperature, are both life-threatening conditions that result when this mechanism fails.

Key Concepts

- Tissues are groups of cells that work together to perform a common function.
- The four main tissue types are: epithelial, connective, muscle, and nervous. Epithelial tissue can be classified by the shape of the cell or the number of cell layers.
- Membranous epithelium covers the organs, and lines the blood vessels and several organs. Glandular epithelium forms the products of glands.
- Most connective tissue binds and supports the other tissues of the body and is characterized by its structure. Specialized connective tissues include blood, bone and cartilage.
- Muscle tissue is composed of cells that contract. The three types of muscle tissue (skeletal, cardiac, and smooth) are characterized by location, structure and control.
- Nervous tissue makes up the brain, spinal cord and nerves. It consists of nerve cells, called neurons, and neuroglia that support, insulate and protect neurons.
- Tight, adhesion and gap junctions hold cells together to form tissues.
- Organs are structures composed of two or more different tissues that work together to perform a specific function. Organs work together to form organ systems.
- Organs are housed in body cavities. The thoracic cavity houses the lungs and heart. The abdominal cavity contains the digestive, urinary and reproductive systems. The cranial cavity encloses the brain and the spinal cavity holds the spinal cord.
- Body cavities and surfaces or organs are covered with membranes. The four types of membranes are mucous, serous, synovial and cutaneous.
- The skin forms our outer covering called the integumentary system. It consists of the epidermis and the dermis.

- Skin color is determined by the quantity and distribution of melanocytes, the cells producing the primary pigment melanin, the amount of carotene in the skin and the blood vessels that come near the surface.
- Hair, nails and glands are derivatives of the skin. Hair and nails are both dead cells hardened by keratin that function in protection. Oil glands, sweat glands and wax glands all secrete materials.
- The skin functions closely with other organ systems.
- Homeostasis is the relative internal consistency that is required for life. It is maintained through a negative feedback mechanism similar to the thermostat in a house.
- Regulation of core body temperature is one example of a negative feedback mechanism.

Study Tips

This chapter is the bridge between the chemical and cellular information of the preceding chapters and the information on systems in the following chapters. A solid knowledge of organelles, cells, and tissues will greatly enhance your understanding of organs and organ systems.

Two reoccurring themes are seen in this chapter. One is the relationship between structure and function. The structure of each of the tissues presented in this chapter enables it to carry out its function. For example the fact that blood is a liquid makes circulation throughout the body much easier. The columnar structure of bones provides them with the strength to support hundreds of pounds. Form the habit of studying both the structure and functions of tissues, organs and systems. Whenever possible, ask the question, "How does this structure relate to the function?"

A second theme that is foundational to human biology is that of homeostasis. The maintenance of a relatively constant internal environment is critical to survival. Feedback loops seen in the controlling mechanisms described in this chapter will reappear. Be sure that you understand the importance of homeostasis and the role of a negative feedback system.

Finally, begin to focus on the interconnectedness found throughout human biology. With each new topic, make connections to other topics. The concepts in this course will continue to build upon one another. Learn each one as you go and develop a flow chart to illustrate how each concept is applied. Ask questions and get clarification as you study the material. To avoid problems later in the semester, understand and be able to apply the material as you learn it.

Review Questions

A. Define the term tissue:

____

____

B. List the four primary tissues discussed in this chapter.

1. ____________________ 2. ____________________

3. ____________________ 4. ____________________

C. List where each of the following tissues might be found.

1. Simple squamous: ____

2. Simple cuboidal: ____

3. Simple columnar: ____

4. Stratified squamous: ____

5. Stratified cuboidal: ____

6. Stratified columnar: ____

7. Structurally what does each type of epithelial tissue have in common? ____________

____

D. Fill in the blank with information concerning connective tissue.

1. Areolar and adipose tissues are ____________ connective tissues.

2. The primary function of loose connective tissue is to ____________.

3. Dense connective tissue is found in ____________, ____________ and the ____________.

4. Compared to loose connective tissue, dense connective tissue contains many more ____________, which are produced by ____________.

5. The three types of specialized connective tissues are ____________, ____________ and ____________.

6. The three types of cartilage in our bodies are structurally very different. The most flexible cartilage is ____________ cartilage found in the outer ear.

7. ____________ forms the outer part of the shock-absorbing disks between the vertebrae of the spine.

8. Bone is strengthened by ____________ and by collagen fibers.

9. ____________ bone is particularly strong because the osteocytes are organized in concentric rings forming a columnar structure.

10. Blood is a liquid, connective tissue made of ____________ and ____________ including red blood cells, white blood cells and platelets.

E. For each muscle type listed below, give the location, type of control and draw a picture of the fibers.

Muscle Type	**Skeletal Muscle**	**Smooth Muscle**	**Cardiac Muscle**
Location			
Control			
Drawing			

F. Number the following types of cell-to-cell junctions so that #1 is the tightest junction and #3 is the loosest. Give the function of each type of junction.

_____ Gap Junction Function: ____

____

_____ Tight Junction Function: ____

____

_____ Adhesion Junction Function: ____

____

G. Match the type of membrane listed below with its location or function. Each membrane type may be used twice.

A. Mucous B. Serous C. Synovial D. Cutaneous

1. __________ lines thoracic and abdominal cavities

2. __________ thick, waterproof and dry

3. __________ lines the cavities such as the knee and finger

4. __________ lines the windpipe and other airways

5. __________ covers the outside of the body

6. __________ lines the lungs and heart

7. __________ secretes mucous into the linings of the respiratory system

8. __________ secretes fluid to lubricate the joints

H. Label the following drawing of the skin and underlying hypodermis using these terms: artery, dermis, epidermis, hair follicle, hair root, hair shaft, hypodermis, oil gland, sweat gland, and vein.

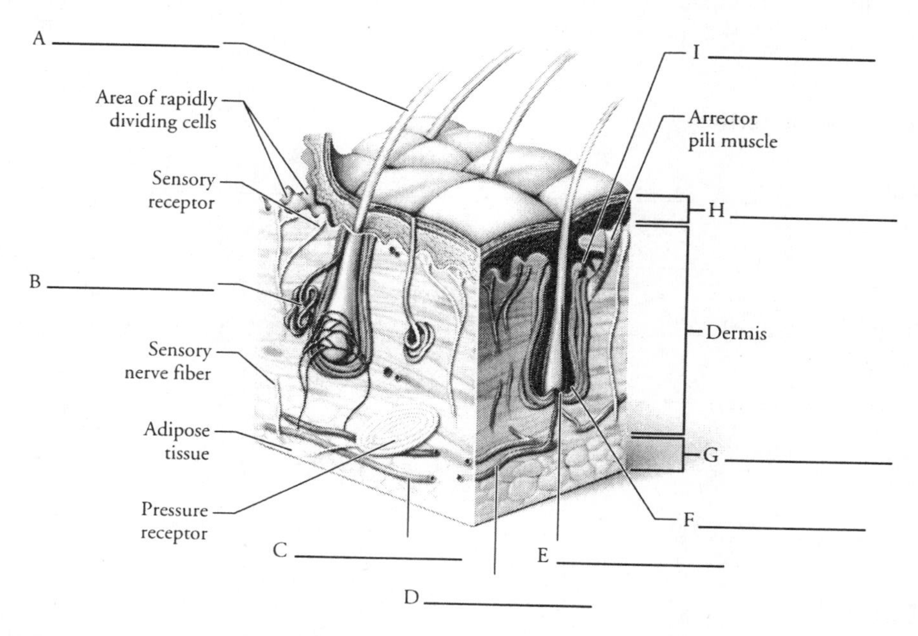

I. Fill in the blanks in the following paragraph describing skin color.

The pigment, __________ is produced by __________. It colors the entire epidermis. All people have about the __________ number of melanocytes. The color of the skin is determined by the distribution and __________ of melanin produced. An orange-yellow pigment of the skin is __________. The pigment found in blood, __________ along with the amount of circulation to the skin can also affect skin color. People with the inherited condition, __________ have melanocytes incapable of producing melanin. The melanocytes disappear from patches of the skin in people with __________. Exposure to UV light can result in the build-up of melanin we call __________.

J. Complete the following table with the missing information about the glands found in skin.

Gland	Location	Secretion
______ glands	All over the body except palms of hands and soles of feet.	
______ glands	All over the body including the forehead, palms, soles of feet, armpits, pubic area and pigmented region around nipples.	
______ glands		Wax
______ glands	Breasts	Milk

K. Define the following terms.

Homeostasis: ____

____

Hyperthermia: ____

____

Hypothermia: ____

____

CRITICAL THINKING

Read each of the following questions carefully. If it helps you, underline or list each component to the question. Use complete sentences in your explanation.

1. Bioengineering is an exciting field of research and creativity with great potential. Researchers have been able to grow tissues in the laboratory. They are especially successful in growing epithethial tissues. Scientists look forward to developing organs made of human cells. Why is it easier to grow a group of cells than a tissue, and a tissue than an organ? Use your knowledge of the levels of biological organization to answer these questions.

2. Think about the functions of the epidermis, specifically the skin, and of muscle tissue. Compare the location of these tissues and how they are controlled. Explain the statement that your face is merely the expression of dead cells. Think about what the text said about Botox and how it paralyzes the muscles. People who are opposed to Botox injections argue that facial exercises that tone facial muscles will improve sagging features and will avoid injections of a poison. What is the relationship between the skin and the underlying muscle structure? What are the positive aspects to this idea?

3. Throughout your study of biology you will constantly be reminded of the relationship between structure and function. The tissues studied in this chapter have characteristic structures, for example bones have columns and epithelial tissue easily forms sheets. Make a list of at least four different tissues discussed in this chapter and their primary function. Next describe the structural characteristics that make that tissue especially well suited to its function.

4. Form a small group to review the concept of homeostasis. Refer to Figure 4-12 in your text and explain how the body regulates temperature. Include the input, the processing, and the output of the information. Next, make an analogy to your home heating system. It is the thermostat that interacts with the fire in the furnace to keep your home at the designated temperature. Explain that system using Figure 4-12.

Practice Test

Choose the one best answer to each question that follows. As you work through these items, explain to yourself why each answer you discard is incorrect and why the answer you choose is correct.

1. A group of cells that work together for a common function.
   a. biosphere
   b. organ
   c. system
   d. tissue

2. Which of the following is NOT a tissue?
   a. connective
   b. epithelial
   c. muscle
   d. respiratory

3. Sometimes swabs are taken of the inside of the mouth for DNA sampling. What type of cell is being collected?
   a. columnar epithelium
   b. compound epithelium
   c. cuboidal epithelium
   d. squamous epithelium

4. Which of the following tissues contains an extracellular matrix?
   a. connective
   b. epithelial
   c. nerve
   d. muscle

5. This tissue contains cells specialized for the storage of fat.
   a. areolar
   b. adipose
   c. cartilage
   d. dense connective tissue

6. Which statement is true concerning bone?
   a. It is alive and growing.
   b. It is dead tissue reinforced by keratin.
   c. Its only function is for support.
   d. It stores vitamin D.

7. Blood is unique as a tissue because it is ____________?
   a. composed of protein
   b. liquid
   c. made of cells
   d. part of the circulatory system

8. Looking through a microscope you saw cells that were striated (striped) and that were interconnected. These cells were taken from the ____________.
   a. calf
   b. heart
   c. stomach
   d. tendon

9. Which tissue responds to stimuli and transmits an electrical impulse?
   a. cardiac
   b. muscle
   c. nerve
   d. squamous

10. The cells that make up organs that must contain fluids, such as the stomach or bladder, are held together by ____________ junctions.
    a. adhesion
    b. gap
    c. interconnecting
    d. tight

11. Where would a kidney be found?
    a. abdominal cavity
    b. cranial cavity
    c. spinal cavity
    d. thoracic cavity

12. The membrane found lining the shoulder joint is the ____________ membrane.
    a. cutaneous
    b. joint
    c. lubricating
    d. synovial

13. The skin serves only to protect us. True or false.
    a. True
    b. False

14. Someone who lost a large amount of skin due to injury or fire would be at risk of which of the following?
    a. dehydration
    b. infection
    c. overheating
    d. all of the above

15. Name the tissue that has no direct circulation and therefore is very difficult to heal.
    a. bone
    b. cartilage
    c. connective
    d. epithelium

16. What is the primary pigment in the skin?
    a. carotene
    b. hemoglobin
    c. melanocytes
    d. melanin

17. What causes the change in skin color when someone blushes?
    a. capillaries of the skin dilate
    b. melanin molecules rush to the surface
    c. melanocytes are undergoing cell division
    d. red is the color of shame

18. Which type of gland is most linked to acne?
    a. adipose
    b oil
    c. sebum
    d. sweat

19. When homeostasis is lost, a person ____________.
    a. blushes
    b. dies
    c. shivers
    d. sweats

20. In a negative feedback mechanism ____________.
    a. body temperature is maintained at exactly 98.6°F.
    b. hypothermia is never a possibility.
    c. systems are turned off when the temperature drops.
    d. the absence of something triggers the system.

Short Answer Questions

Read each of the following questions carefully. Jot down the main points you want to include in your answer. Then write a well-organized explanation. Use a table if it will save words and clearly make comparisons.

1. The relative speed with which people heal after an injury or surgery depends in part on the structure of the tissues. This is especially true of the connective tissues. Why does it take so long to heal an injury to a tendon or ligament? Why does an injury to a bone heal faster than an injury to cartilage? Explain your answer in terms of the structure of these tissues.

2. Blood is the most unusual tissue that you studied in this chapter. What are the characteristics that make blood a tissue? How is it similar to other tissues and how is it different? Are there other tissues that have unusual qualities?

3. Recall the factors that affect the color of someone's skin. A primary factor that gives skin color is melanin. Even albinos have some coloring. They often have a pinkish coloring to their skin, but must stay out of the sun. What causes albinism? Why would someone with no melanin in his or her skin have a pinkish color? Why must they stay out of the sun?

4. List the body cavities and the organs or systems contained in them. What is the function of the body cavities? That is, why not have just one large area for all of the organs?

Answer Key

Review Questions

A. A tissue is a group of cells that work together to accomplish a common function.

B. 1. Epithelial 2. Connective
   3. Muscle 4. Nervous

C. 1. Simple squamous: air sacs of lungs, blood vessel lining
   2. Simple cuboidal: linings of kidney tubules and glands
   3. Simple columnar: lining of gut and respiratory tract
   4. Stratified squamous: surface of skin, lining of mouth and vagina
   5. Stratified cuboidal: ducts of mammary glands, sweat glands, and salivary glands
   6. Stratified columnar: urethra
   7. Each type of epithelial tissue has a basement membrane and connective tissue.

D. 1. loose
   2. bind the skin to the underlying tissues and organs
   3. tendons, ligaments, and dermis
   4. fibers, fibroblasts
   5. cartilage, bone and blood
   6. elastic
   7. fibrocartilage
   8. calcium
   9. compact
   10. plasma, formed elements

E.

Muscle Type	Skeletal Muscle	Smooth Muscle	Cardiac Muscle
Location	attached to skeletal system	blood vessels, airways, organs	heart
Control	voluntary control	involuntary	involuntary
Drawing should be similar to Figure 4-3.			

F. #3 Gap Junction — Function: link adjoining cells by small holes allowing for communication as in cardiac muscle.

   #1 Tight Junction — Function: form a leak-proof seal as needed in the urinary system.
   #2 Adhesion Junction — Function: held together by filaments allowing for some movement as needed in cells that slide along one another such as the muscles.

G. 1. b
   2. d
   3. c
   4. a
   5. d
   6. b
   7. a
   8. c

H. A. hair shaft
   B. sweat gland
   C. vein
   D. artery

E. hair root
F. hair follicle
G. hypodermis
H. epidermis
I. oil gland

I. The pigment melanin is produced by melanocytes. It colors the entire epidermis. All people have about the same number of melanocytes. The color of the skin is determined by the distribution and amount of melanin produced. An orange-yellow pigment of the skin is carotene. The pigment found in blood, hemoglobin, along with the amount of circulation to the skin can also affect skin color. People with the inherited condition, albinism, have melanocytes incapable of producing melanin. The melanocytes disappear from patches of the skin in people with vitiligo. Exposure to UV light can result in the build-up of melanin we call tanning.

J.

Gland	Location	Secretion
Oil glands	All over the body except palms of hands and soles of feet.	Sebum
Sweat glands	All over the body including the forehead, palms, soles of feet, armpits, pubic area and pigmented region around nipples.	Sweat and also a mix of sweat and fatty substances and proteins
Wax glands	External ear canal	Wax
Mammary glands	Breasts	Milk

K. Homeostasis: relatively constant internal environment.
Hyperthermia: abnormally elevated body temperature.
Hypothermia: abnormally low body temperature.

Critical Thinking

1. Tissues are groups of cells that work together to perform a common function. There are only four types of tissues. Organs, on the other hand, are composed of two or more different tissues with a specialized function. Tissues are not as complex as organs and can be grown as a sheet of cells.
2. The outermost layer of the epidermis is the stratum corneum, 25 to 30 rows of flat, dead cells. So, what people see when they look at you are dead cells! Skin functions as a covering that protects the underlying tissues. Muscle tissue is responsible for movement. It is attached to the dermis by connective tissue. Some wrinkles are caused by the contraction of muscle or the weakening of the underlying muscle. Some of these wrinkles could be reduced through exercises that strengthen the facial muscles. This process although not as effective as Botox injections is not as invasive.
3.

Tissue	Primary Function	Structural Characteristics
Simple Epithelium	Provides lining, functions in diffusion, secretion, absorption or filtering.	Loose structure allows for diffusion and the passage of other substances.
Stratified Epithelium	Provides covering, protection.	Tight, cohesive sheets held together by tight junctions. Keratinized cells waterproof and protect underlying cells.

Loose Connective	Support for epithelium	Contains fibroblasts for secretion and elastic and collagenous fibers for structure.
Bone	Support, linkages for muscles, protection	Compact bone arranged in columns hardened with calcium
Cardiac Muscle	Contractions of the heart	Cells are basically striated muscle cells with specialized junctions to hold them together and allow them to contract as a unit.

4. Homeostasis is maintained through a negative feedback mechanism in which the substance that is produced goes back into the system to shut down production. In our body, too much heat stimulates the production of sweat that results in cooling. Once cooled, we no longer need to sweat. In a house, the thermostat is set at a specific temperature. When it senses that the temperature has fallen, it starts the furnace. The thermostat senses the rise in temperature and shuts down the furnace. Thus an increase in the product, heat, causes the system to shut down.

Practice Test

1. d
2. d
3. d
4. a
5. b
6. a
7. b
8. b
9. c
10. d
11. a
12. d
13. b
14. d
15. b
16. d
17. a
18. b
19. b
20. d

Short Answer Questions

1. Tendons and ligaments are made of tightly woven fibers. It does not have an extensive circulatory system and growth and repair is slow. Bone, however, is an active, ever-changing tissue with a good blood supply that promotes prompt healing.
2. All tissues are groups of cells that serve a common function. Blood is made of several types of cells suspended in plasma. The most unusual quality about blood is that it is a liquid. Nearly all the tissues have unique characteristics – muscles contract, nerves are long and conduct impulses.
3. Skin color is the result of the distribution of pigment, melanin, and blood flow. Although albinos cannot produce melanin, they do have skin circulation that gives them a pinkish coloring. Melanin absorbs the harmful UV rays of the sun before it reaches the underlying tissues. Albinos do not have the melanin in their skin to protect them from the sun and will repeatedly burn.
4. Internal organs are located in two main body cavities. The posterior cavity is subdivided into the cranial cavity, in which the brain is located, and the spinal cavity, in which the spinal cord is located. The anterior cavity is subdivided into the thoracic (chest) cavity and the abdominal cavity. The thoracic cavity is further divided into the pleural cavities, which contain the lungs, and the pericardial cavity, which contains the heart. These cavities have two important functions: 1) they help protect the vital organs from damage due to the bumps and thumps that occur as we walk and jump and 2) they allow organs to slide past one another and change shape.

Chapter 5

The Skeletal System

Objectives

After reading the text and studying the material in this chapter, you should be able to:

- List the functions of bone.
- Compare the structure of compact and spongy bone.
- Label a diagram of bone structure.
- Describe the development of bone beginning with the formation of cartilage in the embryo.
- Explain the process of bone growth and what controls it.
- Describe how bones heal after a fracture or break.
- Compare the function of osteocytes, osteoblasts and osteoclasts.
- Explain what is meant by the continual remodeling of bone.
- List the components of the axial skeleton and the appendicular skeleton.
- Describe the structure of the vertebral column as a series of vertebrae separated by cartilaginous disks.
- List the components of the pectoral and pelvic girdles.
- Compare the three types of joints in terms of structure and motion.
- Explain the difference between a ligament and a tendon.
- Describe the structure of a synovial joint.
- Explain how a bone fracture or break is healed.
- Describe the cause of the following conditions: sprain, bursitis, osteoarthritis and rheumatoid arthritis.

Chapter Summary

The **skeleton** is a framework of bones and cartilage that functions in movement and the protection of internal organs. The skeleton provides **support** for soft tissues, gives a place of attachment for muscles, **protects** internal organs, **stores** minerals and fat, and **produces blood cells** in the marrow of certain bones.

Bones have a hard outer layer of **compact bone** surrounding **spongy bone**. The compact bone is covered by a periosteum containing blood vessels, nerves, and cells involved in bone growth and repair. Spongy bone is formed from the thin layers of bone with open spaces inside. It is found in small, flat bones, and in the head and near the ends of the shafts of long bones. In adults, the spaces of some spongy bones are filled with **red marrow**, which generates red blood cells, and the shaft is filled with **yellow marrow**, a fatty tissue for energy storage.

The structural unit of compact bone is called an **osteon**. Living bone cells, **osteocytes** are found in small spaces within the hard matrix. Woven throughout the matrix are strands of elastic protein collagen for strength. Bones are hardened with calcium and phosphorus salts.

During development bone is first formed of cartilage. Long bones begin when **osteoblasts** form a collar of bone around the shaft of cartilage. Then cartilage within the shaft begins to break down forming the marrow cavity. Osteoblasts migrate to this cavity to form spongy bone. After birth, two regions of cartilage remain at each end of the long bone. One is the cap that covers the surfaces that rub against other bones and the second is a plate of cartilage, called the epiphyseal or growth plate. Cartilage cells within the epiphyseal plate divide, forcing the bone away from the shaft. This new cartilage is replaced by bone. Bone growth is stimulated by **growth hormone (GH)**. Thyroid hormones ensure that the skeleton grows with the proper proportions.

Bone fractures are healed by fibroblasts and osteoblasts. When a break occurs, there is bleeding followed by a clot. Fibroblasts secrete collagen fibers that form a **callus** linking the two parts of the bone. This cartilage is later replaced by bone.

Bones continually go through remodeling where new bone is deposited by osteoblasts and old bone is broken down by osteoblasts. The body reabsorbs these minerals. The rate and extent of bone remodeling is in response to the stress felt by the bone. That is why weight-bearing exercise leads to stronger bones and weightlessness can quickly reduce the mass of bones. If bone is broken down faster than it is built, a condition called **osteoporosis** results.

The bones of the human body are arranged into two groups: the first is the **axial skeleton**, including the skull, the vertebral column and the bones of the chest. Those bones protect and support the inner organs. The bones of the vertebral column, called **vertebrae**, are cushioned with **intervertebral disks**. The second is the **appendicular skeleton** including the pectoral girdle, the pelvic girdle and the limbs, allows you to move and interact with the environment. The pectoral girdle connects the arms to the rib cage and the pelvic girdle connects the legs to the vertebral column. The femur is the largest and strongest bone in the body. The structure of the wrists and hands parallel that of the feet and ankles.

Joints are the places where bones meet. They are classified according to their structure or the degree of movement they permit. **Fibrous joints** are held together by fibrous connective tissue. They have no joint cavity and do not permit movement as in the case of the skull. **Cartilaginous joints** allow very little movement as in the case of the vertebral column, the joint between the rib and sternum and the joint holding the two halves of the pubic bone together.

Most joints in the body are freely movable, **synovial joints**. The surfaces of these joints move past one another on a thin layer of hyaline cartilage. They are surrounded by a thin capsule containing synovial fluid, a lubricant. The entire joint is reinforced with **ligaments** that hold bones together and direct movement. **Hinge joints** and **ball-and-socket joints** are so named for the movement they allow. **Sprains** are injuries to ligaments. Bursitis is an inflammation of the **bursae**, the sacks that surround and cushion joints. Arthritis is joint inflammation. **Osteoarthritis** is a degeneration of the surfaces of a joint over time while **rheumatoid arthritis** is an autoimmune condition marked by an inflammation of the synovial membrane.

Key Concepts

- The skeleton is a framework of bone and cartilage that supports and protects the internal organs, allows for movement, stores minerals and fat, and produces red blood cells.
- Compact bone is dense bone found on the outside of all bones and covered by the periosteum.
- Spongy bone is an open latticework of bone found in flat bones and near the ends of long bones.
- The shaft of long bones is filled with a fatty yellow marrow. The spongy bone of some long bones in adults is filled with red marrow.
- The structural unit of the bone is called an osteon. It consists of a central canal with bone cells, called osteocytes, arranged in concentric circles. Bone matrix is hardened by calcium salts and strengthened by fibers of collagen.
- The embryonic skeleton forms as cartilage that is gradually replaced by bone during a process called endochondrial ossification. Osteoplasts form a bony collar around the shaft of the bone. Then cartilage within the shaft breaks down and is replaced by bone. Epiphyseal plates of cartilage remain and allow for bone growth of the long bones.
- Bone growth is stimulated by growth hormone from the anterior pituitary gland.
- Repair of bones begins with a blood clot at the site of the fracture. A cartilaginous callus links the broken ends of the bone. This cartilage is gradually replaced by bone.
- Bone is constantly being remodeled. Osteoblasts deposit new bone and osteoclasts break down existing bone. Weight-bearing stress fosters strong bones and forestalls osteoporosis.
- Joints are places where bones meet. They are classified according to the movement they permit.
- Bones are connected to other bones by ligaments and to muscles by tendons.
- Fibrous joints usually allow no movement and cartilaginous joints allow very limited movement.
- Synovial joints are the most common and allow for the most movement. They have cartilage on the adjoining surfaces and a lubricant. The bones are held together by ligaments.
- Arthritis is an inflammation of a joint. Osteoarthritis occurs when the surface of a joint degenerates. Rheumatoid arthritis is an autoimmune disease causing irritation of the synovial membrane.

Study Tips

As you look through this chapter, you will immediately notice that there is a lot of information about the skeletal system. First note the kinds of information presented. Some of it is explanatory of structure such as the structure of bone tissue or of the synovial joint. Some of it is explanatory of processes such as the formation of bone from cartilage and the process of remodeling. And some of it presents the names of the bones. It is important that you clarify with your instructor what is the most important information for you to know and understand and whether you are expected to memorize the names of all or some of the bones.

Develop a series of note cards. First draw and label a long bone to show its structure including each cell type. Reduce and copy the figures from the text that contain the bones you must know and glue them to note cards so that you can repeatedly study the names of the bones throughout the day. Do this for the different types of joints including example locations. Concentrate on the figure as you study the note cards. Sometimes students are thinking about other things as they turn their cards and later wonder why there was no increased learning. Think about what the terms mean – for example the axial skeleton is found on the main axis of the body. Some students find it helpful to develop mnemonics, jingles or raps to help remember terms that must be memorized.

This chapter will require an understanding of structure and processes. It may also require a great deal of memory work if you must know the names of the bones. It is critical that you ask your instructor what you will need to know.

Review Questions

A. List the functions of the skeletal system.

1. ____

2. ____

3. ____

4. ____

5. ____

6. ____

B. Use the following terms to label the figure below: blood vessel, central cavity, compact bone, osteon, periosteum, shaft, spongy bone, and yellow bone marrow.

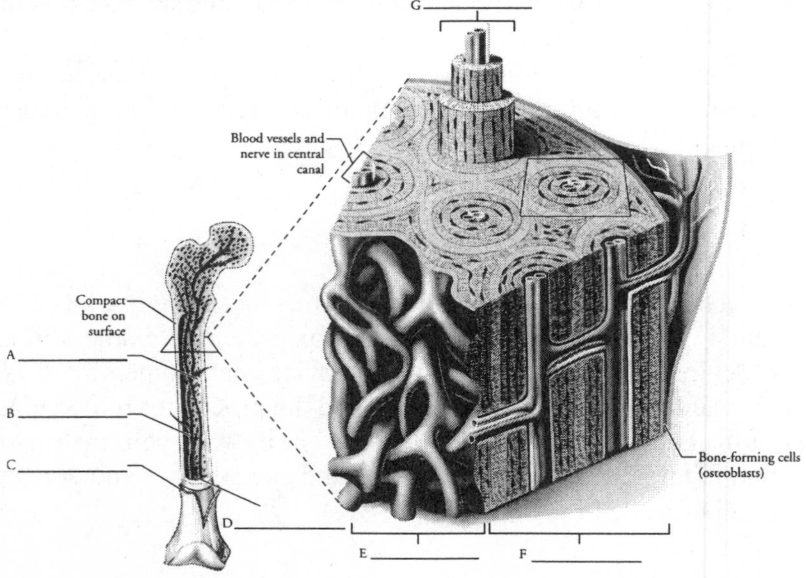

C. Match the term in Column A with the function or specialized structure in Column B.

Column A	Column B
1. compact bone	a. site of fat storage
2. epiphysis	b. bone-forming cells
3. osteoblasts	c. the membrane covering bone that nourishes it
4. osteoclast	d. latticework of thin layers of bone with open areas between
5. osteocyte	e. dense bone with few internal spaces
6. osteons	f. place where blood cells form
7. periosteum	g. bone-dissolving cell
8. red marrow	h. mature bone cell
9. spongy bone	i. enlarged end of a long bone
10. yellow marrow	j. structural unit of compact bone

D. Complete the following paragraph describing the formation of new bone.

During embryonic development, most of the skeleton is formed as ____________. Because cartilage cells are able to divide and form new cartilage cells, this model allows for ____________. Early in development, the cartilage begins to be replaced by ____________. First bone-forming cells, ____________, form a collar of bone around the shaft of the cartilaginous model. The osteoblasts form from ____________ cells. The bony collar supports the shaft as the cartilage within breaks down forming the ____________. Once a cavity is formed within the shaft ____________ migrate into the space and form ____________ bone. Osteoblasts secrete the collagen of the matrix and ____________ salts are deposited on the collagen. These cells form ____________ but bone cannot undergo ____________ ____________. Once the matrix is formed, these cells are called ____________, which are the principal cells in bone tissue. The growth plate, also called the ____________ plate, remains as a site of growth through the end of puberty. Bone growth is stimulated by ____________

____________.

E. List the major parts of the axial and appendicular skeleton below.

Axial Skeleton	Appendicular Skeleton
1. ______________________	1. ______________________
2. ______________________	2. ______________________
3. ______________________	3. ______________________
4. ______________________	4. ______________________

F. Refer to the figures in your text to identify the scientific name for each of the following bones.

1. backbone	1. ______________________
2. collarbone	2. ______________________
3. fingers	3. ______________________
4. thigh bone	4. ______________________
5. head and face	5. ______________________
6. shinbone	6. ______________________
7. hip	7. ______________________
8. shoulder blade	8. ______________________
9. breastbone	9. ______________________
10. cheek bones	10. ______________________

G. Complete the following table with the missing information about the joints.

Joint	**Description/Movement**	**Example**
Fibrous joints		
Cartilaginous joints		
Synovial joints		

H. Circle the correct word to complete the following statements.

1. Osteoporosis is a condition where there is a(n) (increase, decrease) in bone density.
2. Weight-bearing exercises (strengthen, weaken) bones.
3. Ligaments connect bone to (bone, muscle).
4. Tendons connect bone to (bone, muscle).
5. A sprain is a stretching of a (ligament, tendon).
6. Because of their structure, tendons and ligaments heal (quickly, slowly).

Critical Thinking

Read each of the following questions carefully. If it helps you, underline or list each component to the question. Use complete sentences in your explanation.

1. Find a study partner. Then describe bone development starting with cartilage in an embryo and concluding with bone calcification in an adult. Review the structure of cartilage and bone and explain why it is essential that bones develop this way.
2. Form a small group. List the injuries to joints and bones that either those in the group have experienced or read about. Choose three of the most common injuries and describe them in anatomical terms. What bones, tendons, and ligaments were involved? What was the structure of the joint? What was done to correct the problem and how long did it take for a full recovery?
3. Your 80-year-old grandmother is in reasonably good health. However, she just got word that she has osteoporosis. She says that she is going to ride her reclining bike even more to stave off the effects of the disease. Reflecting on what you know about bone structure and the development of bone strength, what advice would you give her?
4. Next time you have turkey or chicken, save the bones. Wash them and then put some in the oven set at 150°F for about two days. Don't leave the oven on when you are not home. What changes occurred in the bone? Put several other bones into a jar with vinegar in it. You will need to change the vinegar every day for about two weeks. What do you notice happened to the bone in the acid solution? What did heat and acid do to the bones to cause these changes? Why weren't the changes in the two bones the same? Put a third group of bones in a plastic bag to leave unchanged as a control. What changes did you notice in these bones?

Practice Test

Choose the one best answer to each question that follows. As you work through these items, explain to yourself why each answer you discard is incorrect and why the answer you choose is correct.

1. Which of the following is not a function of the skeletal system?
   a. manufacture of red blood cells
   b. protection
   c. support
   d. synthesis of growth hormones

2. The bone marrow that produced blood cells is found in ____________.
   a. compact bone
   b. developing bone
   c. spongy bone
   d. strong bones

3. The nonliving portion of the bone is mostly ____________.
   a. chitin, magnesium and nitrate
   b. keratin, sodium and sulfate
   c. matrix fibers, calcium and phosphates
   d. spongy bone, calcium and magnesium

4. The periosteum ____________.
   a. covers the ends of a long bone to cushion it
   b. covers the outer surface of the bone and contains osteoplasts and nerve endings
   c. lines the inner cavity of the long bone and contains blood vessels
   d. is found woven within spongy bone

5. The name *spongy bone* refers to the ____________.
   a. ability to absorb nutrients
   b. lattice work of air spaces
   c. soft touch
   d. wet outer surface of the bone

6. Bone-forming cells are called ____________.
   a. osteoblasts
   b. osteoclasts
   c. osteocytes
   d. osteons

7. Which of the following hormones control growth?
   a. growth hormone
   b. sex hormones
   c. thyroid hormone
   d. all of the above

8. The healing of a fracture or break is dependent upon which of the following tissues?
   a. bone
   b. connective
   c. epithelial
   d. sarcoplasma

9. Which of the following describes bone remodeling?
   a. bone growth through adolescence
   b. deposition and absorption of bone minerals
   c. joint replacement
   d. plastic surgery after an accident

10. Which of the following activities would lead to the strongest bones?
    a. bicycling
    b. canoeing
    c. resting
    d. weight lifting

11. Which of the following is part of the pectoral girdle?
    a. femur
    b. pelvis
    c. scapula
    d. skull

12. The axial skeleton is so named because it ___________.
    a. holds up the axis, another word for Earth
    b. is composed of joints that do not move
    c. is found in the vertical axis of the body
    d. is the centermost part of the body

13. When people have a hip replaced, what is the name of the leg bone involved in the replacement?
    a. femur
    b. fibula
    c. pelvis
    d. tibia

14. Pain in the lower back is associated with which of the following?
    a. cervical vertebrae
    b. lumbar vertebrae
    c. sacrum
    d. thoracic vertebrae

15. The movement in what type of joint in the mother facilitates the delivery of a baby?
    a. cartilaginous
    b. fibrous
    c. pelvic
    d. synovial

16. A pitcher is most likely to injure which type of joint?
    a. ball and socket
    b. cartilaginous
    c. fibrous
    d. hyaline

17. Adequate calcium intake is important for strong bones. What might happen if there is inadequate calcium intake?
    a. calcium would not be available to deposit during remodeling
    b. low blood calcium levels would stimulate the activity of osteoblasts
    c. low calcium levels would force the development of more cartilage
    d. osteocytes would die

18. Ligaments connect ____________.
    a. bone to bone
    b. bone to muscle
    c. bone to organ
    d. muscle to muscle

19. Which type of joint is most likely to develop arthritis?
    a. cartilaginous
    b. fibrous
    c. synovial
    d. they are all equally likely to develop arthritis

20. Which of the following disorders is caused by an attack of the joint tissues by the body's immune system?
    a. bursitis
    b. osteoarthritis
    c. osteoporosis
    d. rheumatoid arthritis

Short Answer Questions

Read each of the following questions carefully. Jot down the main points you want to include in your answer. Then write a well-organized explanation.

1. Draw and label a long bone, including the features of an osteon.

2. Explain what is meant by the constant remodeling of bones. What cells are involved and what do they do? What happens later in life when this process is no longer in balance? What can be done to strengthen bones?

3. When babies are first born, they have a "soft spot" on their heads. What is this? What type of joint is involved? How will this place change over the next several years?

4. Why might it take longer for a knee injury to heal than a broken bone in the femur or tibia? What tissues are involved and what characteristics affect their healing?

ANSWER KEY

REVIEW QUESTIONS

A. 1. support
2. movement
3. protection
4. mineral storage
5. fat storage
6. blood cell production

B. A. yellow bone marrow
B. blood vessel
C. periosteum
D. central cavity
E. spongy bone
F. compact bone

C. 1. e
2. i
3. b
4. g
5. h
6. j
7. c
8. f
9. d
10. a

D. During embryonic development, most of the skeleton is formed as cartilage. Because cartilage cells are able to divide and form new cartilage cells, this model allows for growth. Early in development, the cartilage begins to be replaced by bone. First bone-forming cells, osteoblasts, form a collar of bone around the shaft of the cartilaginous model. The osteoblasts form from cartilage cells. The bony collar supports the shaft as the cartilage within breaks down forming the marrow cavity. Once a cavity is formed within the shaft osteoblasts migrate into the space and form spongy bone. Osteoblasts secrete the collagen of the matrix and calcium salts are deposited on the collagen. These cells form bone but bone cannot undergo cell division. Once the matrix is formed, these cells are called osteocytes, which are the principal cells in bone tissue. The growth plate, also called the epiphyseal plate remains as a site of growth through the end of puberty. Bone growth is stimulated by growth hormone.

E. Axial Skeleton: skull, vertebral column, sternum, rib cage
Appendicular Skeleton: pectoral girdle, pelvic girdle, bones of the arms, bones of the legs

F. 1. vertebral column
2. clavicle
3. phalanges
4. femur
5. skull
6. tibia
7. pelvic girdle
8. scapula
9. sternum
10. zygomatic bones

G.

Joint	Description/Movement	Example
Fibrous joints	no joint cavity, made of fibrous connective tissue, no movement	joints of the skull
Cartilaginous joints	made of cartilage, no joint cavity, allow for little or no movement	joints at pubic bones, vertebral column, ribs and sternum
Synovial joints	surface of the joint has a thick layer of hyaline cartilage to reduce friction, surrounded by a capsule that secretes synovial fluid to lubricate and cushion the joint	shoulder, hip, arms, hands, wrists, feet, etc

H. 1. decrease
2. strengthen
3. bone
4. muscle
5. ligament
6. slowly

Critical Thinking

1. During fetal development, the skeleton is first laid down in cartilage. The cartilaginous model is gradually replaced by bone. Osteoblasts form from cartilage cells and secrete both the salts and the collagen of the matrix that gives bones their strength. These cells form bone, but they cannot undergo cell division; therefore they could not participate in the formation of the underlying structure of the skeleton or bone growth.
2. Many people experience bone fractures and breaks. The repair of a bone fracture involves the formation of a blood clot and then a cartilaginous callus that links the broken ends of the bone. The cartilage is gradually replaced by bone. Many students may have experienced damage to a joint, ligament or tendon usually due to an athletic injury. The joint may have healed when given rest or it may have required surgery and follow-up physical therapy. Use Figure 5-6 showing the major bones of the body for reference.
3. Bone strength is enhanced by weight-bearing exercise and good nutrition including calcium. Encourage your grandmother to eat well, take calcium supplements, and to walk or engage in an aerobics class at the senior center. Sitting on a bike will not strengthen her bones, but it will improve her cardiovascular conditioning.
4. Bones put in the oven will dry. The water will be removed and the living tissue will die due to the heat. These bones will be very brittle as the living tissue gives bones flexibility. The bones put into the vinegar solution will become rubbery in time. The acid dissolves the salts from the bones leaving the remains of the living, soft tissue. The bones in the plastic bag are there for a comparison. They should remain moist, but strong. They will not be as brittle as the bones in the oven or as rubbery as the bones in the acid solution.

Practice Test

1. d
2. c
3. c
4. b
5. b
6. a
7. d
8. b

9. b
10. d
11. d
12. d
13. a
14. b
15. a
16. a
17. a
18. a
19. c
20. d

Short Answer Questions

1. Refer to Figure 5-1 in your text and the study materials in the study guide. Your drawing should include the structures of the long bone including the shaft, the heads, spongy bone found in the shaft, and the structure of the osteon.
2. This period bone remodeling occurs after the initial growth process is completed. It is the ongoing process of bone deposition and absorption. Bone is deposited by osteoblasts. Osteoclasts break down bone and the minerals are reabsorbed by the body. An important factor in determining the rate and extent of bone remodeling is the degree of stress to which a bone is exposed. Bone forms in response to stress and is destroyed when it is not stressed.
3. In an adult the bones of the skull are joined by suture joints that are largely immovable. However, these joints form first as cartilage in the fetus and take about two years to fully calcify into bone and form the fixed joints of an adult. During delivery these joints allow the bones of the skull to move over one another, compressing the skull as the baby moves out the birth canal.
4. Most knee injuries are to the ligaments of the knee. These fibrous connective tissues have less blood supply than the bone itself. Thus it takes them much longer to heal.

Chapter 6

The Muscular System

Objectives

After reading the text and studying the material in this chapter, you should be able to:

- State the function of the muscular system.
- Name the three kinds of muscles.
- List the four traits that all muscles have in common.
- Describe the attachment of muscle to bone and how that leads to movement.
- Demonstrate and explain the use of antagonistic muscles.
- Describe the banded appearance of striated muscles using the terms myofibril, myofilaments, actin and myosin.
- Explain muscle contraction at the molecular level of the actin and myosin filaments.
- Explain why ATP is essential to muscle contraction and relaxation.
- Describe the role of the tropomyosin-troponin complex in muscle contraction.
- Draw a neuromuscular junction and explain the role of acetylcholine in muscle contraction and the release of calcium ions.
- Differentiate between a muscle twitch, a stronger contraction resulting from wave summation, tetanus and fatigue.
- List the sources of ATP for muscle contraction and describe in detail where and how the ATP is generated.
- Compare and contrast slow-twitch and fast-twitch muscles, including where they are located in the body and when they are utilized in different physical activities.
- Describe the best way to build muscle endurance and the requirements for building larger muscle mass.

Chapter Summary

The muscular system produces movement and maintains posture. There are three types of muscles: skeletal, cardiac and smooth. All **muscles are excitable** which means that they respond to stimuli. They have the ability to contract, extend and to return to their original length after being shortened or stretched.

Skeletal muscles are voluntary muscles responsible for moving our body. Each muscle is attached to a bone by a **tendon**. The **origin** of the muscle remains stationary during movement while the **insertion** is attached to the bone that moves. Most muscles are arranged in pairs that work in opposition to one another. They are called **antagonistic pairs**, since while one muscle contracts the opposite muscle must relax.

When skeletal muscles are viewed under a microscope, they have distinct bands. These **striations** are formed by the arrangement of **myofibrils** within the muscle cell. Each myofibril contains groups of long myofilaments. Each **myofilament** is composed of **myosin** and **actin** filaments. Sarcomeres are the contractile units of muscle. They shorten as the actin filaments slide along the myosin filaments.

Muscle contraction occurs at the molecular level. According to the sliding filament model, muscle contracts when actin filaments slide past myosin filaments. The myosin head attaches to the actin filament forming a **cross bridge**. Then it bends and swivels, pulling the actin filament toward the midline of the cell. The movement of the myosin filament is powered by ATP. The **tropomyosin-troponin complex** and calcium ions regulate muscle contraction at the actin-myosin binding sites.

Contraction is triggered when a nerve impulse travels down a motor neuron until it reaches the **neuromuscular junction**. Here it causes the release of acetylcholine from vesicles in the motor neuron. The acetylcholine causes changes in the permeability of the muscle cell, resulting in an electrochemical message similar to a nerve impulse.

A motor neuron and all the muscle cells it stimulates are called a **motor unit**. The strength of muscle contraction depends on the number of motor units that are stimulated. The muscle cells of a motor unit are spread throughout the muscle resulting in even, whole muscle contraction.

All of the muscle cells innervated by a single neuron contract at once causing a **muscle twitch**. If a second stimulus is received before the muscle is fully relaxed, the second twitch will be stronger than the first due to **wave summation**. Taken to the extreme, a sustained powerful contraction is called **tetanus**. **Fatigue** sets in when a muscle is unable to contract even when stimulated.

The ATP for muscle contraction comes from many sources. The initial source is the ATP stored in the muscle cells and then the ATP formed from the creatine phosphate reserves. When those sources are depleted, the muscles depend upon stored glycogen, which is converted to glucose and then to ATP through aerobic respiration or lactic acid fermentation.

Slow-twitch fibers are loaded with mitochondria and therefore deliver prolonged, strong contractions. By contrast, **fast-twitch fibers** contract rapidly and powerfully but with much less endurance. They rely on lactic acid fermentation as their source of energy and therefore tire quickly.

Aerobic exercise increases endurance and coordination, while **resistance exercise** builds strength.

Key Concepts

- The muscular system produces movement and maintains posture.
- There are three kinds of muscles: skeletal, cardiac and smooth.
- All muscles are excitable, contractile, extensible and elastic.
- Muscles are attached to bone by tendons and are arranged in opposing, or antagonistic pairs.
- Sarcomeres are the contractile units of muscle.
- The striated appearance of muscles is caused by the arrangement of myofibrils within the muscle cell. Each myofibril contains groups of myofilaments composed of actin and myosin proteins.

- A muscle contracts when myosin binds to actin, causing the filaments to slide past one another. Contraction is powered by ATP and controlled by two regulatory proteins and calcium ions.
- Nerves stimulate muscle contraction at the neuromuscular junction. Acetylcholine, released from the motor neuron, causes a change in muscle cell permeability resulting in the release of calcium ions and contraction.
- A motor neuron and all the muscle cells it stimulates are called a motor unit. The strength of a muscle contraction depends on the number of motor units involved.
- A muscle twitch is the contraction caused by a stimulus to the muscle. Successive contractions, prior to complete relaxation, build upon the first by wave summation. A sustained contraction is called tetanus.
- ATP is required for muscle contraction. A small amount of ATP is stored in the muscle cells and more can be generated from the energy stored in creatine phosphate. The rest comes from aerobic respiration and lactic acid fermentation.
- Slow-twitch muscles, found in the abdomen and back, contract slowly, powerfully and with endurance. Fast-twitch muscles, found in the arms and legs, contract rapidly and powerfully, but with less endurance.
- Aerobic exercise increases endurance while resistance exercise builds muscle mass.

Study Tips

The focus of this chapter is the relationship between the structure and function of whole muscles and their components. Once again it is critical that you study each figure and be able to explain its purpose and content to a friend. Illustrations provide a clear representation of a process occurring inside the body that may not be well represented by a photograph. As you study each illustration, be sure that you understand the main point of the illustration as presented in the caption. Read the caption first, then study the illustration. Understand each label, symbol and arrow. If an illustration is used to simplify an electron micrograph, as in Figure 6-3, the structure of a muscle, be sure that you can match the labels shown in the illustration to the photograph. If the illustration presents the steps in an ordered process as in Figures 6-4 and 6-5, then practice explaining how each drawing relates to the previous and following drawings.

Ask your instructor to identify the most important learning objectives for this chapter, then highlight them in this study guide. Review the text material, but refer to the figure each time it is referenced in the text. Figure 6-3 shows the increasing detail of a muscle. Highlight the most important terms from the figure and put them in order from largest to smallest. Find a study partner to review the mechanism of muscle contraction with you. Refer to Figures 6-4 and 6-5 to explain the function of each filament and how they cause the muscle to shorten or contract. Redraw a neuromuscular junction using the terms in your notes. Explain what happens as a nerve impulse reaches the muscle. What causes the muscle to contract? Muscle contraction is represented by the graph shown in Figure 6-9. First read the caption, then identify what is represented by each axis of the graph. Explain to your study partner where the stimulus is applied and the resulting effect on the muscle. Identify differences between a single twitch, a wave summation and tetanus.

Finally, reflect on your own activities and use of slow-twitch and fast-twitch fibers. How can one build endurance? What must happen to build more muscle mass? Discuss the various activities that go on at your student activity center and how they utilize different strategies to build fitness and strength.

Review Questions

A. Complete the following statements.

1. The functions of the muscular system are to produce ____________ and maintain ____________.

2. The three kinds of muscles are ____________, ____________, and ____________.

3. Muscles are ____________, which means that they have the ability to stretch.

4. Muscles are ____________, which means that they have the ability to shorten.

5. Muscles are ____________, which means that they can return to their original length after being shortened or lengthened.

6. Muscles are ____________, which means that they respond to stimuli.

B. Use the following terms to label the figure below: biceps, extension, flexion, insertion, origin, and triceps.

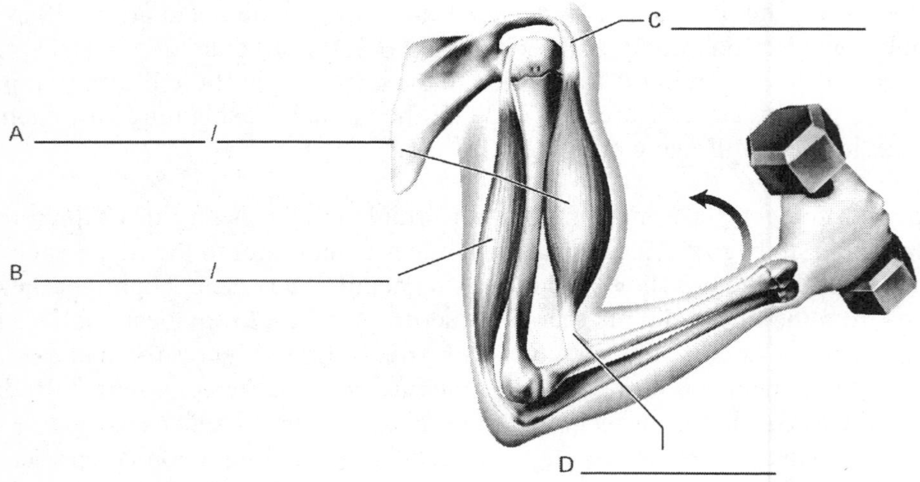

C ____________

A ____________ / ____________

B ____________ / ____________

D ____________

C. Refer to Figure 6-3 in your text to complete the following paragraph about muscle structure.

Skeletal muscles have pronounced bands visible under the microscope called ____________. This banded appearance is caused by the orderly arrangement of ____________. Each myofibril is composed of many, many ____________. The myofilaments are composed of two

filaments. The thicker filament is made of the protein ____________, while the more

numerous, thinner filament is made of ____________.

D. The drawings below show the mechanism of muscle contraction. First study Figure 6-5 in your text. Next, in the space provided, write an explanation of what is happening in each drawing below.

a.

Actin
Myosin heads
ADP
P_i

Resting sarcomere

a. ____

____

____

b.

ADP
P_i

Cross-bridge attachment

b. ____

____

____

c.

ADP
P_i

Pivoting of myosin head

c. ____

____

____

d.

ATP

Cross-bridge detachment

d. ____

____

____

e.

ADP
P_i

Myosin reactivation

e. ____

____

____

E. Match the term in Column A with the function or specialized structure in Column B.

Column A	Column B
1. acetylcholine	a. forms the parallel striations of skeletal muscle
2. actin filament	b. thick filament with heads
3. ATP	c. make up 80% of muscle cell volume and are composed of two protein filaments
4. cross bridges	d. regulatory proteins found on the actin filaments
5. myofibrils	e. storage area for calcium ions
6. myofilaments	f. thin filament
7. myosin filament	g. attachment of myosin filaments to actin filaments
8. sarcomere	h. smallest contractile unit
9. sarcoplasmic reticulum	i. causes a permeability change in the muscle cell
10. tropomyosin-troponin complex	j. source of energy for muscle contraction

F. Complete the following summary of neural control of muscle contraction.

A motor unit includes a ____________ and the ____________ fibers it stimulates. A ____________ ____________ is the connection between the nerve and the muscle. ____________ is released when the nerve impulse arrives at the neuromuscular junction. The vesicles at the terminal end of the motor neuron release a neurotransmitter that changes the ____________ of the muscle cell. The muscle cell now releases ____________ ions that change the shape of ____________ and expose the myosin-binding sites of the actin filament. Myosin attaches to ____________, forming a ____________ ____________. The ____________ head swivels and the filaments slide past one another. The muscle is now ____________. This is the description of the ____________ ____________ model of muscle contraction.

G. Use the following terms to label the drawing below: fatigue, single twitch, stimulus, tetanus, and wave summation. When you have finished, explain the graph to a study partner.

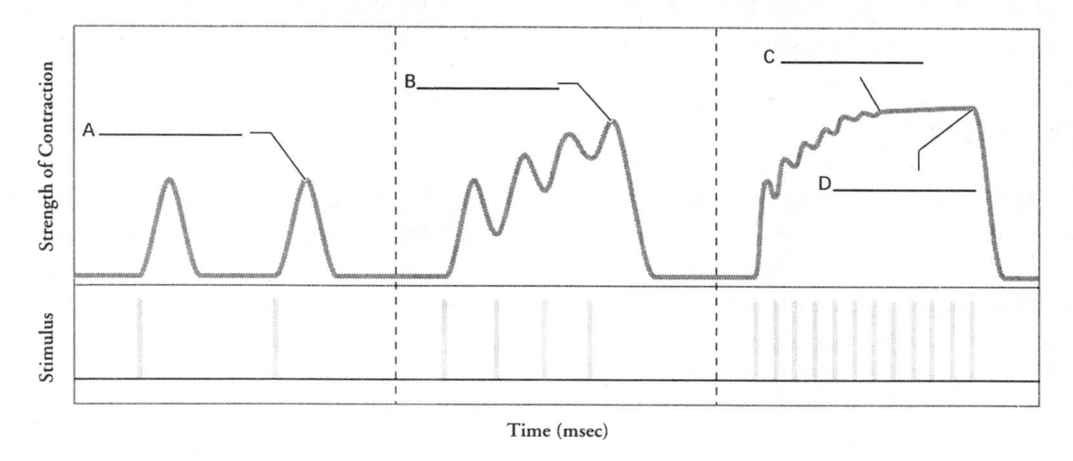

H. Circle the correct word to complete the following statements.

1. The first source of energy for muscle contraction is (ATP, creatine phosphate).
2. (Fast-twitch, slow-twitch) muscle fibers are most useful in long-term, sustained activities.
3. (Fast-twitch, slow-twitch) muscle fibers are most valuable to sprinters and power lifters.
4. (Aerobic, resistance) exercise builds the cardiovascular system and delivers oxygen to the tissues.
5. (Aerobic, resistance) exercise builds muscle mass by increasing the diameter of existing muscle cells.

Critical Thinking Questions

Read each of the following questions carefully. Circle the most important words in the question before formulating your answer.

1. Draw the sliding-filament model of muscle contraction and explain the drawing to a study partner. Explain each attachment and what physically causes the muscle to shorten. Using your drawing explain why this is called the *sliding-filament* model.
2. Review the discussion of the various sources of ATP as energy for muscular contraction. Summarize your recommendations regarding how an athlete might increase the amount of ATP available for a marathon race. Include a rationale for your suggestions.
3. A weight lifter accidentally grabbed a live electric wire. The news reported that he died. First, from your knowledge of the additive effect of wave stimulation on muscle contraction and the relative strength of antagonistic pairs, explain why the weight lifter who was of great strength could not let go of the electric wire. Next, explain why when the rescue team found the body stiff, they knew he had been dead at least 4 hours.

4. Skeletal muscles are characterized by a banded appearance. What causes this appearance and what changes would occur as the muscle contracts? The actin and myosin filaments are of different thicknesses. Draw what a cross section through a muscle fiber might look like at the area between the dark bands and through one of the dark bands. Justify your prediction.

PRACTICE TEST

Choose the one best answer to each question that follows. As you work through these items, explain to yourself why each answer you discard is incorrect and why the answer you choose is correct.

1. Which of the following is a function of the muscular system?
   a. production of red blood cells
   b. produces movement
   c. removes lactic acid
   d. synthesis of steroids

2. Which of the following is not a trait of muscle cells?
   a. contractile
   b. excitable
   c. extensible
   d. motile

3. Muscles are attached to bones by _______________.
   a. insertion couplers
   b. ligaments
   c. other muscles
   d. tendons

4. Choose the term that describes muscles that oppose one another.
   a. antagonistic
   b. contrary
   c. pairing
   d. synergistic

5. The end of the muscle that is attached to the bone that moves.
   a. insertion
   b. ligament
   c. origin
   d. proximal

6. The contractile unit of a skeletal muscle.
   a. actin filaments
   b. myofibrils
   c. myosin filaments
   d. sarcomere

7. The striated, or striped, appearance of skeletal muscles is caused by which of the following.
   a. only actin filaments
   b. myofibrils
   c. only myosin filaments
   d. sarcomere

8. According to the sliding filament model of muscle contraction, __________.
   a. actin forms a covalent bond with myosin
   b. actin slides along myosin
   c. myosin slides along actin
   d. both actin and myosin slide toward the center of the cell

9. Which of the following is smallest?
   a. actin filament
   b. myofibril
   c. myosin filament
   d. sarcomere

10. What is the first energy source for muscle contraction at the molecular level?
    a. ATP
    b. creatine phosphate
    c. myosin head molecules
    d. respiration

11. Tropomyosin-troponin complex controls muscle contraction by ___________.
    a. accelerating actin contraction
    b. contracting the myofilaments
    c. covering the myosin heads
    d. splitting the ATP into ADP and P

12. Where are the calcium ions needed for muscle contraction stored?
    a. ribs
    b. sarcomere
    c. sarcoplasmic reticulum
    d. transverse tubules

13. The junction between a motor neuron and a muscle cell is called a ___________.
    a. motor nerve union
    b. neuromuscular junction
    c. neurotransmitter
    d. vesicle

14. What chemical causes the permeability of the muscle cell membrane to change?
    a. acetylcholine
    b. actin activator
    c. ATP
    d. calcium

15. A motor unit is a ___________.
    a. motor neuron and all the muscle cells it stimulates
    b. motor neuron and the spinal nerves it stimulates
    c. sensory neuron and the muscle cells it stimulates
    d. the bound actin and myosin filaments

16. The strength of a contraction depends upon the number of ___________ involved.
    a. motor neurons
    b. motor units
    c. neuromuscular junctions
    d. neurotransmitter molecules

17. When does wave summation occur?
    a. when a second stimulus is received before the muscle relaxes
    b. when the muscle has a powerful and sustained contraction
    c. when the muscle is unable to contract in spite of continued stimulation
    d. when the relaxation phase is longer than the contraction phase

18. Endurance sports, as opposed to sports requiring fast-twitch muscles, require stores of ___________ found in the muscles.
    a. ATP
    b. creatine phosphate
    c. glucose
    d. glycogen

19. Which of the following is more powerful?
    a. fast-twitch muscles
    b. mixed muscles
    c. slow-twitch muscles
    d. slow-twitch and fast-twitch muscles have the same strength

20. Which of the following would build the largest muscles?
    a. aerobic exercise
    b. anaerobic exercise
    c. electrically stimulated exercise
    d. resistance exercise

SHORT ANSWER QUESTIONS

Read each of the following questions carefully. If it helps you, underline or list each component to the question. Formulate your answer and jot down the main points. Then use complete sentences in your explanation.

1. Draw and label a sarcomere showing the relationship between actin and myosin filaments.

2. Demonstrate the action of the actin and myosin filaments either through a drawing or a model using strips of paper. Include in your explanation the exact mechanism of movement

including the tropomyosin-troponin complex, calcium ions, formation of cross bridges, and the disengaging of the cross bridges.

3. Differentiate between a single muscle twitch, the summation of several twitches, tetanus and muscle fatigue.

4. Two friends join the track team. Danielle is a sprinter and she repetitively runs wind sprints during practice and engages in a challenging weight-training program. Tanisha is a distance runner. She runs long distances each practice with a weekly run that takes her to near exhaustion. She does not have a prescribed weight-training program. Discuss why these two very different training programs are appropriate for each athlete. Include the energy sources for each activity and why more muscle mass will be an advantage to Danielle while a well developed cardiovascular system will be an advantage to Tanisha.

Answer Key

Review Questions

A. 1. movement, posture
2. skeletal, cardiac, smooth
3. extensible
4. contractile
5. elastic
6. excitable

B. A. biceps/flexion
B. triceps/extension
C. origin
D. insertion

C. Skeletal muscles have pronounced bands visible under the microscope called striations. This banded appearance is caused by the orderly arrangement of myofibrils. Each myofibril is composed of many, many myofilaments. The myofilaments are composed of two filaments. The thicker filament is made of the protein myosin while the more numerous, thinner filament is made of actin.

D. 1. In the resting sarcomere, the myosin and actin filaments are parallel but not connected. ATP has been split into ADP and P_i.
2. The actin and myosin filaments connect forming cross bridges. Phosphate is released from the myosin head.
3. Using the stored energy from ATP, the myosin head swivels and moves the actin filament ahead. This is the actual muscle contraction.
4. Another ATP attaches to the myosin head releasing the cross bridge. The two filaments are no longer attached.
5. The myosin head splits ATP into ADP and P_i and stores the energy. The myosin head returns to the original resting position.

E. 1. i
2. f
3. j
4. g
5. a
6. c
7. b
8. h
9. e
10. d

F. A motor unit includes a motor neuron and the muscle fibers it stimulates. A neuromuscular junction is the connection between the nerve and the muscle. Acetylcholine is released when the nerve impulse arrives at the neuromuscular junction. The vesicles at the terminal end of the motor neuron release a neurotransmitter that changes the permeability of the muscle cell. The muscle cell now releases calcium ions that change the shape of troponin and expose the myosin-binding sites of the actin filament. Myosin attaches to actin, forming a cross bridge. The myosin head swivels and the filaments slide past one another. The muscle is now shorter. This is the description of the sliding filament model of muscle contraction.

G. A. single twitch
B. summation
C. tetanus
D. fatigue

H.
1. ATP
2. Slow-twitch
3. Fast-twitch
4. Aerobic
5. Resistance

Critical Thinking Questions

1. Drawing should be similar to Figure 6-4 showing thicker myosin filaments between thin actin filaments. During muscle contraction, the actin filaments slide along the myosin filaments, shortening the sarcomere. The myosin head attaches to the actin molecules, then bends, causing the actin filament to slide across the myosin filament. Then the myosin head is freed and returns to its original resting position.
2. There are several sources of ATP available to muscle cells. First, the muscle uses the ATP stored in it, then it uses ATP formed from creatine phosphate and ADP, and finally it uses ATP generated from glycogen stored in muscles and broken down to form glucose. Training for a long-distance race will improve the amount of oxygen that reaches the cells and will also train the muscles to store more glycogen.
3. Because the muscles were continually stimulated by the electric current and because the muscles that cause a hand to grip are much stronger than the antagonistic muscles, the weight lifter only continued to grab the electric wire tighter and tighter. When a person dies, they can no longer make the ATP required for muscle contraction. Specifically, the ATP allows the myosin head to detach and return to the resting state. Without fresh ATP, the cross bridges cannot be broken and the muscle becomes stiff.
4. The darkest bands of the sarcomere are where the actin and myosin filaments overlap. They shorten during contraction. A cross section would include thick myosin filaments regularly spaced between thin actin filaments. The lighter areas of the sarcomere are where the actin filaments occur alone. This cross section would show only thin actin filaments and no myosin filaments.

Practice Test

1. b
2. d
3. d
4. a
5. a
6. d
7. b
8. b
9. a
10. d
11. c
12. c
13. b
14. a
15. a
16. b
17. a
18. d
19. a
20. d

Short Answer Questions

1. The drawing should be similar to Figure 6-4 showing the Z line, the actin and myosin filaments overlapping.
2. The answer to this question includes information from Figure 6-5 and Figure 6-6. A critical feature is that the myosin head bends, moving the actin filament over it. The two filaments do not move in opposing directions. The myosin head forms cross bridges with the actin filament to a binding site. The binding sites are blocked by the troponin-tropomyosin complex until the complex changes shape after coming in contact with a calcium ion. The cross bridges are broken when a new ATP molecule binds to the myosin head, allowing it to return to its more relaxed position.

3. The best answer would include a graph similar to Figure 6-9. A single twitch is isolated from others; summation shows increasing strength with each stimulus; tetanus is a continuous contraction; and fatigue occurs when there is no more contraction although there may be more stimulation.
4. Sprinters have more fast-twitch muscles than endurance athletes. The fast-twitch cells are rich in glycogen deposits but depend more heavily on anaerobic means of producing ATP. Thus they can benefit from a weight-training program. However, they will tire more quickly than the slow-twitch cells that are more aerobic.

Chapter 7

Neurons: The Matter of the Mind

Objectives

After reading the text and studying the material in this chapter, you should be able to:

- Identify the two main parts of the nervous system.
- Differentiate between a neuron and neuroglial cells in structure and function.
- Label a generalized neuron.
- Explain the role of the myelin sheath.
- Describe how a nerve cell maintains a resting state using the sodium-potassium pump and what changes occur in the membrane upon nerve stimulation.
- Explain how a nerve impulse is a bioelectrical signal including how ions move across the membrane.
- Describe an action potential as a wave of activity along an axon.
- Draw and label a synapse showing the movement of neurotransmitters.
- Describe the specific role of common neurotransmitters in the maintenance of a stable mental condition and in disease situations.

Chapter Summary

The **nervous system** integrates and coordinates many of the body's activities. It is divided into the central nervous system (CNS), the brain and spinal cord, and the peripheral nervous system (PNS). Two types of specialized cells carry out the functions of the nervous system. **Neurons** are excitable cells that generate and transmit messages. **Neuroglial cells** (also called glial cells) are ten times as numerous and provide structural support, growth factors, and insulating sheaths around the nerves. Glial cells are able to reproduce, unlike neurons.

Neurons are the functional unit of the nervous system. **Sensory** (or **afferent**) **neurons** carry information toward the CNS from **sensory receptors**. **Motor** (or **efferent**) **neurons** carry information away from the CNS to an **effector**. **Association neurons** (or **interneurons**) are located between sensory and motor neurons within the CNS where they integrate and interpret sensory signals.

Dendrites carry information toward the **cell body** of a neuron and a single long **axon** carries information away from the cell body. A **nerve** consists of parallel axons, dendrites, or both from many neurons that are covered with tough connective tissue. Most axons not found in the CNS are electrically insulated by a **myelin sheath**. **Schwann cells** form the myelin sheath, insulating it and allowing messages to travel faster as they jump from **one node of Ranvier** to the next. The myelin sheath also facilitates nerve repair outside the CNS. **Multiple sclerosis (MS)** results from the destruction of the myelin sheath that surrounds axons found in the CNS.

A nerve impulse, or action potential, is a bioelectrical signal involving sodium ions (Na+) and potassium ions (K+) that cross the cell membrane. They cross through the **ion channels** without a direct need for an energy source. They are also transported across the membrane by the **sodium-potassium pump**. This requires an energy source, ATP, but allows for concentration of the ions so that the concentration of Na^+ ions is higher outside and K^+ ions is higher inside the cell. When a neuron is not conducting a nerve impulse, it is in a **resting state**. Since the ions are not evenly distributed on either side of the membrane, there is a slight difference in charge across the membrane called the **resting potential**. When the neuron is stimulated, there is a sudden reversal of charge across the membrane because the sodium gates open and sodium ions enter the cell. The minimum charge that causes the sodium gates to open is called the **threshold**. Next the potassium gates open and potassium ions rush out of the cell causing the cell to return to the original state, or **repolarize**. An **action potential** is the sudden reversal of the charge across the membrane followed immediately by its restoration. These changes occur in a wave along the axon. For a very brief period following an action potential, the nerve cannot be stimulated again. This is called the **refractory period**. Since the ion gates are either open or closed, action potentials are all-or-none-events.

Communication between neurons is by **neurotransmitters**, chemicals that cross the gap between two neurons. A **synapse** is the junction between the **presynaptic neuron**, which sends a message to the **postsynaptic neuron**. When the impulse reaches the **synaptic knob** of the sending neuron, calcium ions cause the membrane of the **synaptic vesicles** to fuse with the plasma membrane and to release the neurotransmitter substances. They diffuse across the **synaptic cleft** to the other side where they bind with receptors on the postsynaptic cell causing the ion channels to open. After the neurotransmitter crosses the membrane, it is quickly broken down or pumped back into the synaptic knob of the presynaptic axon. If this occurs at an **excitatory synapse** and enough receptor sites bind with neurotransmitter substances to cause depolarization to threshold value, an action potential is generated in the postsynaptic cell. However, in an **inhibitory synapse**, the postsynaptic cell becomes more negatively charged and there is no action potential generated. It is the summation of the excitatory and inhibitory effects that determine whether the cell will generate an action potential.

The chemical balance of our neurotransmitters affects our moods, behavior and health. **Acetylcholine**, the neurotransmitter released by motor neurons at the neuromuscular junction can also be released into the synaptic cleft between two neurons. Dopamine, norepinephrine and serotonin affect our emotional state. **Alzheimer's disease** and **Parkinson's disease** are caused by deficiencies of these chemicals in the brain.

Key Concepts

- The nervous system integrates and coordinates the body's activities.
- It consists of two parts: the central nervous system (brain and spinal cord) and the peripheral nervous system.
- Neurons are nerve cells that are supported and nourished by neuroglial cells. Neuroglial cells form myelin sheaths and supply nerve growth factors determining whether a nerve can be repaired.

- Sensory (afferent), motor (efferent) and association (interneurons) are the three categories of neurons.
- A typical neuron has a cell body with a nucleus and other organelles, numerous branched dendrites that carry impulses toward the cell body and a single axon that conducts impulses away from the cell body.
- An insulating layer called the myelin sheath made of Schwann cells surrounds many axons. The myelin sheath insulates the axon, speeds the nerve impulse, and allows for regeneration.
- A nerve impulse (action potential) is caused by the change in charge as sodium and potassium ions cross the cell membrane. The ions may passively cross the cell membrane using ion channels or be concentrated by the sodium-potassium pump, which requires an energy source. The sodium-potassium pump maintains the resting state of the neuron.
- When a stimulus reaches a neuron it causes sodium ions to rush into the cell, resulting in a change in the charge across the membrane, or depolarization. Next the cell is repolarized by potassium ions entering the cell. The wave of depolarization and repolarization is called an action potential.
- Neurotransmitter chemicals are released from presynaptic vesicles, diffuse across the synaptic cleft and are received by the postsynaptic membrane. Postsynaptic cells integrate excitatory and inhibitory input. If a threshold is reached, an action potential is generated in the postsynaptic cell. Neurotransmitter substances are quickly removed from the synapse.
- Different neurotransmitters are involved in many behavioral systems. Brain chemistry affects our moods, behavior and the onset of various diseases including Alzheimer's and Parkinson's.

Study Tips

This chapter lays the foundation for the next chapter on the nervous system. It focuses on the structure and function of neurons. The structure of a neuron is fairly straightforward and it will not be difficult to learn the types of neurons and their functions. However, the details of an action potential and the role of neurotransmitters are more complex and depend upon your understanding of the plasma membrane. If you are not confident that you understand how materials cross the plasma membrane, it will be a wise investment of your time to review that material.

As in the last chapter, the figures in this chapter showing the mechanisms of the membrane activity are truly worth a thousand words. Apply the study tips from the last chapter as you study this chapter. First, be sure that you understand the structure of a neuron and the function of each part. Then, focus on the processes that are critical to the transmission of nerve impulses. Find a study partner and explain the mechanisms depicted in the figures. Focus on Figure 7-4, the passage of ions across the plasma membrane, and Figure 7-5, the propagation of an action potential along the axon, Figure 7-6, the structure of a synapse.

REVIEW QUESTIONS

A. Draw a typical neuron and label it using these terms: axon, cell body, dendrites, interneuron, motor neuron, myelin sheath, nodes of Ranvier, nucleus, Schwann cell, sensory neuron, and synapse. Draw arrows to show the direction of the impulse.

B. Summarize the function of each of the following.

1. Axon: ______________________________

____

2. Dendrite: ______________________________

____

3. Motor Neuron: ______________________________

____

4. Interneuron: ______________________________

____

5. Sensory Neuron: ______________________________

____

C. Fill in the blanks in the following sentences describing the role of the myelin sheath.

Most axons in the peripheral nervous system are enclosed with an insulating layer called the

___________ ___________. In the PNS, the type of cell that forms the myelin sheath is the

___________ cell. The space between Schwann cells is called a ___________ of

___________. This anatomy allows for very fast transmission of an impulse because the

impulse jumps from node to node. The myelin sheath is necessary for the ___________ of

neurons after an injury. ___________ ___________ is the disease that involves progressive

destruction of the myelin sheath.

D. Match the organ, structure or event in Column A with the function in Column B.

Column A	Column B
1. action potential	a. Na^+
2. depolarization	b. used to concentrate Na^+ and K^+ on opposite sides of the membrane
3. ion channels	c. when sodium ions enter the cell
4. potassium	d. nerve impulse
5. refractory period	e. charge necessary to start an action potential
6. repolarization	f. period when a neuron cannot generate an action potential when stimulated
7. resting potential	g. K^+
8. sodium	h places where ions can cross the plasma membrane without using energy
9. sodium-potassium pump	i. return to resting potential
10. threshold	j charge difference across a membrane not conducting an action potential

E. Complete the following statements about an action potential by circling the correct words.

1. The plasma of a resting neuron is charged with the inside (negative, positive) compared to the outside.

2. The loss of charge difference across the membrane is (depolarization, repolarization).

3. During depolarization, (potassium, sodium) ions (enter, exit) the axon. The inside of the membrane becomes (negatively, positively) charged.

4. (Depolarization, Repolarization) occurs when the membrane potential returns to close to its resting value. This is caused by (potassium, sodium) ions leaving the axon.

5. The original distribution of ions is restored by the (potassium-nitrate, sodium-potassium) pump.

F. Label the following diagram using these terms: axon, cell body, dendrites, neurotransmitter molecules, presynaptic membrane, postsynaptic membrane, receptors, synaptic cleft, and synaptic vesicle.

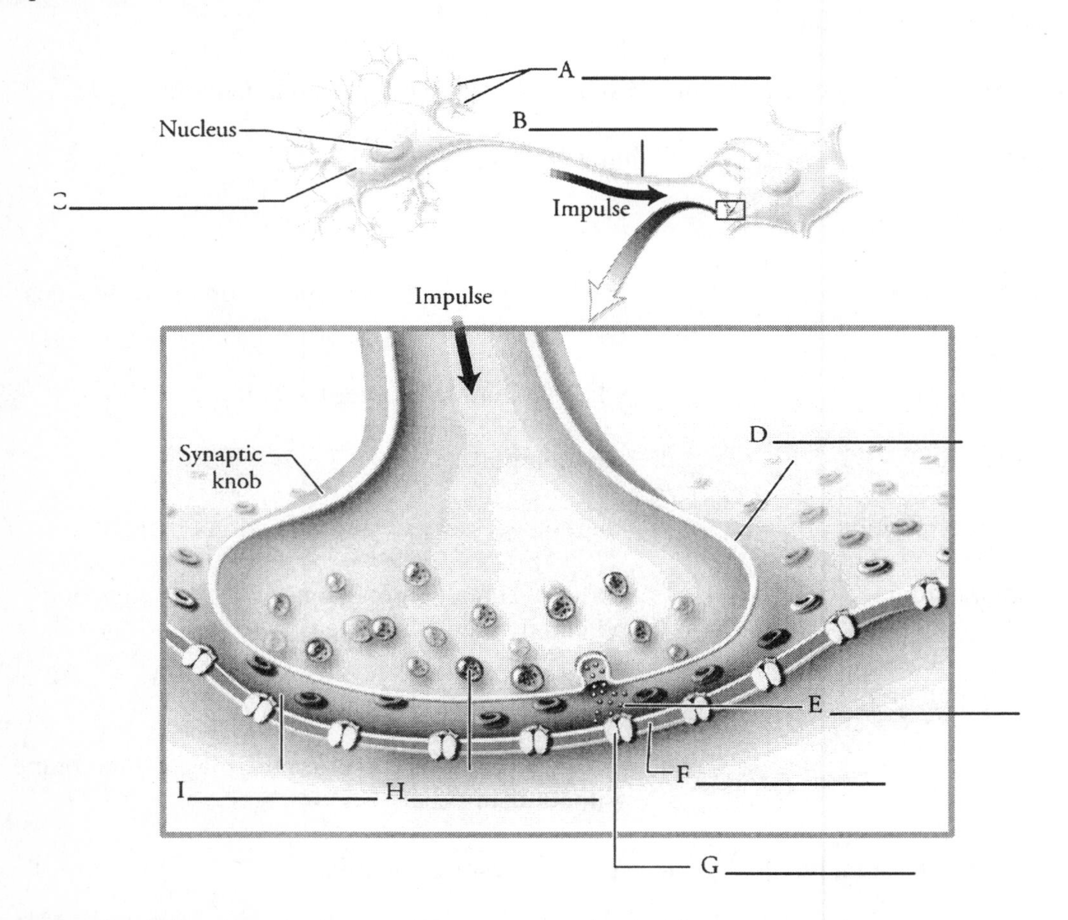

G. Complete the following statements about communication between neurons.

1. A ____________ diffuses across the gap between two neurons.

2. The junction between one neuron and the next is called a ____________.

3. The gap between two cells is called the ____________ ____________.

4. The small bulb-like swelling at the end of an axon is the ____________ ____________.
5. The ____________ neuron sends the message while the ____________ neuron receives the message.
6. The synaptic knobs of the presynaptic neuron contain synaptic vesicles loaded with neurotransmitters.
7. Neurotransmitters may be broken down by enzymes as is the case for acetylcholine that is broken down by acetylcholinesterase.
8. The loss of acetylcholine may be the cause of some of the symptoms of Alzheimer's disease.
9. An increase in the level of serotonin seems to reduce depression.
10. The death of dopamine producing cells results in Parkinson's disease.

Critical Thinking Questions

Read each of the following questions carefully. If it helps you, underline or list each component to the question. Use complete sentences in your explanation.

1. Use Figure 7-3a to explain why the myelin sheath and especially the nodes of Ranvier help a nerve impulse to travel faster. Draw an analogy to another mechanism outside the body that allows faster travel by "skipping" along a pathway.

2. Explain to a study partner how a nerve impulse travels along an axon. Include the movement of ions across the membrane. Next explain how a nerve impulse crosses a synapse. How are these two means of neural communication different and similar? Consider their structure, mechanism and outcome.

3. Demonstrate the refractory period with your study partner by doing the following activity. Have your partner sit with his or her back to you. Touch the back of your partner's neck with a sharp pencil or pin. Record whether your partner could feel the prick. Continue to touch the back of the neck with a sharp object. Record when your partner can no longer feel the prick. This is the refractory period when a neuron can no longer be stimulated.

4. Construct a chart of common neurotransmitters including the name and action. Indicate if an abnormally high or low concentration results in a physical or psychological disorder. Choose one pharmaceutical designed to mimic a neurotransmitter. Use the Internet to investigate the mechanism of the prescription drug and add that action to your chart.

5. If possible, visit with or observe people who have experienced nervous system damage through a stroke, disease, or injury. Note their struggles in movement, communication and understanding. Use your knowledge of neurons to explain the basis for these problems. What is the likelihood that their situation will change, either for the better or worse?

Practice Test

Choose the one best answer to each question that follows. As you work through these items, explain to yourself why each answer you discard is incorrect and why the answer you choose is correct.

1. Which of the following carry impulses toward the cell body?
   a. axon
   b. dendrite
   c. glial cells
   d. Schwann cells

2. Neurons with a myelin sheath carry impulses ________________ than unmyelinated cells.
   a. faster
   b. slower
   c. there is no difference

3. Which cells are responsible for nerve growth and regeneration?
   a. axons
   b. glial
   c. neurons
   d. interneurons

4. A muscle twitch results from the impulse carried by a ____________ neuron.
   a. associative
   b. motor
   c. musculoneural
   d. sensory

5. By definition a nerve consists of ____________.
   a. a neuron and the support cells making up the myelin sheath
   b. a single dendrite, cell body and axon
   c. several neurons bundled together
   d. the entire neuromuscular junction

6. The transmission of a nerve impulse along a neuron is ____________.
   a. chemical
   b. electrical
   c. electrochemical
   d. neurological

7. The transmission of a nerve impulse between two neurons across a synapse is ____________.
   a. chemical
   b. electrical
   c. electrochemical
   d. neurological

8. Which of the following mechanisms are used as ions cross the membranes of neurons?
   a. active transport using ATP
   b. passive diffusion
   c. sodium-potassium pump
   d. all of the above

9. If the inside of the neuron is negatively charged compared to the outside, the neuron is in the ____________.
   a. action potential
   b. excitatory phase
   c. retraction period
   d. resting state

10. What is the result of the sudden reversal in charge that occurs when a neuron is stimulated?
    a. action potential
    b. excitatory phase
    c. refractory period
    d. resting state

11. Action potentials are all-or-none. Which of the following summarizes this principle?
    a. an action potential can vary in both strength and duration
    b. if a threshold stimulus is reached, the action potential will always be of the same strength
    c. several stimuli will cause a stronger response
    d. the effects of several stimuli are combined until the neuron fires

12. Which of the following is true of the refractory period?
    a. the action potential is approaching the synapse
    b. the charge inside the neuron is becoming more positive
    c. the nerve impulse can no longer increase in intensity
    d. the neuron cannot be stimulated

13. As an action potential travels along an axon, depolarization means that ____________.
    a. acetylcholine is released from the presynaptic vesicles
    b. potassium ions enter the cell quickly
    c. sodium ions enter the cell quickly
    d. sodium and potassium move in and out equally

14. What conclusions can you draw from the following observation? An axon was stimulated with .075 volts and there was no response. Later it was stimulated with .15 volts and there was a response.
   a. Acetylcholinesterase levels were too high.
   b. Summation was occurring.
   c. The threshold stimulus was between .075 volts and .15 volts.
   d. None of the above can be concluded given this information.

15. What would be the result if excitatory neurotransmitters were not broken down?
   a. A migraine headache would result.
   b. The neurotransmitter would decompose on its own.
   c. There would be constant stimulation.
   d. There would be strong high followed by severe depression.

16. What would be the result if calcium could not enter the synaptic knob?
   a. neurotransmitters would break down faster
   b. neurotransmitters would not be released
   c. the cell could regenerate
   d. the response would be inhibitory

17. Where are synaptic vesicles located?
   a. along the entire length of the axon
   b. at the end of the axons
   c. at the end of the dendrites
   d. at the end of both axons and dendrites

18. Which of the following is NOT a neurotransmitter?
   a. acetylcholine
   b. acetylcholinesterase
   c. dopamine
   d. serotonin

19. Which of the following is the cause of Alzheimer's disease?
   a. low levels of dopamine
   b mercury contamination of drinking water
   c. pesticides accumulating in the CNS
   d. protein clusters in and between neurons

20. Which of the following most likely causes Parkinson's disease?
   a. excessive acetylcholine production
   b. high aluminum concentrations
   c. low dopamine production
   d. low serotonin levels

Short Answer Questions

Read each of the following questions carefully. If it helps you, underline or list each component to the question. Formulate your answer and jot down the main points. Then use complete sentences in your explanation.

1. Draw a nerve cell including the cell body, dendrites and an axon. Next draw in detail the axon showing the myelin sheath and Schwann cells. Using this diagram, explain why regeneration of nerves is more likely in peripheral nerves than the CNS.

2. Choose any stimulus and draw the pathway it takes to get to the brain and to effect a change in a muscle or gland. Include interneurons, motor neurons, sensory neurons, a sense organ and a muscle or gland in your drawing.

3. Explain the movement of ions across the membrane during an action potential. Begin with the resting state, continue with what happens when the stimulus arrives and threshold is reached, and the processes of depolarization and repolarization.

4. Draw a synapse and explain how an impulse travels from one neuron to the next. Include the role of the presynaptic vesicles, calcium, neurotransmitter chemicals, and postsynaptic receptor sites.

Answer Key

Review Questions

A. Drawing should include the following terms and might look similar to text Figure 7.3: axon, cell body, dendrites, interneuron, motor neuron, myelin sheath, nodes of Ranvier, nucleus, Schwann cell, sensory neuron, and synapse. The direction of impulse should go from sensory neuron → interneuron → motor neuron and within each neuron it should be dendrites → cell body → axon.

B.
1. Axons carry impulses away from the cell body to the synapse.
2. Dendrites carry impulses from the synapse toward the cell body.
3. Motor neurons conduct information away from the central nervous system to an effector.
4. Interneurons are located within the CNS; they integrate information.
5. Sensory neurons conduct information from the sensory receptors toward the central nervous system.

C. Most axons in the peripheral nervous system are enclosed with an insulating layer called the myelin sheath. In the PNS, the type of cell that forms the myelin sheath is the Schwann cell. The space between Schwann cells is called a node of Ranvier. This anatomy allows for very fast transmission of an impulse because the impulse jumps from node to node. The myelin sheath is necessary for the repair of neurons after an injury. Multiple sclerosis is the disease that involves progressive destruction of the myelin sheath.

D.
1. d
2. c
3. h
4. g
5. f
6. i
7. j
8. a
9. b
10. e

E.
1. The plasma of a resting neuron is charged with the inside negative compared to the outside.
2. The loss of charge difference across the membrane is depolarization.
3. During depolarization, sodium ions enter the axon. The inside of the membrane becomes positively charged.
4. Repolarization occurs when the membrane potential returns to close to its resting value. This is caused by potassium ions leaving the axon.
5. The original distribution of ions is restored by the sodium-potassium pump.

F.
A. dendrites
B. axon
C. cell body
D. presynaptic membrane
E. neurotransmitter molecules
F. postsynaptic membrane
G. receptors
H. synaptic vesicle
I. synaptic cleft

G.
1. A neurotransmitter diffuses across the gap between two neurons.
2. The junction between one neuron and the next is called a synapse.
3. The gap between two cells is called the synaptic cleft.
4. The small bulb-like swelling at the end of an axon is the synaptic knob.

5. The presynaptic neuron sends the message while the postsynaptic neuron receives the message.
6. The synaptic knobs of the presynaptic neuron contain synaptic vesicles loaded with neurotransmitters.
7. Neurotransmitters may be broken down by enzymes as is the case for acetylcholine that is broken down by acetylcholinesterase.
8. The loss of acetylcholine may be the cause of some of the symptoms of Alzheimer's disease.
9. An increase in the level of serotonin seems to reduce depression.
10. The death of dopamine producing cells results in Parkinson's disease.

Critical Thinking Questions

1. The myelin sheath serves as an electrical insulator for the axon. It also allows the signal to move along the axon faster. Because the axon is insulated by the myelin sheath, the impulse must jump along from one node of Ranvier to the next. This jumping action is much faster than if the impulse had to travel the entire length of the axon. This is analogous to skipping vs. walking, and to throwing a basketball from person to person across a basketball court rather than dribbling it the entire way.
2. A nerve impulse is carried along an axon as a bioelectrical signal caused by Na+ and K+ crossing the neuron's membrane. The action potential is a sudden reversal in the charge difference across the membrane (depolarization) followed by its restoration (repolarization.) Communication occurs between neuron by chemical messengers. A neurotransmitter is released from vesicles at the presynaptic membrane, diffuses across a gap between the two neurons, and is received by receptors on the postsynaptic membrane where it can be either excitatory or inhibitory.
3. Be careful not to hurt your partner. Stimulate the same place with increasingly quick pricks. Soon, your partner will not be able to distinguish between touches.
4. Usually a search of the Internet by the name of the neurotransmitter or pharmaceutical will give the web sites with information. The clinical information will most likely be associated with the manufacturer's web site. You might try these drugs: Accept®, Exelon®, Reminyl®, Zoloft®, Prozac®, and Paxil®.

Neurotransmitter	Action	Clinical Effect
Acetylcholine	neurostimulator	low levels may lead to Alzheimer's disease
Serotonin (also dopamine and norepinephrine)	produced feeling of general well-being	low levels can lead to depression
Acetylcholinesterase	breaks down acetylcholine	causes muscle spasms by inhibiting acetylcholinesterase
Dopamine	control of complex muscular movements	low levels lead to Parkinson's disease

5. Damage to the PNS is most likely to show gradual healing. Damage to the CNS will not. Therefore, stroke patients and those with spinal cord injuries are unlikely to see improvement. However, damage to the nerves in limbs due to an accident may show improvement. In general, problems arise if the neurotransmitters have been damaged, if the neurons themselves have been damaged as in the case of Alzheimer's disease, or if the nerve pathway has been severed. Diseases such as Alzheimer's and Parkinson's will progressively get worse.

Practice Test

1. b	11. b
2. a	12. d
3. a	13. c
4. b	14. c
5. b	15. c
6. c	16. b
7. a	17. b
8. d	18. b
9. d	19. d
10. a	20. c

Short Answer Questions

1. The drawing of the neuron should be similar to Figure 7-2a with multiple dendrites and one long axon. The next drawing should resemble Figure 7-3a showing the myelin sheath and Schwann cells. The myelin sheath plays a role in helping to repair a neuron in a cut or crushed nerve. Cells of the CNS are not covered in the myelin sheath and do not have the regenerative advantage of the Schwann cells.
2. Your drawing should look similar to Figure 7-1.
3. Refer to Figure 7-5 in the text. During the resting state, the plasma membrane is charged with the inside negative compared to the outside. When a stimulus arrives, the loss of the charge difference across the membrane occurs as the sodium channels open and sodium ions enter the axon. The inside of the axon becomes positively charged. This is depolarization. Repolarization occurs as the potassium channels open and potassium ions leave the neuron. Finally the sodium-potassium pump restores the original distribution of ions and the charge is restored to its original resting potential.
4. The drawing should resemble something similar to Figure 7-7. First packets of neurotransmitters are released from the synaptic knobs of the presynaptic neuron. When the nerve impulse reaches the synaptic knob, it causes the calcium ion channels to open. This causes the membranes of the synaptic vesicles to fuse with the plasma membrane and release the neurotransmitter substance into the synaptic cleft. The neurotransmitter diffuses across the synaptic cleft and binds to receptors on the membrane of the postsynaptic neuron. When the neurotransmitter binds to its receptor, an ion channel is opened. The opening of the channel can excite or inhibit the postsynaptic neuron. The postsynaptic cells integrate excitatory and inhibitory synaptic input. The effects of the neurotransmitters are temporary as they are either deactivated by enzymes or actively pumped back into the axon tip.

Chapter 8

The Nervous System

Objectives

After reading the text and studying the material in this chapter, you should be able to:

- Describe the function of the central and peripheral nervous systems.
- List the components of both the central and peripheral nervous systems.
- Describe the structure of the cerebrum.
- List the three functional areas of the cerebral cortex.
- Explain the location and functions of the primary sensory and motor areas of the cerebral cortex.
- Compare and contrast the left and right hemispheres.
- List the functions of the thalamus and hypothalamus.
- Describe the functions of the cerebellum.
- Describe the function of the medulla.
- Describe what causes/controls emotions.
- Describe two functions of the reticular activating system.
- List the functions of the spinal cord and relate its structure to efficiently performing these functions.
- Describe the reflex arc that occurs when you have stepped on something sharp.
- List the structures that provide protection for the central nervous system.
- Explain why meningitis is so very serious.
- Describe the function of the somatic and autonomic nervous systems.
- Compare the sympathetic and parasympathetic nervous systems.
- Describe the role of the sympathetic nervous system in an emergency.
- Differentiate between most common headaches and migraine headaches both in terms of cause and severity.
- Describe what causes a stroke and what determines the resulting damage.
- Explain what causes a coma.
- Explain why, based on the structure and function of the spinal cord, an injury results in the lack of function below the injury.

Chapter Summary

The nervous system is divided into the **central nervous system** and the **peripheral nervous system**. The central nervous system is the brain and spinal cord. It is concerned with integrating incoming information and coordinating all voluntary and involuntary responses. Large clusters of nerve cells, called **ganglia**, are located outside the central nervous system. The peripheral nervous system consists of the nerves branching from the central nervous system and the ganglia. The peripheral nervous system can be further subdivided on the basis of function. **The somatic nervous system** receives sensory information and directs voluntary movements and the **autonomic nervous system** regulates involuntary reactions.

The **cerebrum** is the thinking, conscious part of the brain. It consists of two hemispheres, which receive sensory information from and direct the movements of the opposite side of the body. The cerebrum consists of the outer layer of **gray matter** called the **cerebral cortex**, the underlying **white matter** made of myelinated nerve tracts that allow for communication between various areas of the brain. The **corpus callosum** connects the two cerebral hemispheres. The sensory, motor and association areas are in the cerebral cortex. The **primary sensory area** receives sensory information from the body while the **primary motor area** controls the skeletal muscles.

The **thalamus** is the relay station of the brain for all sensory information except smell. It also directs motor activity, cortical arousal and memory. The **hypothalamus** maintains homeostasis by regulating blood pressure, heart rate, breathing rate, digestion and body temperature. The hypothalamus coordinates the nervous and endocrine systems through its connection to the pituitary gland. As part of the limbic system, the hypothalamus is a center for emotions. It also serves as the master biological clock.

The **cerebellum** controls sensory-motor coordination. It integrates information from the motor cortex and sensory pathways to produce movements. The cerebellum also stores memories of learned motor skills. The **medulla oblongata** contains reflex centers to regulate the rhythm of breathing, force and rate of the heartbeat, and blood pressure. It serves as the pathway for all sensory messages to the higher brain centers and motor messages leaving the brain. The **limbic system**, which includes several brain structures, is largely responsible for emotions. The **reticular activating system** filters sensory input and keeps the cerebral cortex in an alert state.

The **spinal cord** is a tube of neural tissue running from the medulla to the bottom of the rib cage. Spinal nerves arise from the cord and exit through the openings between the stacked vertebrae of the vertebral column. The spinal cord conducts messages between the brain and the body and serves as a reflex center. A reflex action is an automatic response to a stimulus in a pre-wired circuit called a **reflex arc**. The brain and spinal cord are protected by the bones of the vertebral column and skull, three connective tissue membranes that form the **meninges**, and **cerebrospinal fluid**. Inflammation of the meninges is called **meningitis**.

The peripheral nervous system includes the sensory receptors, the peripheral nerves and ganglia, and specialized motor endings that stimulate the effectors. The **peripheral nervous system** is divided into the **somatic nervous system**, which governs conscious sensations and voluntary movements, and the **autonomic nervous system**, which is concerned with our unconscious, involuntary internal activities. The autonomic nervous system can be divided into the **sympathetic** and **parasympathetic nervous systems**, two branches with antagonistic actions. The sympathetic nervous system gears up the body for stressful or emergency situations while the parasympathetic nervous system adjusts body functioning so that energy is conserved during nonstressful times.

Health problems with the nervous system vary in significance. Headaches are usually caused by tension in the neck or by dilation of the blood vessels of the head. Migraine headaches are caused by an imbalance in the brain's chemistry. A stroke results when the brain is deprived of blood and nerve cells die. The extent and location of the damage caused by a stroke depends on the affected region of the brain. A coma is caused by trauma to neurons in regions of the brain responsible for stimulating the cerebrum. A comatose person is totally unresponsive to all sensory input and cannot be awakened. A spinal cord injury results in loss of function below the site of injury.

Key Concepts

- The nervous system is divided into the central nervous system and the peripheral nervous system.
- The central nervous system consists of the brain and spinal cord and is responsible for integrating incoming information and coordinating all voluntary and involuntary responses.
- The peripheral nervous system consists of all neural tissue outside the central nervous system including ganglia.
- Bone, membranes and cerebrospinal fluid protect the central nervous system.
- The brain is the primary organ of the nervous system and consists of many parts with specific functions.
    - ✓ The cerebrum is the largest and most prominent part of the brain and is responsible for consciousness.
    - ✓ The thalamus receives sensory input, sorts information and relays it to the appropriate regions of the brain for processing.
    - ✓ The hypothalamus maintains homeostasis and is the bridge to the endocrine system.
    - ✓ The cerebellum is responsible for coordination of sensory input and motor output.
    - ✓ The medulla oblongata contains the reflex centers that keep us alive.
    - ✓ The pons connects the brain to the medulla oblongata, and thus to the spinal cord.
- The limbic system allows us to experience emotions and to maintain short- and long-term memory.
- The reticular activating system filters sensory input and stimulates the cerebral cortex to maintain consciousness.
- The spinal cord transmits messages to and from the brain and is a reflex center.
- The peripheral nervous system consists of the somatic nervous system and the autonomic nervous system.
- The somatic nervous system carries information to and from the CNS resulting in sensations and voluntary movement.
- The autonomic nervous system governs involuntary, unconscious activities and maintains homeostasis.
- The autonomic nervous system consists of the sympathetic nervous system that takes over during emergency conditions and the parasympathetic nervous system that conserves energy during non-stressful times.
- Health issues and injuries of the nervous system vary in seriousness depending upon the nature of the problem and the location.

Study Tips

Take a moment to look at the list of topics on the very first page of the chapter. This shows you the different divisions of the nervous system and some possible problems associated with it. You will need to be able to recall the components of each facet of the nervous system and the function of each organ/structure. In addition, you will need to understand how these organs and systems work together.

Some students are linear thinkers and study best from notes organized one point after another such as in a jot list or outline. If you are such a student, then you may benefit from outlining the chapter and making a list of each organ and structure. If you do not learn best from studying one point at a time, but rather learn by "seeing" how all the parts relate to one another, then you may benefit from developing a concept map for the content in this chapter. If concept mapping is new to you, you may want to learn more and to view example concept maps at the web site http://cmap.coginst.uwf.edu/info/printer.html, The Theory Underlying Concept Maps and How to Construct Them.

To begin your study of the nervous system, first identify the major divisions. You may list them, if you are a linear thinker, or write them on sticky notes so that you can move them around in a concept map. Next determine the function of each system and the components that make it up. If you are making a concept map, draw lines to connect these structures. Finally, determine the interconnections between all of the components. Draw lines that connect any parts of the nervous system that interact with one another. At this point, your concept map will begin to look like a spider web. It is these interconnections that make the nervous system so unique and that also make injuries so debilitating.

Review Questions

A. Use the definition of the central nervous system, peripheral nervous system, somatic nervous system and autonomic nervous system to complete the following sentences. One system may be used twice. Each sentence gives an example of the function of that system.

1. Movement of food through the intestine is controlled by the ____________.

2. You will rely on your memory, located in the ____________, as you study for the next test.

3. The heat receptors in your fingertips are part of the ____________.

4. A quick reflex, such as when you step on a sharp stone with your bare foot, is a result of

   both the ____________ and the ____________.

5. Everything neural that does not involve the brain and spinal cord involves the

   ____________.

B. Fill in the blanks in the following sentences.

1. The deepest groove in the brain is called the ____________ ____________, which separates the brain into the ____________ and ____________ ____________.

2. The cerebral cortex is made of ____________ ____________ while the layer just below this is called white matter and is made of ____________ ____________.

3. Communication between the two hemispheres of the brain is accomplished by the ____________ ____________ and the ____________ ____________.

4. One important area of the cerebrum is the ____________ area that is responsible for the analysis of information and communication between areas of the brain.

5. When doing mathematics or other analytical tasks, one primarily uses the ____________ hemisphere, but the parts of the brain used for spatial tasks and visual memory are located in the ____________ hemisphere.

C. Match the organ or structure in Column A with the function in Column B.

Column A	Column B
1. cerebellum	a. emotional brain
2. hypothalamus	b. relay station
3. limbic system	c. alert system
4. pons	d. connects nervous and endocrine systems
5. reticular activating system	e. balance and muscle memory
6. thalamus	f. connects spinal cord to brain

D. Draw a reflex arc that would result if someone stepped on a piece of glass and cut his or her foot. Use the following terms to label your drawing: stimulus, sensory neuron, interneuron, and motor neuron.

E. Fill in the blank.

As you relax watching a television show, the ____________ nervous system is busy conserving energy for a time of action. Your relaxation is interrupted by the sound of a car accident. Immediately the ____________ nervous system kicks in, making your heart beat quickly and allowing you to run to the road to be of assistance. As you approach the scene of the accident, you count the people in the car, take note of injuries and begin to be of assistance. The ____________ nervous system allows you to perceive the victim's injuries and call for help.

F. Match the term in Column A with the symptom in Column B.

Column A	Column B
1. coma	a. Suddenly Grandpa fell to the ground unable to talk or stand up.
2. migraine	b. They had been swimming at the quarry when one person did not come to the surface after diving into the water. After pulling the victim out of the water, they realized that he could not walk.
3. spinal cord injury	c. She clutched her head, complaining of pain on the left side right behind her eye.
4. stroke	d. Tanya had been hit by a car while riding her bike. Her friends visited her in the hospital commenting that she looked asleep. But they knew she was unconscious and could not be awakened.

Critical Thinking Questions

Read each of the following questions carefully. If it helps you, underline or list each component to the question. Use complete sentences in your explanation.

1. Studying the material in this chapter and preparing for your exam require many functions of the nervous system. First list the areas of the nervous system that are involved in your learning and study. Then choose at least three of those areas and explain how they interact with other systems to help you succeed in this course.

2. Animals lower on the evolutionary tree have less well-developed nervous systems than animals higher on the evolutionary tree. In general, the cerebrum of lower animals is much smaller than that of higher animals, including humans, but in animals requiring exceptional balance, the cerebellum is proportionally larger in lower animals than in humans. Explain

the differences in the function the cerebrum and the cerebellum and why greater cerebral capacity is required in more highly developed animals.

3. Explain the statement, "The nervous system can be subdivided into many component parts, yet all parts must work together in order for the organism to function effectively." Include examples in your explanation and address the apparent contradiction between the clear subdivisions of the nervous system and their necessary interconnectedness.

4. Choose either a stroke or a spinal injury and explain, based on the nature of neurons, the severity of these injuries and why victims don't "get better."

Practice Test

Choose the one best answer to each question that follows. As you work through these items, explain to yourself why each answer you discard is incorrect and why the answer you choose is correct.

1. What are the main parts of the central nervous system?
   a. brain and cranial nerves
   b. brain and vagus nerve
   c. spinal cord and brain
   d. spinal nerves and spinal cord.

2. A ganglion is a ____________.
   a. clusters of nerve cell bodies
   b. node of the spinal cord
   c. subdivision of the brain stem
   d. very large neuron

3. Which part of the brain is most important to your recall of information for success on this test?
   a. cerebellum
   b. cerebrum
   c. medulla
   d. thalamus

4. What is the primary function of the cerebellum?
   a. center of consciousness
   b. control of digestion, circulation, and breathing movements
   c. coordination of complex muscular movements
   d. coordination of endocrine and nervous responses

5. You just finished running and your heart rate is fast as is your breathing. The part of the brain controlling basic body functions (heart rate and blood pressure) is the ____________.
   a. cerebellum
   b. cerebral cortex
   c. medulla
   d. spinal cord

6. The thalamus is responsible for which of the following functions?
   a. connection between speech and hearing
   b. control of many of the endocrine glands
   c. integration and relay of information
   d. motor coordination

7. A major portion of the motor area of the brain is devoted to the ____________.
   a. foot and toes
   b. hand and fingers
   c. senses
   d. all of the above have equal area

8. As Tonya read the obituary of her neighbor, she felt sad. What system is considered our emotional brain?
   a. central nervous system
   b. frontal lobes
   c. limbic system
   d. reticular activating system

9. Joseph's heart was racing as he read the concluding chapter to the mystery. What controlled the physiological response to his emotions?
   a. cerebral cortex
   b. hypothalamus
   c. reticular activating system
   d. thalamus

10. What part of our brain takes the information you just heard in lecture and converts it from short-term memory to long-term memory?
    a. corpus callosum
    b. hippocampus
    c. hypothalamus
    d. thalamus

11. What area of the brain filters sensory information and helps to keep us alert?
    a. cerebral cortex
    b. hippocampus
    c. medulla
    d. reticular activating system

12. The spinal cord is similar to the brain in all of the following ways except one. In which way are they different?
    a. contains both gray and white matter
    b. controls motor responses
    c. controls thinking
    d. protected by bone

13. The primary functions of the spinal cord involve ____________.
    a. controlling muscle activity and maintaining balance
    b. intelligence and memory
    c. reflex actions and communication between the brain and spinal nerves
    d. speech, smell, taste, hearing and vision

14. What is the primary advantage to a spinal reflex?
    a. It depends upon the decision-making abilities of the brain.
    b. It efficiently controls regions farthest from the brain.
    c. It involves both gray and white matter.
    d. It is very fast.

15. What is the greatest disadvantage/weakness of nerve cells?
    a. They are constantly expanding to cover new areas with greater detail.
    b. They can only conduct information.
    c. They cannot divide and produce new cells.
    d. They are protected by both bone and tissues.

16. Cerebrospinal fluid functions to ____________ the brain.
    a. cushion
    b. nourish
    c. support
    d. all of the above

17. Spinal nerves are able to carry which of the following types of information?
    a. intelligence carried in the interneurons
    b. motor only
    c. sensory only
    d. both motor and sensory

18. The reaction of the body to an emergency is controlled by the ____________ nervous system.
    a. central
    b. parasympathetic
    c. somatic
    d. sympathetic

19. What is the cause of the most common type of headache?
    a. bright lights
    b. constricted cranial vessels
    c. tension in the muscles of the face, head, and neck
    d. usual brain chemistry imbalances

20. Which of the following might be the result of a spinal injury, such as that from a diving accident?
    a. aneurism
    b. coma
    c. paralysis
    d. stroke

Short Answer Questions

Read each of the following questions carefully. If it helps you, underline or list each component to the question. Formulate your answer and jot down the main points. Then use complete sentences in your explanation.

1. List the components of the central and the peripheral nervous systems. Then compare the functions of the two systems.

2. Describe the structure and function of the cerebral cortex. Include any differences in the two hemispheres.

3. Identify the parts of the brain that are involved in an emotional response to a stimulus and describe how they interact.

4. Compare the parasympathetic and sympathetic nervous systems. Include the level of activity, the anatomy of the two systems, the effects they each have on various organ systems, and when each system is used.

ANSWER KEY

REVIEW QUESTIONS

A. 1. Movement of food through the intestine is controlled by the autonomic nervous system.
   2. You will rely on your memory, located in the central nervous system, as you study for the next test.
   3. The heat receptors in your fingertips are part of the somatic nervous system.
   4. A quick reflex, such as when you step on a sharp stone with your bare foot, is a result of both the central nervous system and the peripheral nervous system.
   5. Everything neural that does not involve the brain and spinal cord involves the peripheral nervous system.

B. 1. The deepest groove in the brain is called the longitudinal fissure, which separates the brain into the left and right hemispheres.
   2. The cerebral cortex is made of gray matter while the layer just below this is called white matter and is made of myelinated axons.
   3. Communication between the two hemispheres of the brain is accomplished by the corpus callosum and the basal ganglia.
   4. One important area of the cerebrum is the association area that is responsible for the analysis of information and communication between areas of the brain.
   5. When doing mathematics or other analytical tasks, one primarily uses the left hemisphere, but the parts of the brain used for spatial tasks and visual memory are located in the right hemisphere.

C. 1. e
   2. d
   3. a
   4. f
   5. c
   6. b

D. The reflex arc should resemble Figure 8.10 in your text showing the perception of pain by sensory neurons being carried to the spinal cord and then to motor neurons that stimulate the appropriate muscles. You might also include the connection of the interneurons as the impulse is carried to the brain as is registered as an injury.

E. As you relax watching a television show, the parasympathetic nervous system is busy conserving energy for a time of action. Your relaxation is interrupted by the sound of a car accident. Immediately the sympathetic nervous system kicks in, making your heart beat quickly and allowing you to run to the road to be of assistance. As you approach the scene of the accident, you count the people in the car, take note of injuries and begin to be of assistance. The somatic nervous system allows you to perceive the victim's injuries and call for help.

F. 1. d
   2. c
   3. b
   4. a

CRITICAL THINKING QUESTIONS

1. The cerebrum is the thinking, conscious center of the brain. It receives the sensory information that brings new knowledge to you. The cerebellum coordinates the writing activity of taking notes and answering questions. The medulla oblongata maintains heart rate and blood pressure so that you are alive. The corpus callosum connects the two hemispheres while the thalamus filters out non-essential information as you sit listening to a lecture. As the instructor looks your way and may call

on you for a response, the hypothalamus interfaces with the endocrine system to heighten your response levels.

2. It is the cerebrum that accounts for 83% of our total brain weight. It is home to our mental functions and those characteristics that make us "human." The primary function of the cerebellum is sensory-motor coordination. It integrates information from the motor cortex and sensory pathways to produce smooth movements. As humans we use language, mathematics and stored knowledge to a far greater degree than other animals. However, other animals rely on a greater degree of balance as they rest on a wire or walk along the ledge of a cliff preparing to pounce on a prey.
3. The nervous system is divided into the central and peripheral nervous systems. The central nervous system consists of the brain and spinal cord. The peripheral nervous system consists of everything else and is divided into the somatic and autonomic nervous systems. The somatic nervous system controls conscious functions and the autonomic nervous system controls internal organs. The autonomic nervous system can be further subdivided into the sympathetic and parasympathetic nervous systems. While divided both anatomically and by function, all aspects of the nervous system must work together for us to accomplish everyday tasks. The CNS gives us our ability to think, process information and recognize friends. The peripheral nervous system, controlling voluntary movement, allows us to receive sensory information and swing a bat or shake hands. The two branches of the autonomic nervous system oppose one another. The parasympathetic nervous system adjusts body functions so that energy is conserved during restful times while the sympathetic nervous system gears the body to face stressful or emergency situations. As we end the day, we can reflect on many instances where the smooth integration of these systems was critical.
4. Both a stroke and a spinal cord injury damage parts of the central nervous system. As discussed in the previous chapter, the nature of the neurons in the CNS and the lack of glial cells restrict the recovery of damaged cells. A stroke is the death of nerve cells in a region of the brain. These cells are permanently unable to function and therefore are unable to carry out their usual cognitive or motor tasks. The spinal cord contains pathways that allow the brain to communicate with the rest of the body. Therefore, damage to the spinal cord can impair sensation and motor control below the site of the injury. If the cord is completely severed, there is a complete loss of sensation and voluntary movement below the level of the cut.

Practice Test

1. c
2. a
3. b
4. c
5. c
6. c
7. b
8. c
9. b
10. b
11. a
12. c
13. c
14. d
15. c
16. d
17. d
18. d
19. c
20. c

Short Answer Questions

1. The central nervous system consists of the brain and spinal cord. The central nervous system integrates and coordinates all voluntary and involuntary nervous functions. The nerves and ganglia located outside of the central nervous system make up the peripheral nervous system. The peripheral nervous system keeps the central nervous system in contact with all parts of the body.
2. The outer layer of each hemisphere of the cerebrum is called the cerebral cortex. It consists of billions of neuroglial cells, nerve cell bodies and unmyelinated axons described as gray matter. It is highly convoluted to increase surface area. The cerebral cortex is divided into two hemispheres. Each hemisphere is divided into the frontal, parietal, temporal and occipital lobes. The cerebrum is

the thinking portion of the brain. Each hemisphere receives sensory impressions from and directs the movements of the opposite side of the body. There are three functional areas of the brain: motor areas, sensory areas, and association areas.

3. The limbic system is our emotional brain. The hypothalamus coordinates the activities of the nervous and endocrine systems. It also connects to lower brain centers such as the medulla. The thoughtful responses of the nervous system are connected to the emotional responses of the endocrine system through the pituitary gland.
4. The sympathetic nervous system gears the body to face an emergency or stressful situation. Its neurons release mostly norephinephrine and respond as a whole. In contrast, the parasympathetic nervous system adjusts body functions so that energy is conserved during relaxation. It secretes acetylcholine as a neurotransmitter. The effects of the parasympathetic nervous system occur more independently of one another. Both systems innervate nearly all internal organs having opposite effects.

Chapter 8A

Drugs and the Mind

Objectives

After reading the text and studying the material in this chapter, you should be able to:

- Differentiate between tolerance, cross-tolerance and physical dependence.
- Describe the mechanisms of action of psychoactive drugs.
- List examples of each class of drugs discussed in this chapter.
- Explain why different individuals may experience different levels of intoxication having had the same amount of alcohol.
- List the effects of alcohol on the various body systems, nutrition, cancer and fetal development.
- Explain the effects of THC.
- Describe the long-term effects of marijuana on the body.
- Describe how stimulants act on the CNS.
- Give the effects of stimulants, including amphetamines, on the body.
- Describe the danger of hallucinogenic drugs.
- Explain the effects of sedatives on the CNS.
- Compare the positive and negative aspects to opiates.

Chapter Summary

A psychoactive drug is one that alters one's mood or emotional state. Psychoactive drugs alter the communication between neurons by stimulating, inhibiting or dampening the release of a neurotransmitter, or by altering the binding site of the neurotransmitter.

When a person's body requires higher and higher doses of a drug to generate a response, we say that they have built up a tolerance to that drug. Cross-tolerance occurs when tolerance to one drug results in a lessened response to another, usually similar, drug. When someone can no longer function normally without the drug, then they have a dependence on it. People tend to use, and abuse, drugs that stimulate pleasure centers.

The effects of alcohol on a person's behavior depend on the blood alcohol level, which is affected by how much alcohol is consumed, how quickly it is consumed, the rate of absorption, distribution of body fat and metabolic rate. The amount of ethanol per ounce of alcoholic beverage can vary from less than 4% to nearly 50%. The intoxicating effects of alcohol begin when it is absorbed into the blood and is delivered to the brain. Some alcohol is absorbed through the stomach lining and some through the intestinal walls. Since alcohol is distributed to all parts of the body, the same amount of alcohol will affect a small person much more than a large person. The liver converts alcohol to carbon dioxide and water at the rate of about one-third of an ounce of pure ethanol per hour. If intake is faster than metabolism, then a person becomes more intoxicated the more alcohol that is consumed. Due to differences in weight, percent body fat and levels of stomach enzymes, alcohol has a much greater effect on women than on men.

Alcohol has a negative effect on nearly every organ in the body. In the brain, it permanently kills nerve cells resulting in a loss of cognitive ability, impairs judgment, slows reflexes and impairs balance and coordination. Finally the brain can be affected by excessive amounts of alcohol to the point of causing unconsciousness, coma and death.

Alcohol has little nutritional value and is very high in calories. It can inhibit the absorption of certain vitamins and may cause the kidneys to eliminate crucial ions. Excessive alcohol consumption damages the liver and causes the build up of fat reducing blood flow through the liver. Hepatitis and cirrhosis of the liver can result in death. A person who drinks heavily is at least twice as likely to develop cancer of the mouth, tongue or esophagus.

The effect of alcohol on the circulatory system is mixed. In moderate amounts, alcohol reduces the effects of stress and releases HDL into the blood. However, when consumed in larger quantities it damages the heart and blood vessels. Alcohol consumption can have devastating effects on the fetus and even limited chronic consumption while pregnant may lead to fetal alcohol syndrome.

Although alcohol is America's number one drug problem, marijuana is the most widely used illegal drug in the United States today. The principal psychoactive ingredient in the leaves, flowers, and stems of *Cannabis sativa* is THC. THC binds to receptor sites in the brain that regulate mood, memory, pain, appetite, and movement. The most harmful long-term result of prolonged smoking of marijuana is marked trauma to the respiratory system. Marijuana causes an increase in heart rate and blood pressure. THC reduces the levels of testosterone and sperm production in men. The effect on the functions of the female reproductive system is unclear. THC has detrimental effects on driving skills and performance.

Stimulants are drugs that excite the CNS. Cocaine augments the neurotransmitters dopamine and norepinephrine bringing about a rush of intense pleasure, a sense of self-confidence and power, clarity of thought, and increased physical vigor. This high is very short lived followed by depression, anxiety and extreme fatigue. Cocaine has negative effects on the cardiovascular system as do all stimulants and can be a fatal depressant to the respiratory system. Amphetamines are another class of stimulants. Tolerance to amphetamines develops along with both physical and psychological dependence. Unfortunately, amphetamines can cross the placenta and affect a developing fetus. Caffeine, another stimulant, blocks the effect of the inhibitory chemical adenosine. Low doses of caffeine increase wakefulness and mental alertness, but higher doses can lead to agitation, anxiety, tremors, rapid breathing, and irregular heartbeats. Caffeine is addicting and can lead to withdrawal symptoms of headache, depression and anxiety when consumption is stopped.

Hallucinogenic drugs alter sensory perception. Psychedelic drugs act by augmenting the action of the neurotransmitters serotonin, norepinephrine or acetylcholine. The real dangers of these drugs are the distortions of reality that may lead to dangerous actions. Tolerance for and cross-tolerance among psychedelic drugs develop quickly.

Sedatives depress the CNS. First they reduce the effect of inhibitory neurons and then excitatory neurons. The effects of sedatives are additive and continued use leads to both tolerance and dependence. "Date rape" drugs fall into this classification.

Opiates are natural or synthetic drugs that reduce pain, produce a sense of euphoria and reduce anxiety. They act by binding to the sites of the body's natural opiate producers.

Key Concepts

- Psychoactive drugs alter communication among neurons.
- Tolerance is a progressive decrease in the effectiveness of a drug while cross-tolerance is the tolerance for one drug while taking another, usually related drug.
- Dependence requires continued use of a drug for normal functioning. Most addictive drugs stimulate the pleasure centers of the brain.
- Alcohol depresses the CNS.
- The concentration of ethanol varies depending upon the drink. The effect it has on the body depends upon the concentration, absorption rate, a person's weight and percent body fat.
- Alcohol has negative effects on nearly all systems of the body, including a developing fetus. Alcohol consumption leads to accidents and poor judgment.
- Marijuana is the most widely used illegal drug in the United States.
- THC is the psychoactive ingredient in marijuana that produces feelings from those of well being to hallucinations and paranoia.
- Continued marijuana use negatively affects the respiratory, circulatory and reproductive systems.
- Stimulants excite the CNS.
- Cocaine and amphetamines augment the neurotransmitter dopamine, bringing about sense of pleasure, and increase the effects of norepinephrine, making one feel alert, energetic and confident.
- Caffeine blocks the effect of the inhibitory chemical adenosine, a neuromodulator, a chemical that is released by one neuron but affects other nearby neurons.
- Hallucinogenic drugs alter sensory perceptions by augmenting the neurotransmitters serotonin, norepinephrine, or acetylcholine.
- Sedatives depress the CNS. The effects of several sedatives are additive and can be fatal.
- Opiates are important medically because they alleviate severe pain, but they also cause a rush of euphoria and physical dependence.

Study Tips

This chapter focuses on the effects of various psychoactive drugs. One study technique that is effective in situations that depend upon memorization is flash cards. Although you may first think of elementary school mathematics, flash cards can help you organize the information and take advantage of many short opportunities for study. Get a package of brightly colored note cards and use one color for each type of drug – alcohol, marijuana, stimulants, sedatives, psychedelic drugs and opiates. As you read the text and attend lecture, focus on the most important concepts to include on the cards. Use the headings in the text to guide your organization. Clarify with your instructor what you will need to know about each drug. You will certainly need to know its classification, its effect, and the method of action. You may also need to know its short- and long-term effects.

REVIEW QUESTIONS

A. Write a short definition of each term listed below and elaborate with an example.

1. Tolerance: ____

____

2. Cross tolerance: ____

____

3. Physical dependence: ____

____

B. Fill in the blanks in the following sentences.

1. Alcohol ____________ the central nervous system.

2. The effect alcohol has on the body is determined by the amount of ____________ in the beverage.

3. The alcohol content of a beverage is measured as ____________.

4. One degree of proof equals ____________ percent alcohol.

5. A standard "drink" contains ____________ ounces of pure ethanol.

6. This is equivalent to about one bottle of beer, a ____________ of wine, or a jigger of ____________ distilled spirits.

7. The intoxicating effects of alcohol begin when it is absorbed from the ____________ system into the blood and is delivered to the ____________.

C. Alcohol absorption varies from person to person. List the factors that affect alcohol absorption.

1. ____

2. ____

3. ____

4. ____

D. Complete the following table showing the effect alcohol has on the various systems/organs of the body.

System/Organ	**Effect**
Nervous System	
Nutrition	
Liver	
Cancer	
Heart and Blood Vessels	
Fetal Development	

E. Write the drug in the space in Column A which matches the effect or example in Column B. Each of the following may be used twice: alcohol, amphetamines, caffeine, cocaine, hallucinogens, marijuana, opiates, and sedatives.

Column A	Column B
1. ____________________	a. depresses CNS and has an additive effect when mixed with alcohol
2. ____________________	b. interferes with re-uptake of dopamine
3. ____________________	c. causes cirrhosis of the liver
4. ____________________	d. alters sensory perception
5. ____________________	e. inhibits adenosine activity
6. ____________________	f. may be used to treat ADHD

7. ______________________ g. THC causes feelings of "bliss" and affects mood

8. ______________________ h. examples are morphine, codeine, heroin

9. ______________________ g. most widely used illegal drug in US

10. ______________________ h. reduces pain

11. ______________________ g. examples are Valium® and GHB (gamma-hydroxybutrate)

12. ______________________ h. causes up to ½ of driving accidents per year

13. ______________________ g. longer acting than cocaine

14. ______________________ h. crack is an extremely potent form

15. ______________________ g. examples are LSD, ecstasy and mescaline

16. ______________________ h. in moderation increases mental alertness; in excess causes anxiety and irregular heartbeats

Critical Thinking Questions

Read each of the following questions carefully. If it helps you, underline or list each component to the question. Use complete sentences in your explanation.

1. Justin, a lineman on the football team, and Amy, a petite and thin dancer, went to a party where they drank alcohol. They consumed the same amount of alcohol, but it seemed to affect Amy much more. Explain why this might be the case including the factors that affect alcohol absorption.

2. Explain why drinking coffee won't help someone recover from a drinking binge or too many sleeping pills.

3. Use the Internet to identify a reputable web site that provides data on drug and alcohol use and abuse. Choose at least three drugs, including alcohol and nicotine, and find the percentage of people in your age group who use the drug. Is there a problem of abuse? What are the physical effects of quitting? What is the cost to the person and to society at large when there is an addiction?

4. If possible, visit with a pediatrician or pediatric nurse about the effects of Fetal Alcohol Syndrome. Use the Internet to find out how many children are born with FAS in your area.

Practice Test

Choose the one best answer to each question that follows. As you work through these items, explain to yourself why each answer you discard is incorrect and why the answer you choose is correct.

1. Drugs that alter one's mood are called ___________.
   a. addictive
   b. neurological
   c. psychoactive
   d. psychedelic

2. When it takes increasing amounts of a drug to have the same effect, someone is said to be ___________.
   a. abusive
   b. cross tolerant
   c. dependent
   d. tolerant

3. Illegal drugs gain their popularity because they primarily affect which portion of the brain?
   a. cerebellum
   b. frontal lobe of the cerebrum
   c. limbic system
   d. psychological system

4. The ingredient in alcoholic beverages that affects the body is ___________.
   a. ethanol
   b. grain alcohol
   c. methanol
   d. all of the above

5. Blood alcohol level is solely dependent upon the number of drinks consumed.
   a. True
   b. False

6. Two students were arguing. One claimed that two glasses (5 oz each) of wine at dinner did not contain nearly the same amount of alcohol as two bottles of beer (12 oz each.) Was this student correct?
   a. Yes
   b. No

7. Why is it a good idea to eat food while drinking alcoholic beverages?
   a. Food dilutes the alcohol.
   b. Food slows the absorption of alcohol through the stomach.
   c. Food covers part of the surface of the stomach, preventing absorption.
   d. all of the above

8. Which organ is responsible for metabolizing alcohol?
   a. liver
   b. pancreas
   c. small intestine
   d. stomach

9. Which of the following is not true about alcohol and its effects on the body?
   a. it affects coordination and judgment
   b. it increases blood pressure and heart rate
   c. the calories help with general nutrition
   d. when combined with smoking, rates of cancer increase

10. Nearly ½ of all traffic fatalities are due to which of the following?
    a. alcohol
    b. amphetamines
    c. caffeine
    d. opiates

11. What is the active drug in marijuana?
    a. ADHD
    b. GHB
    c. PCB
    d. THC

12. What system is most negatively impacted by long-term smoking of marijuana?
    a. circulatory
    b. nervous
    c. respiratory
    d. reproductive

13. Which of the following is not similar to the others in effect?
    a. amphetamines
    b. caffeine
    c. cocaine
    d. codeine

14. Amphetamines are used to treat ____________.
    a. attention deficit hyperactivity disorder
    b. fetal alcohol syndrome
    c. hangovers
    d. psychosis

15. What is the mechanism of action of caffeine?
    a. blocks adenosine action
    b. blocks pain
    c. increases dopamine levels
    d. increases serotonin levels

16. It has been proven that using marijuana leads to use of more serious drugs.
    a. True
    b. False

17. What is the effect of hallucinogens on the body?
    a. alter heart rate and respiration rate
    b. alter sensory perception
    c. decrease use of frontal lobe of cerebrum
    d. increase norepinephrine levels

18. Which of the following is not a hallucinogen?
    a. angel dust
    b. crack
    c. ecstasy
    d. Special K

19. Barbiturates are a ____________.
    a. depressant
    b. narcotic
    c. psychedelic
    d. stimulant

20. The primary effect of opiates is to ____________.
    a. cause sedation
    b. create an altered state of consciousness
    c. reduce pain
    d. relieve anxiety

Short Answer Questions

Read each of the following questions carefully. If it helps you, underline or list each component to the question. Formulate your answer and jot down the main points. Then use complete sentences in your explanation.

1. Explain why someone may remain drunk long after he or she has stopped drinking.

2. When do the intoxicating effects of alcohol begin and what are they?

3. Describe the physiological effects of stimulants. When can they become deadly?

4. What is unique about the effects of psychedelic drugs compared to the other classes of drugs studied in this chapter?

5. Opiates are of great value to those suffering from cancer, yet they can lead to destruction of a life when misused. What is the effect of opiates and how is this used medically? What is their attraction as an illegal drug?

Answer Key

Review Questions

A. 1. The progressive decrease in the effectiveness of a drug. For example, when more and more of a drug is needed to produce a high.
  2. Development of a tolerance for a second drug, one not taken, by taking a similar drug as is the case when someone takes codeine and develops a tolerance for other opiates.
  3. When a drug is needed to maintain normal functioning as in the case of sleeping pills and caffeine.

B. 1. Alcohol depresses the central nervous system.
  2. The effect alcohol has on the body is determined by the amount of ethanol in the beverage.
  3. The alcohol content of a beverage is measured as proof.
  4. One degree of proof equals 0.5% percent alcohol.
  5. A standard "drink" contains 0.5 ounces of pure ethanol.
  6. This is equivalent to about one bottle of beer, a glass of wine, or a jigger of 80 proof, 40% distilled spirits.
  7. The intoxicating effects of alcohol begin when it is absorbed from the digestive system into the blood and is delivered to the brain.

C. 1. The rate of alcohol absorption depends on its concentration.
  2. Overall body size/weight
  3. Percentage of body fat
  4. Rate of metabolism in the liver

D.

System/Organ	Effect
Nervous System	Depresses inhibitory neurons. Slows activity of all neurons making thinking and reflexes much slower. Affects balance and coordination causing accidents. Kills nerve cells.
Nutrition	High in calories but little nutritional value. Metabolized for energy before body fat.
Liver	Fat accumulates reducing blood flow and possible cyst formation. Can lead to alcoholic hepatitis and cirrhosis.
Cancer	Drinkers are twice as likely to develop cancer of the mouth, tongue, or esophagus than a nondrinker. When combined with smoking, risk of cancer is multiplied. That means that the risk of cancer is greater than the sum of the risks for cancer from either drinking or smoking alone.
Heart and Blood Vessels	Moderate amounts relieve stress and raise HDL levels. Reduces the heart's ability to pump blood. Raises blood pressure, causes the heart to enlarge.
Fetal Development	Low birth weight. Increased risk of miscarriage and stillbirth. Cause of birth defects and mental retardation. Fetal alcohol syndrome.

E.
1. sedatives
2. cocaine
3. alcohol
4. hallucinogens
5. caffeine
6. amphetamines
7. marijuana
8. opiates
9. marijuana
10. opiates
11. sedatives
12. alcohol
13. amphetamines
14. cocaine
15. hallucinogens
16. caffeine

Critical Thinking Questions

1. If Justin and Amy consume the same amount of alcohol, Amy will feel the effects before Justin. This is because Amy is much smaller than Justin and her body contains a higher percentage of fat than does Justin's. Alcohol dissolves more slowly in fat than in water, which prolongs the effects of the alcohol in women. In addition, women have less of the enzyme found in the stomach that metabolizes alcohol and this means that they absorb more alcohol into their bloodstream than a man. Finally, the same amount of alcohol will have a much greater effect on someone who weighs less.
2. Alcohol must be metabolized in the liver to lose its effect. Because it cannot be stored in the body, it circulates in the blood stream until it is metabolized. A cup of coffee may counter the drowsiness caused by alcohol, but it does not reduce the level of intoxication. Although the effects of a stimulant (coffee) and a depressant (sleeping pills) are opposite, they act on different receptors and do not "undo" one another.
3. It is sometimes difficult to identify reliable web sites. However, web sites that are not involved with advertising or selling drugs would be the least biased. Generally web sites sponsored by the government (ending in gov), educational institutions including medical schools (ending in edu), and other non-profit organizations (ending in org) are most reliable.
4. The effects of Fetal Alcohol Syndrome are associated with behavioral disorders, mental retardation, growth deficiency, and characteristic facial features. It is the third leading cause of birth defects associated with mental retardation. The rates of FAS can be over 50% is geographic areas where there is high alcohol abuse among pregnant women.

Practice Test

1. c
2. d
3. c
4. a
5. b
6. b
7. d
8. a
9. c
10. a
11. d
12. c
13. d
14. a
15. a
16. b
17. b
18. b
19. a
20. c

Short Answer Questions

1. The liver metabolizes alcohol and there is no way to speed up that process. The liver can only process about 1/3 of an ounce of pure ethanol per hour. One will remain intoxicated until the blood alcohol level is reduced.
2. The intoxicating effects of alcohol begin when it is absorbed from the digestive system into the blood and is delivered to the brain. Alcohol is a depressant. It affects the brain centers that are involved in discrimination, fine motor control, memory and concentration. Alcohol releases the brain from inhibitory controls so that it reduces anxiety and often creates a sense of well-being.
3. Stimulants excite the central nervous system. However, when they wear off, there is always depression to the CNS. If breathing becomes depressed or inhibited it may cause respiratory failure. In addition, stimulants can cause heart attacks by restricting blood flow to the heart.
4. Psychedelic drugs are hallucinogens. These drugs mimic the effect of naturally occurring drugs. Therefore the normal physiological reactions are not especially harmful, but the distortions of reality that are created by the drug experience may lead to some actions that are dangerous.
5. Opiates are natural or synthetic drugs that affect the body in ways similar to the major pain-relieving agent in opium. They are medically important because they alleviate severe pain, such as that following surgery or caused by cancer. However, they have a very high potential for abuse because tolerance and physical dependence occur. The effect on the CNS is very quick and produces a rush of euphoria.

Chapter 9

Sensory Systems

OBJECTIVES

After reading the text and studying the material in this chapter, you should be able to:

- List the sensory receptors and the stimuli they respond to.
- Differentiate between the general senses and the special senses.
- Describe the mechanism of touch and the two types of receptors that are involved.
- Explain how we perceive body position.
- Label the parts of the eye and explain the function of each part in the perception of sight.
- Explain how we visualize an object using both eyes and describe common visual impairments.
- Describe the chemical reactions in the rods and cones that cause neural impulses to reach the brain.
- Label the parts of the ear and describe the role of each part in hearing.
- Describe how pitch and loudness are perceived.
- Differentiate between the two types of hearing loss.
- Explain how we perceive and maintain balance.
- Describe how odor is perceived by the chemoreceptors in the nasal cavity.
- Explain how taste is perceived and identify the four primary taste regions of taste on the tongue.

CHAPTER SUMMARY

Sensory receptors respond to stimuli by generating electrochemical messages. All sensory receptors are selective, responding best to one form of energy. When the **receptor potential** reaches threshold level, an action potential is generated. The body contains many specialized receptors including **mechanoreceptors**, **thermoreceptors**, **photoreceptors**, **chemoreceptors**, and **pain receptors**. **Exteroceptors** are located near the surface and respond to external stimuli while **interoceptors** monitor conditions inside the body.

Receptors for the **general senses** are located throughout the body and perceive touch, pressure, temperature, sense of body position and pain. These receptors rely on either free nerve endings or encapsulated nerve endings. Touch and pressure are detected by receptors in the skin. Free nerve endings in Merkel disks receive touch as do the encapsulated nerve endings in Meissner's corpuscles. Pacinian corpuscles respond to pressure. Heat and cold receptors respond to changes in temperature. Body and limb position are detected by **muscle spindles** responding to the stretch of a muscle and **Golgi tendon organs** measuring muscle tension. Tissue damage causes the release of inflammatory chemicals that stimulate sensory neurons. Pain receptors are found in all tissues of the body. **Referred pain**, pain felt somewhere besides the site of the injury, is common with damage to internal organs.

The outer layer of the eye is made of the **sclera** that protects and shapes the eye and provides attachment for muscles and the **cornea**, which allows light to enter. The **choroids**, **ciliary body**, and **iris** make up the middle layer. The **pupil** allows light to enter the eye and reach the innermost layer is the **retina**, which contains the photoreceptors, **rods** and **cones**. The **fovea** contains cones for focused vision. The **optic nerve** carries visual information to the brain and forms a **blind spot** where it leaves the retina. The two fluid-filled areas of the eye are the **vitreous humor**, the main cavity of the eye, and the aqueous humor, the cavity in the front of the eye between the cornea and the lens. **Glaucoma** results when pressure of the aqueous humor reaches dangerous levels.

The cornea bends light as it enters the eye. The ciliary muscle can change the shape of the lens allowing the image to be focused on the retina. The elasticity of the lens provides for the process of **accommodation**. A **cataract** is a lens that has become cloudy usually due to aging. Depth perception and a focused image are accomplished by **convergence**, which keeps both eyes focused on the midline of an object. **Farsightedness**, **nearsightedness**, and **astigmatism** are the three most common visual problems. These focusing problems may be caused by discrepancies in the lens or the shape of the eye.

The function of all photoreceptors is to respond to light with a neural message sent to the brain. The light-absorbing portion of the pigment molecules in all photoreceptors is **retinal** which is bound to a protein called an **opsin**. Rods allow us to see in dim light. Three types of cones—red, blue and green—allow us to see color.

We hear sound waves produced by vibrations. The **outer ear**, consisting of the **pinna** and **external auditory canal**, receives the waves. The **middle ear**, including the **tympanic membrane**, the **malleus**, **incus**, and **stapes**, and the **Eustachian tube**, amplify the sound. The **inner ear** is a transmitter and consists of the **cochlea** and **vestibular apparatus**. The **organ of Corti** is most directly responsible for the sense of hearing. Vibrations transmitted from the middle ear to the fluid within the cochlea activate **hair cells** that stimulate the nerves that carry the impulse to the brain. The more hair cells stimulated, the louder the sound. Pitch is interpreted by the frequency of impulses in the auditory nerve. There are two types of hearing loss: conductive and sensorineural. Ear infections vary in severity depending upon what portion of the ear is affected.

The two components of the **vestibular apparatus** are the **semicircular canals** and the **vestibule**. The semicircular canals contain sensory receptors that monitor movement. The vestibule monitors balance when we are not moving. When the brain receives conflicting sensory information regarding movement, one tends to feel motion sick.

Olfactory receptors are located in the roof of the nasal cavity. They are neurons with long cilia covered by a coat of mucus. Odor molecules dissolve in the mucus and bind to the receptors causing a stimulation that is relayed to the olfactory bulb in the brain.

Taste is perceived by taste buds located on the tongue and inner surfaces of the mouth. Taste cells have **taste hairs** that project into a pore at the tip of the **taste bud**. When food molecules are dissolved in water, they enter the pore and stimulate the taste hairs. Taste buds sense the four basic tastes of sweet, salty, sour and bitter.

KEY CONCEPTS

- Sensory receptors generate electrochemical messages in response to external (exteroceptors) and internal (interoceptors) stimuli.
- The general senses include touch, pressure, temperature, sense of body and limb position and pain.
- Mechanoreceptors are responsible for touch, pressure, hearing, equilibrium, blood pressure and body position.
- Light touch is perceived by free nerve endings called Merkel disks. Meissner's corpuscles are encapsulated nerve endings that indicate where the touch occurred.
- Pacinian corpuscles respond to pressure.
- Thermoreceptors detect changes in temperature.
- Muscle spindles and Golgi tendon organs monitor body position.
- Photoreceptors, the rods and cones of the eye, detect light.
- Chemoreceptors monitor chemical levels and function in smell and taste.
- Pain receptors respond to stimuli caused by physical or chemical damage to tissues.
- The eye consists of three layers. The outer layer (sclera and cornea), the middle layer (choroids, ciliary body, and iris) and the inner layer (retina and photoreceptors).
- The cornea and lens focus the image on the retina.
- Cataracts occur when the lens becomes cloudy and can only be corrected by surgery. Other focusing problems can be corrected with glasses, contacts or corneal surgery.
- When light hits the rods or cones, a chemical change occurs in the pigment molecule that causes a neural message to travel to the brain via the optic nerve.
- Color vision depends upon the cones that are most sensitive to different wavelengths of light.
- Sound enters the outer ear as vibrations in the air that are transferred to the eardrum, the bones of the middle ear, and finally the cochlea and vestibular apparatus of the inner ear.
- The organ of Corti, located in the cochlea, is lined with hair cells and is responsible for our sense of hearing.
- Loud sounds are perceived because more hair cells are stimulated while different pitches are perceived by the activation of hair cells at different places along the basilar membrane of the cochlea.
- Conductive hearing loss occurs when sounds are prevented from reaching the inner ear. Sensorineural hearing loss occurs when there is damage to the hair cells or nerve supply of the inner ear.
- Balance is perceived by the three semicircular canals, which register sudden movement, and the vestibule, which indicates the position of the head with respect to gravity.
- Smell receptors, lined with cilia and coated by mucus, are located in the nasal cavity. Odorous molecules dissolve in the mucus and stimulate receptor hairs. Impulses are transferred to the olfactory bulb and then the brain.
- Taste buds are located on the tongue and inside of the mouth and throat. They have taste hairs that perceive sweet, salty, sour and bitter.

STUDY TIPS

The key to understanding the senses is to focus on the relationship between structure and function. First list the different classes of receptors and the stimuli they receive. Next concentrate on the general senses of touch, pressure, temperature, body position and pain. Know how each sense is perceived and identify the special organs involved in that perception. A set of flash cards might help you learn this information.

The special senses are covered in more detail in the chapter and therefore should be given more of your time and attention as you study. The figures in this chapter will be the focal point (pardon the pun) for your study of vision, hearing, balance, smell and taste. Some students use a copier to make an enlargement of the figure showing structure of an organ. They then add a short summary of the function of each structure to that copy, putting both structure and function in one place for further review.

Study Figure 9-4, the structure of the eye. As you identify each structure, refer to Table 9-1 and review the function of that part of the eye. Review both the structure and function of the parts of the eye as you explain to a study partner how one sees clearly in three dimensions. Next, focus on the structure of the retina. Identify the rods and cones and explain the chemical reactions that occur to cause a neural stimulus. Explain to someone not in the class how we perceive color.

Hearing involves the mechanical transfer of sound vibrations. Study Figure 9-11, the structure of the ear. Identify each part of the ear and describe what happens to the vibrations as they come to and pass beyond that structure. Review Table 9-4 to reinforce the relationship between structure and function. The ear is also used to perceive movement and balance. Review the description of these processes in the text and then use Figure 9-14 to explain these perceptions to a study partner.

Smell and taste function together to allow us to enjoy a great meal. Use Figures 9-15 and 9-16 to explain how chemicals are perceived and generate neural impulses to the brain. As a last step to your review, identify the commonalities among the senses and the transmission of an external stimulus into a nerve impulse.

REVIEW QUESTIONS

A. Write the receptor, from the following list, that receives the stimuli given below: chemoreceptor, mechanoreceptor, pain receptor, photoreceptor and thermoreceptor.

1. ______________________ Detects changes in light intensity in the eye.

2. ______________________ Responds to stimuli caused by physical or chemical damage to tissues.

3. ______________________ Detects changes in temperature.

4. ______________________ Responds to chemical stimulation, found in the nose and mouth.

5. ______________________ Senses touch, pressure, hearing, and equilibrium.

B. Use the following terms to label the drawing shown below: aqueous humor, cornea, fovea, lens, optic nerve, pupil, retina, sclera, and vitreous humor. After each term include a brief statement summarizing the function of that structure. Use Table 9-1 as a resource.

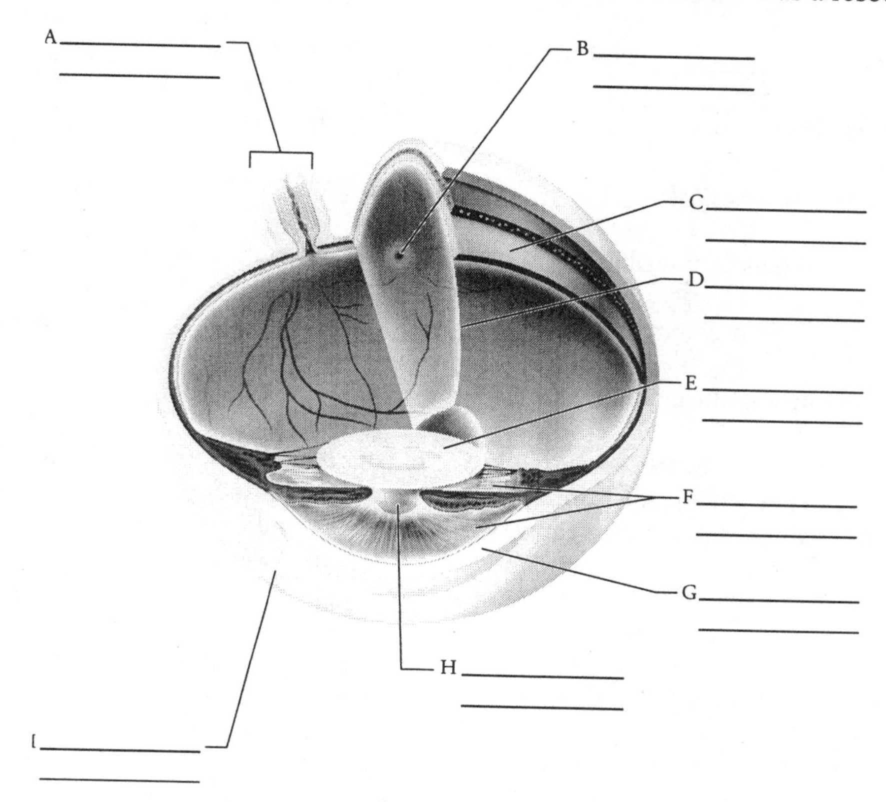

C. Complete the following table summarizing common visual problems.

Condition	Where the image focuses	Images seen clearly	Corrected with
Farsightedness	Behind the retina		
Nearsightedness		Up close, reading	
	NA	All images are asymmetrically blurred	Lenses to correct asymmetrical nature of cornea or lens

D. Refer to Figure 9-5 in your text to complete the following paragraph on how light is converted to nerve signals.

Light waves of the visible spectrum strike an object and are ____________. The light enters the eye through the ____________ and strikes the photoreceptors on the ____________. The pigment molecules are made of ____________ and the protein ____________. We are able to see in dim light because of the ____________ located on the retina. Color is perceived by the ____________. When light strikes a photoreceptor, the molecule changes shape, changing the ____________ of the membrane to ____________. This causes changes in the neural cells, which send an impulse through the ____________ nerve to the brain.

E. Use the following terms to label the drawing of the ear shown below: anvil (incus), auditory canal, auditory nerve, cochlea, eardrum, Eustachian tube, hammer (maleus), oval window, pinna, round window, semicircular canal, stirrup (stapes), and vestibule.

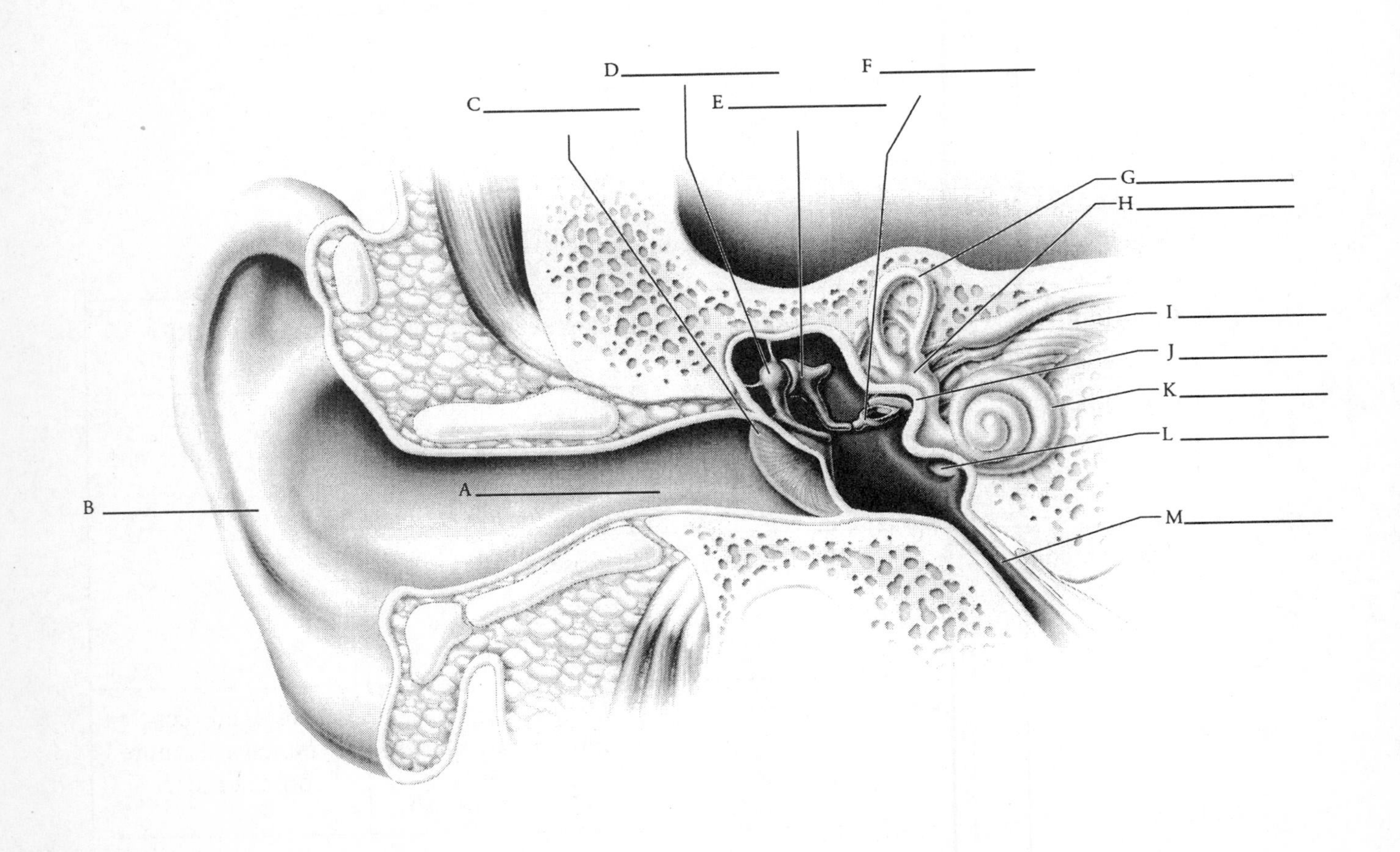

F. Match the structure in Column A with the function in Column B.

Column A	Column B
1. cochlea	a. perceives static equilibrium
2. eardrum	b. organ of hearing
3. Eustachian tube	c. area that amplifies sound
4. external ear	d. perceives dynamic equilibrium (movement)
5. inner ear	e. equalizes pressure
6. middle ear	f. snail-shaped organ housing organ of Corti
7. organ of Corti	g. transfers vibrations to the oval window
8. semicircular canals	h. transfers sound waves to bones of middle ear
9. stirrup (stapes)	i. area that converts sound to nerve impulses
10. vestibule	j. receives sound

G. Put the following events in order from the perception of sound to nerve impulse.

______ a. Sound waves reach the outer ear, travel down the ear canal.

______ b. Nerve stimulus going to the auditory nerve.

______ c. Sound waves travel through the bones of the middle ear.

______ d. Hair cells receive vibrations.

______ e. The tympanic membrane vibrates.

______ f. Basilar membrane vibrates.

______ g. Oval window vibrates.

H. Label the following diagram and then write a short description of the perception of smell. Use the following terms: bony plate, mucous layer, odor particles, olfactory bulb, olfactory hair, and receptor cell body.

Smell is perceived when ____

____

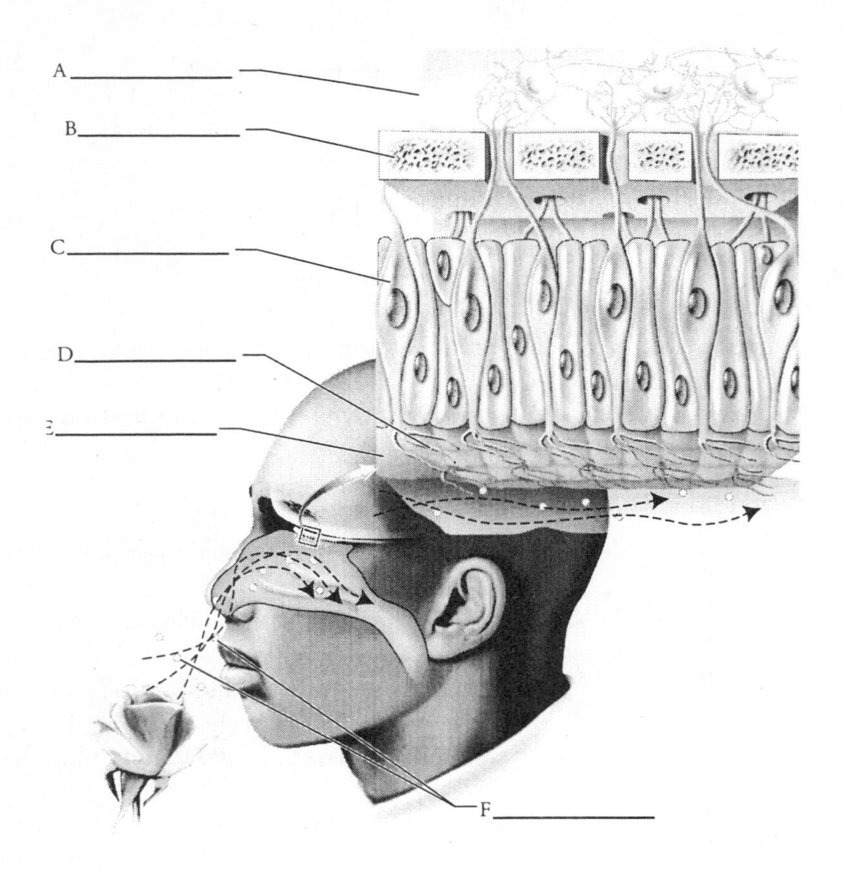

Critical Thinking Questions

Read each of the following questions carefully. If it helps you, underline or list each component to the question. Use complete sentences in your explanation.

1. When someone has cataracts, they may be able to receive a lens implant. This is a surgical replacement of their cloudy lens with an artificial lens. Based on your knowledge of vision, explain why they may need a new prescription for their glasses following this surgery.

2. You and a friend are enjoying a concert of classical music. Describe how sound enters your ear and how both volume and pitch are perceived in the ear.

3. Two people are dizzy. One is a teenager who has been reading in the car and feels motion sick; the other is a young child who has been spinning in circles and just fell down. What are the similarities and differences in their perception of motion? How are motion and equilibrium perceived?

4. Suppose you just came into the house and realized that there were freshly baked cookies waiting for you. What senses would be involved in your enjoyment of the cookies? Include as many senses as you can and describe their interaction.

5. Suppose a nerve transmitted a stimulus received by the ear to the eye. What might be the result?

Practice Test

Choose the one best answer to each question that follows. As you work through these items, explain to yourself why each answer you discard is incorrect and why the answer you choose is correct.

1. Which of the following is true regarding sensory receptors?
   a. demonstrate sensory adaptation
   b. respond best to one form of energy
   c. response varies with the strength of the stimulus
   d. all of the above

2. Merkel disks perceive ____________.
   a. light touch
   b. pain
   c. temperature
   d. where the touch occurs

3. Which of the following is true regarding thermoreceptors that detect heat?
   a. They adapt very quickly.
   b. They perceive both internal and external temperatures.
   c. They perceive temperatures from room temperature to boiling.
   d. They react to hot and cold with the same precision.

4. Muscle spindles and Golgi tendon organs are ____________.
   a. chemoreceptors
   b. mechanoreceptors
   c. photoreceptors
   d. thermoreceptors

5. Referred pain is most common among ____________.
   a. athletes
   b. broken limbs
   c. internal organs
   d. skin abrasions

6. The photoreceptors are located on the ____________.
   a. cornea
   b. lens
   c. retina
   d. sclera

7. The point of focus on the retina is the ____________.
   a. blind spot
   b. fovea
   c. lens
   d. optic nerve

8. Abnormal pressure within the eye is called ____________.
   a. cataracts
   b. farsightedness
   c. glaucoma
   d. myopia

9. What focuses the image on the retina?
   a. cornea
   b. fovea
   c. lens
   d. a and c only
   e. a, b and c

10. Which of the following statements is/are true regarding cataracts?
    a. common with aging
    b. lens becomes cloudy
    c. twice as common in smokers
    d. all of the above are true

11. When the curvature of the cornea or lens is irregular, someone is said to have ____________.
    a. astigmatism
    b. convergence
    c. farsightedness
    d. nearsightedness

12. Rhodopsin is a(n) ____________.
    a. chemoreceptor in the eye
    b. enzyme found in cones
    c. photopigment found in the rods
    d. pigment that colors the iris

13. Which of the following is true concerning sight?
    a. Color-blind people have ineffective rods.
    b. Lens accommodation increases with age.
    c. Sight depends only on the perception in the brain and not the optic nerve.
    d. Sight requires a chemical reaction in the retina.

14. What organ in the ear first responds to vibrations in the air?
    a. cochlea
    b. eardrum
    c. middle ear bones
    d. semicircular canals

15. Which of the following is not involved in the transmission of sound waves?
    a. cochlea
    b. semicircular canals
    c. stapes
    d. tympanic membrane

16. To relieve the pressure on the eardrum when scuba diving or flying in an airplane, one must force air into the ___________.
    a. auditory canal
    b. Eustachian tube
    c. nasal cavity
    d. vestibular apparatus

17. Hearing and balance both are sensed by ___________.
    a. chemoreceptors
    b. equilibrium sensors
    c. mechanoreceptors
    d. vibrations

18. Dynamic equilibrium is dependent upon ___________.
    a. balance
    b. chemoreceptors
    c. inner ear
    d. semicircular canals

19. Which of the following senses are most connected to the limbic system?
    a. hearing
    b. sight
    c. smell
    d. taste

20. Which of the following tastes is most protective – that is, which is most common of spoiled or poisonous food?
    a. bitter
    b. salty
    c. sour
    d. sweet

Short Answer Questions

Read each of the following questions carefully. If it helps you, underline or list each component to the question. Formulate your answer and jot down the main points. Then use complete sentences in your explanation.

1. Compare and contrast the location and mechanisms of the general senses and the special senses.

2. Use a drawing or figure to trace the transfer of vibrations from a speaker to the auditory nerve.

3. Compare and contrast the function and location of the rods and cones of the eye.

4. Pure oxygen damages the optic nerve. Before this was known, premature babies were sometimes given too much oxygen and were blinded by it. From your knowledge of vision, why would an eye transplant not restore vision to these individuals?

5. Describe the similarities between the perception of taste and smell.

Answer Key

Review Questions

A. 1. photoreceptor
   2. pain receptor
   3. thermorceptor
   4. chemoreceptor
   5. mechanoreceptor

B. A. optic nerve: sends impulse to brain
   B. fovea: color vision
   C. retina: generates neural messages
   D. vitreous humor: reflects light, shape
   E. lens: fine focus
   F. aqueous humor: refracts light, shape
   G. cornea: reflects light, focus
   H. pupil: lets light in
   I. sclera: outer layer, protection

C.

Condition	Where the image focuses	Images seen clearly	Corrected with
Farsightedness	Behind the retina	In the distance	Convex lens in glasses or contacts, or corneal surgery
Nearsightedness	Focus is in front of the retina	Up close, reading	Concave lens in glasses or contacts, or corneal surgery
Astigmatism	NA	All images are asymmetrically blurred	Lenses to correct asymmetrical nature of cornea or lens

D. Light waves of the visible spectrum strike an object and are reflected. The light enters the eye through the pupil and strikes the photoreceptors on the retina. The pigment molecules are made of retinal and the protein opsin. We are able to see in dim light because of the rods located on the retina. Color is perceived by the cones. When light strikes a photoreceptor, the molecule changes shape, changing the permeability of the membrane to sodium ions. This causes changes in the neural cells, which send an impulse through the optic nerve to the brain.

E. A. external auditory canal
   B. pinna
   C. tympanic membrane
   D. hammer (malleus)
   E. anvil (incus)
   F. stirrup (stapes)
   G. semicircular canal
   H. vestibule
   I. auditory nerve
   J. oval window
   K. cochlea

L. round window
M. eustachian tube

F. 1. f
2. h
3. e
4. j
5. i
6. c
7. b
8. d
9. g
10. a

G. a. 1
b. 7
c. 3
d. 6
e. 2
f. 5
g. 4

H. Smell is perceived when odor molecules are carried into the nasal cavity by the air. They dissolve in the mucus and bind to receptors on the olfactory hairs. Olfactory hairs are bent, stimulating a response in the receptor cell body, which transfers the impulse to the olfactory bulb. Impulses are grouped as they travel to the brain along the olfactory nerve tract.
A. olfactory bulb
B. bony plate
C. receptor cell body
D. olfactory hair
E. mucous layer

Critical Thinking Questions

1. Cataracts result when the lens becomes cloudy or opaque usually because of aging. The faulty lens can be replaced by an artificial lens. Since a natural lens changes shape to keep an image focused, a new lens will have slightly different properties and may require a new prescription.
2. Amplitude of the wave determines the loudness of the sound and the frequency of the wave determines the pitch of the sound. Sound waves travel through the air and enter the ear through the auditory canal. The tympanic membrane (eardrum) vibrates at the same frequency as the sound wave and transfers those vibrations to the middle ear and on to the oval window. Movement of the oval window causes the basilar membrane to vibrate the hair cells, which transfer information regarding volume to the brain. However, pitch is perceived differently. Different regions of the basilar membrane are activated by different pitches. Therefore pitch is interpreted based on the different regions of the basilar membrane that are stimulated by sounds of different frequency.
3. The child who has been spinning has set the fluid in the semicircular canals into motion. There is a time lag between when the child stops moving and the fluid stops moving. Therefore, the child still feels the sensation of motion. However, motion sickness is thought to be caused by a mismatch of sensory input from the vestibular apparatus and the eyes. Your eyes are telling the brain that you are stationary while your vestibular system detects motion.
4. You would first perceive the cookies through your sense of smell. It is closely linked to the limbic system in the brain. When you tasted the cookies, the high sugar content would indicate calories are on the way and the warmth would bring an emotional response of caring. The sense of taste and smell would combine to let you know what you are eating.

5. Since the stimulus would be received by the ear, a sound would be picked up by the nerve. However, since the stimulus would be interpreted by the eye, which perceived light, a sound would be interpreted as a flash of light.

Practice Test

1. d
2. a
3. a
4. b
5. c
6. c
7. b
8. c
9. d
10. d
11. a
12. c
13. d
14. b
15. b
16. b
17. c
18. d
19. c
20. a

Short Answer Questions

1. The general senses include touch, pressure, vibration, temperature, pain, and a sense of body and limb position. They arise from receptors in the skin, muscles, joints, bones and internal organs. They are widespread through the body, and, in all cases except temperature, are mechanoreceptors. The special senses are smell, taste, vision, hearing, and the sense of balance. The receptors of these senses are located in the head, usually within specific structures. Photoreceptors and chemoreceptors are found only in the special senses.
2. Your drawing should be based on the information in Figure 9-11. Vibrations should leave the speaker and enter the ear causing the eardrum to vibrate. Vibrations are then transferred through the bones of the middle ear to the inner ear, where the cochlea converts the waves to neural messages that are sent to the brain. Some students may find it easier to use a simple flow chart rather than a drawing.
3. Use the information in Figure 9-5 to help answer this question. The rods are more numerous than the cones and are responsible for our black and white vision. They are very sensitive to light and allow us to see in dimly lit rooms. The cones are responsible for color vision. They produce sharp images and are concentrated in the center of area of the retina called the fovea.
4. The optic nerve conducts information from the eye to the brain. It is responsible for the final transmission of an "image." Even if a perfect image is perceived by the anatomy of the eye and it is not transmitted to the brain, there can be no recognition of sight. To date, we are not able to replace the optic nerve like we can other parts of the eye involved with focus.
5. Both taste and smell depend upon chemoreceptors. They both depend upon the chemical dissolving in mucus or saliva before it can be perceived by either taste buds or olfactory receptors. In both cases, the molecules bend hairs in the receptors, which ultimately send information to the brain.

Chapter 10

The Endocrine System

Objectives

After reading the text and studying the material in this chapter, you should be able to:

- Differentiate between endocrine and exocrine glands.
- Describe the action of lipid-soluble and water-soluble hormones.
- Explain and give an example of hormone regulation by negative and positive feedback mechanisms.
- Describe the role of the hypothalamus in the regulation of the anterior pituitary gland.
- List the effect of each of the following anterior pituitary hormones including problems with oversecretion and undersecretion: growth hormone (GH), prolactin (PRL), thyroid-stimulating hormone (TSH), adrenocorticotropic hormone (ACTH), follicle-stimulating hormone (FSH), and luteinizing hormone (LH).
- Define and give examples of tropic hormones.
- Explain the role of the hypothalamus in the manufacture of hormones released by the posterior pituitary gland.
- Explain the effect of antidiuretic hormone (ADH) and oxytocin (OT).
- Describe the effect of thyroid hormone (TH) including oversecretion and undersecretion.
- Explain the regulation of blood calcium by calcitonin (CT) and parathyroid hormone (PTH).
- List the hormones produced by the adrenal cortex and give the hormone, its effect and the results of oversecretion and undersecretion.
- Describe the effects of the hormones produced by the adrenal medulla.
- Compare the effects of the two pancreatic hormones, glucagon and insulin, and describe the various forms of diabetes.
- Describe the effect of the thymus gland on the health of our immune system.
- Relate the production of melatonin to seasonal affective disorder.
- Define prostaglandins and describe their mechanism of action as compared with endocrine hormones.

Chapter Summary

The endocrine system consists of endocrine glands and organs with some endocrine tissue. **Endocrine glands** are made up of secretory cells that release their products directly into the adjacent extracellular fluids. The products, called **hormones**, then diffuse into the blood stream so that they reach all cells in the body. Hormones are one of the chemical messengers of the body. They circulate throughout the body until they reach **target cells,** those that respond to the hormone, influencing growth, development, metabolism and behavior.

Hormones are either lipid-soluble or water-soluble. Steroids are lipid-soluble hormones derived from cholesterol. The ovaries, testes, and adrenal glands are the main organs that secrete steroid hormones. Once they enter the cell, they activate the synthesis of specific proteins. Water-soluble hormones are made of amino acids and cannot pass through the lipid bilayer of the plasma membrane. They exert their effects indirectly by binding to receptors on the surface of the target cell, which stimulates **second messengers** within the cell that carry out the effect of the hormone.

Feedback mechanisms regulate the secretion of hormones. Usually control is by a **negative feedback mechanism** whereby the increased blood level of the hormone inhibits its further release. On the other hand, some hormones are regulated by a **positive feedback mechanism** in which the outcome of a process, further stimulates the process. During times of severe stress, the nervous system can override the feedback mechanisms of the endocrine system.

The **pituitary gland** is located at the base of the brain and consists of two lobes. The anterior pituitary is connected to the **hypothalamus** by a direct venous system called the **pituitary portal system**. The **anterior pituitary** responds to releasing and inhibiting hormones from the hypothalamus by modifying its synthesis and secretion of the following six hormones. Four of these hormones (TSH, ACTH, FSH, and LH) are **tropic hormones**, influencing the secretion of hormones by other glands.

- **Growth hormone (GH)** stimulates the liver to produce somatomedins, which stimulate an increase in cell size and the rate of cell division in target cells. Abnormally high production of GH in childhood results in abnormal growth in height called **giantism**. High levels of GH in adulthood cause **acromegaly**, resulting in the enlargement of the tongue and thickening of the bones of the hands, feet and face. Insufficient production of GH in childhood results in **pituitary dwarfism**.
- **Prolactin (PRL)** stimulates the mammary glands to produce milk and is produced during lactation. Excess PRL may cause infertility and lactation when there is no pregnancy. In men, PRL is involved with the production of mature sperm in the testes and overproduction can lead to sterility.
- **Thyroid-stimulating hormone (TSH)** acts on the thyroid gland to stimulate the release of thyroid hormones.
- **Adrenocorticotropic hormone (ACTH)** controls the synthesis and secretion of glucocorticoid hormones from the adrenal cortex.
- **Follicle-stimulating hormone (FSH)** promotes development of egg cells and secretion of estrogen in females. In males, FSH promotes the production of sperm.
- **Luteinizing hormone (LH)** causes ovulation and the secretion of estrogen and progesterone. In males, LH stimulates the production and secretion of testosterone.

Cells within the **posterior lobe of the pituitary** do not produce any hormones. Rather, neurons of the hypothalamus manufacture antidiuretic hormone (ADH) and oxytocin (OT), which travel down the nerve cells into the posterior pituitary where they are stored and released.

- **Antidiuretic Hormone (ADH)** causes the kidneys to remove water from the fluid destined to become urine. ADH is also called **vasopressin** because it constricts blood vessels and raises blood pressure during times of severe blood loss. A deficiency of ADH results in **diabetes insipidus** characterized by excessive urine production and dehydration.
- **Oxytocin (OT)** stimulates uterine contractions of childbirth and milk ejection

from the mammary glands. In men, OT may facilitate the transport of sperm and be involved in male sexual behavior.

Hormones secreted by the **thyroid gland** regulate metabolism and decrease blood calcium.

- **Thyroid hormone (TH)** stimulates the protein synthesis, breakdown of lipids and the use of glucose for the production of ATP. Too little TH in early development causes **cretinism** characterized by dwarfism, mental retardation and slowed sexual development. Underproduction of TH in adulthood causes **myxedema**, a condition in which fluid accumulates in facial tissues. Oversecretion of TH results in **Graves' disease**, an autoimmune disorder causing increased metabolic rate and heart rate. Iodine is needed for the production of TH. A diet deficient in iodine can produce a **simple goiter**.
- **Calcitonin (CT)** helps regulate calcium concentration in the blood by either stimulating the absorption of calcium by bone or by increasing the excretion of calcium in the urine.

The **parathyroid glands** secrete **parathyroid hormone (PTH)**, or **parathormone**. Low levels of blood calcium stimulate the secretion of PTH, which causes calcium to move from the bone and urine to the blood. CT and PTH work together to maintain proper levels of blood calcium. Undersecretion of PTH can result in nervousness and muscle spasms while oversecretion pulls calcium from bone tissue causing weakened bones and elevated blood calcium levels that might lead to kidney stones, calcium deposits in the soft tissue and deceased activity of the nervous system.

The two **adrenal glands** are located at the top of the kidneys. There are two distinct parts to the adrenal glands, the adrenal cortex and the adrenal medulla. The **adrenal cortex** secretes the following hormones.

- **Gonadocorticoids**, including **androgens** and **estrogens**, are secreted by the adrenal cortex in both males and females. However, their influence is insignificant compared to that of the testes in males and ovaries and placenta in females.
- **Mineralocorticoids** affect mineral homeostasis and water balance. **Aldosterone** acts on cells of the kidneys to increase reabsorption of sodium ions into the blood and promote the excretion of potassium ions in the urine. Addison's disease is caused by the undersecretion of cortisol and aldosterone. This autoimmune disorder causes the body to destroy its own adrenal cortex.
- **Glucocorticoids** affect glucose homeostasis. They act on the liver to promote the conversion of fat and protein to intermediate substances available to the body's cells. They also conserve glucose by inhibiting the uptake of glucose by muscle and fat tissue. They can also inhibit the inflammatory response by slowing the movement of white blood cells to the site of injury. Oversecretion of glucocorticoids can result in **Cushing's syndrome**.

The **adrenal medulla** produces **epinephrine** (adrenaline) and **norepinephrine** (noradrenaline). Both of these hormones are used in our response to danger. These hormones prepare us for the **fight-or-flight response** necessary for survival.

Hormones of the **pancreas**, secreted from the **pancreatic islets**, regulate blood glucose levels through glucagon and insulin.

- **Glucagon** increases glucose in the blood by converting glycogen to glucose in the liver. Amino acids are taken into the liver to make more glucose.
- **Insulin** decreases blood glucose levels. Insulin stimulates transport of glucose into muscle cells, white blood cells, and connective tissue cells. It also inhibits the breakdown of glycogen to glucose and

prevents conversion of amino and fatty acids to glucose. Therefore, insulin promotes protein synthesis, fat storage and the use of glucose for energy.

Diabetes mellitus is a group of metabolic disorders characterized by an abnormally high level of glucose in the blood. **Type 1 diabetes mellitus** is an autoimmune disease whereby a person's own immune system attacks the cells of the pancreas responsible for insulin production. **Type 2 diabetes mellitus** is characterized by a decreased sensitivity to insulin. Gestational diabetes develops in women during pregnancy but resolves after delivery. Other types of diabetes result from insulin deficiencies due to damage to the pancreas. Too much insulin and the resultant severe depletion of blood glucose cause **insulin shock**.

The **thymus gland** secretes hormones such as **thymopoietin** and **thymosin** which are involved in the maturation of T lymphocytes. These hormones and the thymus gland play a very important role in our immunity.

The **pineal gland** contains secretory cells that produce the hormone, **melatonin**. Melatonin is linked to sleep, fertility and aging. Too much melatonin can result in **seasonal affective disorder (SAD)**.

Prostaglandins are lipid molecules continually released by the plasma membranes of most cells. Prostaglandins are not true hormones because they act locally. However, their effects are very diverse, including the reproductive system, fertility, blood clotting and body temperature.

Key Concepts

- The endocrine system communicates via chemical messengers called hormones.
- Endocrine glands secrete hormones directly into the fluid just outside the cells where they diffuse directly into the bloodstream to be transported throughout the body.
- Lipid-soluble hormones move through the plasma membrane and combine with receptor molecules in the cytoplasm.
- Water-soluble hormones are modified amino acids. They cannot enter the cells and instead bind to receptors on the surface of the target cell.
- Feedback mechanisms regulate the secretion of hormones. In a negative feedback mechanism, the outcome of a process causes the system to shut down. In a positive feedback mechanism the outcome of a process stimulates the process.
- Nervous control sometimes overrides feedback mechanisms.
- Hormones influence growth, development, metabolism, and behavior.
- The hypothalamus produces releasing hormones and inhibiting hormones that control secretions of the anterior pituitary gland.
- The pituitary gland consists of two lobes. The anterior pituitary produces growth hormone (GH), prolactin (PRL), thyroid-stimulating hormone (TSH), adrenocorticotropic hormone (ACTH), follicle-stimulating hormone (FSH), and luteinizing hormone (LH). The posterior pituitary releases oxytocin (OT) and antidiuretic hormone (ADH) produced in the hypothalamus.
- Thyroid hormones regulate metabolism by thyroid hormone (TH) and decrease blood calcium by calcitonin (CT).
- Parathyroid hormone increases blood calcium levels.

- The adrenal gland consists of the adrenal cortex, which secretes gonadocorticoids, mineralocorticoids, and glucocorticoids, and the adrenal medulla which secretes epinephrine and norephinephrine effective in our fight-or-flight response.
- Hormones of the pancreas, glucagon and insulin, regulate blood glucose.
- Hormones of the thymus gland promote maturation of white blood cells involved in our defense response.
- The pineal gland secretes melatonin, which influences our daily rhythms.
- Prostaglandins are lipid molecules released by the plasma membranes of most cells that are chemical messengers with only local effects.

Study Tips

This chapter contains a wealth of information that can be overwhelming if not organized properly. The material can be divided into two major sections: the mechanism of action and information on the specific glands. First organize the information on the way hormones regulate the body. The chapter begins with a discussion of endocrine and exocrine glands; be sure you understand the difference. Write a sentence explaining how each type of gland releases its products. Next, the text explains the mechanism of hormone action at the cellular level and the two types of feedback mechanisms. Write an explanation of how water-soluble and lipid-soluble hormones interact with the cell. Next write a general description of the feedback mechanism and differentiate between positive and negative feedback. Draw a simple flow chart of these two systems. As you read about each hormone to follow, there are a few regulatory mechanisms imbedded in the discussion that you should note, including the role of the regulation of anterior pituitary hormones by the hypothalamus and the role of the hypothalamus in the production of hormones of the posterior pituitary. Finally, note the difference between the action of hormones and prostaglandins.

The remaining information in the chapter deals with specific glands, their products, and the effects of the hormones including any diseases or conditions resulting from over or undersecretion. This material will require memorization and lends itself well to flash cards. Organize this information by where the hormone is produced. You might find Table 10.1 summarizing the endocrine glands and tissue helpful, but read the text and add to the information in the table as it is only a barebones summary. Make one flash card for each hormone. Begin with the six anterior and two posterior pituitary hormones. Then move to the thyroid and parathyroid hormones and include how we maintain blood calcium levels. Next identify the hormones of the adrenal cortex and the adrenal medulla. Finish your flash cards with a description of the action of the thymus and pineal glands. Choose a few glands/hormones to study each day. It will be impossible to learn all of this information at the last minute.

Review Questions

A. Complete the following sentences about endocrine and exocrine glands. Try this on your own first, then refer to Figure 10-1 in your text and the accompanying explanation if you need help.

   Our bodies contain two types of glands, ____________ and ____________. Exocrine glands

   secrete their products into ____________ that open into the ____________ of the body, into the spaces within ____________, or into ____________ within the body. An example of an exocrine gland is ____________. Endocrine glands release their products into ____________ just outside the cells. These products then ____________ into the ____________ to be transported throughout the ____________. The products of endocrine glands are called ____________. They are one of the ____________ messengers of the body. The endocrine system consists of the endocrine glands and ____________ tissue. Two organs with at least some endocrine tissue are ____________ and ____________.

B. The mode of action of lipid-soluble and water-soluble hormones is very different. Refer to the text description and to Figures 10-3 and 10-4 to complete the following sentences comparing the action of the two types of hormones.

1. ____________ hormones are able to diffuse through the cell membrane, while ____________ hormones must remain outside the cell.

2. ____________ hormones bind to a receptor in the cytoplasm while ____________ hormones bind dot a receptor on the surface of the membrane.

3. In the case of ____________ hormones, the hormone-receptor complex enters the nucleus, binds to the DNA and activates certain genes. In the case of ____________ hormones, intermediates, called G proteins, link the hormone to the second messenger.

4. Within the cell, ____________ hormones cause proteins to be synthesized, which ultimately alter the activity enzymes within the cell. In the case of ____________ hormones, the second messenger cAMP activates the enzymes within the cell, which may also affect protein synthesis.

C. Write a brief description of a negative and positive feedback mechanism. Give an example of each type of feedback mechanism.

1. In a negative feedback mechanism ____

____

2. In a positive feedback mechanism ____

____

D. Label the following figure with the name of each gland. Choose from the following list of glands: adrenal, pancreas, parathyroid, pineal, pituitary, thymus, and thyroid.

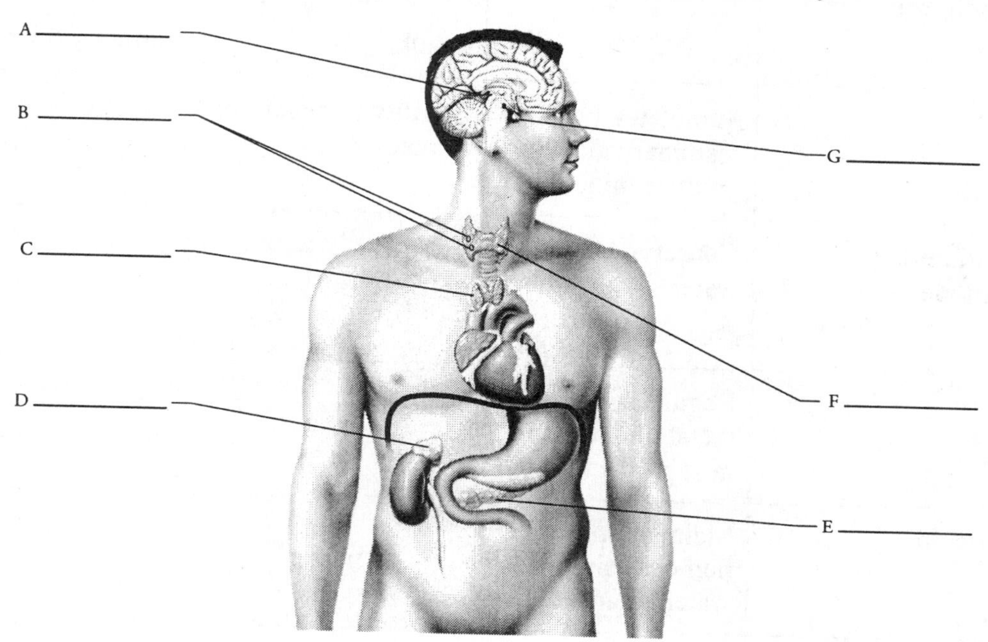

E. Match the gland/structure in Column A with the hormone in Column B.

Column A	Column B
1. adrenal cortex	a. glucagon
2. adrenal medulla	b. calcitonin
3. anterior pituitary	c. thymopoietin
4. hypothalamus	d. parathyroid hormone
5. pancreas	e. epinephrine
6. parathyroid	f. melatonin
7. pineal	g. oxytocin
8. posterior pituitary	h. prolactin
9. thymus	i. releasing hormones
10. thyroid	j. mineralocorticoids

F. Complete the following table listing the gland, hormone, effect, and result of oversecretion and undersecretion.

Hormone	Primary Effect	Oversecretion	Undersecretion
Growth Hormone		Children: Adults:	Children: pituitary dwarfism
	Stimulates mammary glands to produce milk	Sterility in males and females	
Antidiuretic Hormone	Conserve body water		
	Regulates metabolic rate and heat production		Fetus: cretinism Adults: [mysedema]
Corticoids	Maintain mineral homeostasis and water balance		
	Influences daily rhythms		

G. Match the following hormones listed in Column A with the correct function listed in Column B.

Column A	Column B
1. adrenocorticotropic hormone	a. promotes development of egg cells
2. calcitonin	b. stimulates milk letdown
3. epinephrine	c. causes ovulation
4. follicle-stimulating hormone	d. stimulates absorption of calcium in the bone
5. glucagon	e. maturation of white blood cells
6. insulin	f. lowers blood sugar levels
7. leutinizing hormone	g. increases blood calcium levels

8. oxytocin — h. causes body to react to stress or fear

9. parathyroid hormone — i. raises blood sugar levels

10. thymopoietin — j. controls secretion of glucocorticoid hormones

H. Complete the following sequence of events describing the maintenance of normal blood calcium levels.

When levels of calcium are ***high***, the thyroid gland releases ____________, which results in

____.

When levels of calcium in the blood are *low*, the ___________ release parathyroid hormone

causing ____.

Critical Thinking

Read each of the following questions carefully. Circle the most important words in the question before formulating your answer.

1. Review the action of water-soluble and lipid-soluble hormones. People with severe allergic reactions and asthma can receive treatment using epinephrine or cortisone. If epinephrine is a water-soluble hormone and cortisone is a steroid, explain why the injection of epinephrine produces immediate relief while the steroid-based treatment can take several days.

2. Form a diverse group and prepare for a panel interview on the pros and cons of the use of growth hormone in the elderly and in children who are below average height. One student should represent the athletically driven parent, another a family of successful, but genetically short individuals, another the elderly and their frailty.

3. The hypothalamus exerts some control over the anterior pituitary. It does this through releasing hormones and inhibiting hormones. First define a portal system and then explain the value of the pituitary portal system.

4. Type 2 diabetes mellitus is the most common form of diabetes. It is increasing in this country and appears to be related to a diet high in sugar and obesity. From your knowledge of insulin, explain the connection to diet, exercise and weight loss.

5. Describe the effect of the hormone melatonin. A disorder associated with too much melatonin is seasonal affective disorder (SAD). Explain why treatment of SAD often involves repeated exposure to bright light.

PRACTICE TEST

Choose the one best answer to each question that follows. As you work through these items, explain to yourself why each answer you discard is incorrect and why the answer you choose is correct.

1. The endocrine system coordinates body activities and maintains homeostasis through _________.
   a. adrenalin
   b. chemical messengers
   c. electrical impulses
   d. neurological impulses

2. Endocrine glands release their products directly into ____________.
   a. bloodstream
   b. connective tissue
   c. ducts
   d. fluid just outside the cells

3. Which of the following is true of peptide hormones?
   a. The hormone enters the cell and directly influences the production of an enzyme.
   b. The hormone-receptor complex binds to the DNA.
   c. They bind to a receptor site on the surface of the cell.
   d. They enter the cell through the plasma membrane.

4. Which of the following hormones are able to enter the cell because of their structure?
   a. estrogen
   b. progesterone
   c. testosterone
   d. all of the above

5. Which of the following is an example of a negative feedback mechanism?
   a. A drop in core body temperature causes an increase in metabolic rate. The increase in body temperature inhibits the production of thyroid stimulating hormone.
   b. A drop in levels of vitamin K cause slowed clotting times. The addition of dietary vitamin K improves clotting.
   c. The intake of growth hormone causes long bone growth.
   d. Uterine contractions stimulate the production of a hormone, which causes more contractions.

6. Which of the following is not produced by the anterior lobe of the pituitary gland?
   a. antidiuretic hormone
   b. growth hormone
   c. luteinizing hormone
   d. prolactin

7. The posterior pituitary and the hypothalamus are connected by which of the following?
   a. a portal system
   b. blood vessels
   c. ducts
   d. nerves

8. Oversecretion of which hormone will result in acromegaly?
   a. acreatin
   b. growth hormone
   c. metabolic increasing hormone
   d. testosterone

9. Which of the following is a tropic hormone?
   a. adrenaline
   b. oxytocin
   c. thyroid-stimulating hormone
   d. thyroxin

10. When the doctor asks you to swallow and feels the sides of your throat, what gland is he or she feeling?
    a. parathyroid
    b. salivary
    c. thyroid
    d. thymus

11. Cretinism and myxedema may be reversed by ________.
    a. administration of simple proteins
    b. administration of thyroxin
    c. injection of stimulating hormone
    d. surgery

12. Why did a man experience nervousness and muscle spasms following surgical removal of the thyroid gland?
    a. adrenaline levels were too high
    b. decreased levels of calcium
    c. low levels of parathormone
    d. removal of the thyroid gland

13. Which of the following raises heart rate, blood sugar levels, and blood pressure and allows one to perform the extraordinary feats at the scenes of accidents reported on the news?
    a. aldosterone
    b. epinephrine
    c. glucocorticoids
    d. testosterone

14. A person with Cushing's syndrome has a pendulous abdomen with thin arms and legs. What causes this?
    a. adrenal cortex offset the effects of the adrenal medulla
    b. cortisol causes the conversion of protein to glucose and then to fat
    c. pancreas secretes insulin
    d. thyroxin increases metabolic rate

15. Late in the afternoon, blood glucose levels fall, stimulating the secretion of which hormone?
    a. glucagon
    b. glycogen
    c. insulin
    d. pancreatisome

16. Which is the most common form of diabetes?
    a. diabetes insipidus
    b. Type 1 diabetes mellitus
    c. Type 2 diabetes mellitus
    d. They are all about the same frequency.

17. What is the most common cause of insulin shock?
    a. too little insulin
    b. too much insulin
    c. too much sugar intake
    d. swift drop in insulin levels

18. The control of our natural biorhythms and day/night cycle is by ____________.
    a. pancreas and insulin
    b. pineal gland and melatonin
    c. retina and dim light
    d. thymus gland and thyroxin

19. Cancer may be more prevalent in the elderly because of a decrease in immune function due to decreased output from the ________ gland.
    a. adrenal cortex
    b. pineal
    c. thyroid
    d. thymus

20. What is the primary difference between prostaglandins and other endocrine hormones?
    a. They act locally.
    b. They affect the reproductive system.
    c. They are fat-soluble.
    d. They cause an increase in metabolism.

Short Answer Questions

Read each of the following questions carefully. Jot down the main points you want to include in your answer. Then write a well-organized explanation.

1. In the past, the anterior lobe of the pituitary gland was referred to as the master gland. Why would this be an appropriate label?

2. Why are thyroid-related disorders difficult to diagnose?

3. Describe the control of lactation.

4. Explain how glucagon and insulin work together to maintain a relatively constant blood sugar level.

Answer Key

Review Questions

A. Our bodies contain two types of glands, exocrine and endocrine. Exocrine glands secrete their products into ducts that open into the surface of the body, into the spaces within organs, or into cavities within the body. An example of an exocrine gland is sebaceous/ salivary. Endocrine glands release their products into fluid just outside the cells. These products then diffuse into the bloodstream to be transported throughout the body. The products of endocrine glands are called hormones. They are one of the chemical messengers of the body. The endocrine system consists of the endocrine glands and endocrine tissue. Two organs with at least some endocrine tissue are hypothalamus, thymus, pancreas, ovaries, testes, heart, placenta, stomach, small intestine, and kidneys.

B. 1. Lipid-soluble, water-soluble.
   2. Lipid-soluble, water-soluble
   3. lipid-soluble, water-soluble
   4. lipid-soluble, water-soluble

C. 1. In a negative feedback mechanism, the outcome of a process feeds back and shuts down the system. Example: The drop in blood glucose levels caused by insulin stop the production of insulin.
   2. In a positive feedback mechanism, the outcome of a process feeds back on the system, further stimulating the process. Example: Oxytocin stimulates uterine contractions, which further stimulate the production of oxytocin.

D. A. pineal gland
   B. parathyroid glands
   C. thymus gland
   D. adrenal gland
   E. pituitary gland
   F. thyroid gland
   G. pancreas

E. 1. j
   2. e
   3. h
   4. i
   5. a
   6. d
   7. f
   8. g
   9. c
   10. b

F.

Hormone	Primary Effect	Oversecretion	Undersecretion
Growth Hormone	Stimulates growth	Children: giantism Adults: acromegaly	Children: pituitary dwarfism
Prolactin	Stimulates mammary glands to produce milk	Sterility in males and females	

Antidiuretic Hormone	Conserve body water		Diabetes insipidus
Thyroid Hormone	Regulates metabolic rate and heat production	Graves' disease	Fetus: cretinism Adults: mysedema
Corticoids	Maintain mineral homeostasis and water balance	Cushing's syndrome	Addison's disease
Melatonin	Influences daily rhythms	Seasonal affective disorder	

G. 1. j
2. d
3. h
4. a
5. i
6. f
7. c
8. b
9. g
10. e

H. When levels of calcium are ***high***, the thyroid gland releases calcitonin, which results in calcium being taken up by the bones. When levels of calcium in the blood are ***low***, the parathyroid glands release parathyroid hormone causing calcium to be released by the bones, reabsorbed by the kidneys, and absorbed by the intestine.

CRITICAL THINKING

1. Water-soluble hormones attach to the surface of the cell and cause an immediate change in the enzymes and resultant protein production. Lipid-soluble hormones must first penetrate the cell membrane, then enter the nucleus, and stimulate protein production from within the cell.
2. Discussion should include the ethical considerations of altering the life of a minor by administering an artificial hormone that has not been thoroughly tested. It should also include the advantage given to the wealthy in terms of the availability of that treatment. The discussion of the elderly should focus on the natural aging process versus an altered aging process whereby some systems may retain vigor while others do not.
3. A portal system provides a direct circulatory link between two organs – in this case between the hypothalamus and the pituitary gland. This portal system provides a direct connection between the hypothalamus and the pituitary for the releasing and inhibiting hormones so that they do not need to travel through the entire body.
4. Most forms of diabetes mellitus are characterized by an abnormally high level of glucose in the blood. Type 2 diabetes is characterized by a decreased sensitivity to insulin resulting from a decrease in insulin receptors on target cells. The exact relationship between chronically high blood glucose levels and Type 2 diabetes is speculative but treatment seems to hinge on restricting the intake of sugars, increase in exercise that metabolizes sugars, and weight control.
5. Melatonin is produced by the pineal gland. When neurons of the retina are stimulated by bright light, an impulse is sent to the hypothalamus that inhibits secretion of melatonin. Since there is not as much light available during the winter months, some people suffer from SAD due to excessive

amounts of melatonin. Exposure to bright light daily, especially in the morning, will inhibit the production of melatonin and relieve the symptoms of SAD.

PRACTICE TEST

1. b
2. d
3. c
4. d
5. a
6. a
7. d
8. b
9. c
10. c
11. b
12. c
13. b
14. b
15. a
16. c
17. b
18. b
19. d
20. a

SHORT ANSWER QUESTIONS

1. The anterior pituitary produces and secretes six major hormones. All except growth hormone act on other glands. Thus the anterior lobe of the pituitary has control over many other glands and resultant body functions.
2. Control of the thyroid occurs in many places. The anterior pituitary secretes thyroid-stimulating hormone, the thyroid gland secretes thyroid hormone, and iodine a dietary factor is needed to produce thyroid hormone.
3. Prolactin stimulates the mammary glands to produce, but not to eject, milk. The sucking of the baby stimulates touch receptors that in turn send nerve impulses to the hypothalamus. The hypothalamus prompts the posterior pituitary to release oxytocin, which travels to the mammary glands, where it stimulates cells to eject milk.
4. When levels of glucose are high, the pancreas secretes insulin, which lowers blood glucose levels. It causes the liver and muscle cells to take up glucose and adipose tissue to use glucose to form fat. When blood glucose levels are low, the pancreas secretes glucagon, which increases glucose in the blood by causing the liver to break down glycogen into glucose.

Chapter 11

Blood

Objectives

After reading the text and studying the material in this chapter, you should be able to:

- List the functions of blood.
- Describe the composition of plasma and the function of plasma proteins.
- List the three main formed elements of blood.
- Describe the function of red blood cells and the cause for various anemias.
- Give the life cycle of a red blood cell.
- Describe the origin and role of platelets.
- Differentiate between the types of white blood cells both in their structure and function.
- Describe the various leukemias and possible treatments.
- Explain how antibodies and antigens determine blood type.
- Diagram the transfusion relations among blood types.
- List the sequence of steps leading to a blood clot.
- Describe various problems associated with abnormal clotting.

Chapter Summary

Blood functions in transportation, protection, and temperature regulation. Blood is made of plasma, the liquid matrix, and formed elements, the cellular component of blood.

Plasma is about 93% water with the remaining 7% consisting of dissolved substances including nutrients, ions, dissolved gases, hormones and waste products. Most of the dissolved substances in the blood are **plasma proteins** that help maintain the water balance between the cells and the blood. They include 1) albumins which are needed for the water-balancing properties of plasma, 2) globulins that transport lipids and fat-soluble vitamins and 3) clotting proteins such as fibrinogen.

Stem cells, undifferentiated cells formed in the red bone marrow, give rise to the **formed elements** including platelets, white blood cells and red blood cells. **Platelets**, sometimes called thrombocytes, are fragments of larger precursor cells called megakaryocytes. They are essential to blood clotting.

White blood cells (**WBC**s) or **leukocytes**, help defend the body against disease and remove wastes, toxins, and damaged and abnormal cells. WBCs are nucleated cells that are produced in the red bone marrow. They have the unique ability to squeeze between the walls of the blood vessels and move to the site of infection, inflammation or tissue damage. Some WBCs are capable of **phagocytosis**, the process of engulfing the invading microbe.

The five types of white blood cells can be classified into two groups based on the presence or absence of granules in their cytoplasm. **Granulocytes** include neutrophils, eosinophils, and basophils. **Neutrophils** are the most abundant of the WBCs. They engulf microbes by phagocytosis, thus curbing the spread of infection. **Eosinophils** defend against parasitic worms and lessen the severity of allergies. **Basophils** release histamine, a chemical that attracts other white blood cells and causes the blood vessels to dilate. Histamine also plays a role in allergic reactions.

Agranulocytes include monocytes and two types of **lymphocytes**. **Monocytes** are the largest of all blood cells. They develop into macrophages, which are the phagocytic cells that engulf invading microbes, dead cells and cellular debris. **B lymphocytes** give rise to plasma cells, which produce antibodies. **T Lymphocytes** are specialized white blood cells that kill cells not recognized as coming from the body or cells that are cancerous.

Red blood cells (**RBC**s), also called **erythrocytes**, transport oxygen to the cells and carry about 23% of the total carbon dioxide. They are the most numerous cells in the blood and make up 45% of the total blood volume. Each RBC is packed with hemoglobin molecules that bind to oxygen, making **oxyhemoglobin**, needed for aerobic respiration. However, the RBCs have a greater attraction for carbon monoxide, which prevents hemoglobin from carrying oxygen. Since RBCs lose their nucleus and organelles as they mature, they do not consume the oxygen they carry but rely on fermentation as their source of energy.

RBCs are produced in the red bone marrow and live about 120 days. Macrophages engulf and destroy dying red blood cells. The liver degrades the hemoglobin into globin that is broken down into amino acids and heme. Iron is salvaged from heme and the rest is degraded into bilirubin, which is secreted as bile. RBC production is usually in response to the body's need for oxygen. Under special circumstances **erythropoeitin**, a hormone, can speed RBC production.

Disorders of red blood cells include **anemia**, which can result from too little hemoglobin, too few red blood cells, or both. The most common cause is iron deficiency leading to inadequate hemoglobin production. Hemolytic anemias occur when red blood cell destruction exceeds production. Sickle-cell anemia is caused by genetically abnormal hemoglobin. Pernicious anemia occurs when there is insufficient production of red blood cells because of a lack of vitamin B_{12}.

White blood cell disorders include infectious mononucleosis, a viral disease of the lymphocytes. **Leukemia** is a cancer of the WBCs that causes the number of WBCs to increase. The very large numbers of these WBCs prevent the development of normal RBCs, WBCs and platelets. Sometimes bone marrow transplants or stem cells from umbilical cord blood can cure leukemia.

Blood types are genetically determined by the glycoproteins found on the surface of RBCs. Blood types are named by the antigen found on the surface of the cell. People do not have antibodies in their plasma against those antigens found on their own RBCs. When someone's antibodies contact a foreign cell it causes clumping, or **agglutination**, of those cells. Therefore, for a safe transfusion, the recipient must not have antibodies for antigens found on the donor's RBCs. Rh is another important antigen that becomes critical during pregnancies of Rh negative women. Anti-Rh antibodies can develop in the mother and cross the placenta, destroying the fetus's RBCs, a condition called **hemolytic disease of the newborn**.

Upon injury, a blood vessel spasms to reduce blood flow. Next collagen fibers form in the wound that snag platelets. When platelets contact collagen fibers, they swell, form extensions and stick together to form a **platelet plug**. Next **prothrombin activator** converts **prothrombin** to **thrombin**, which causes **fibrinogen** to form long strands of **fibrin**. The fibrin strands form a web that traps blood cells and forms a clot. Vitamin K is required for proper blood clotting. When the wound is healed, **plasmin** formed from **plasminogen,** digests the fibrin strands of the clot. **Hemophilia** is an inherited condition that causes one or more of the clotting factors not to be made. A clot that continues to circulate is called an **embolus**, but when it lodges in a vessel it is called a **thrombus** and can cause a heart attack or stroke.

Key Concepts

- Blood transports nutrients to cells and waste products from cells. It also defends against disease and regulates body temperature.
- Blood consists of plasma, the liquid matrix, and formed elements.
- There are three types of plasma proteins. Albumins aid in the balance of water flow between blood and cells. Globulins transport lipids and fat-soluble vitamins. Clotting proteins stop blood flow at the site of an injury.
- Formed elements of the blood are platelets, leukocytes (WBCs) and erythrocytes (RBCs). Undifferentiated cells called stem cells give rise to the formed elements.
- Leukocytes fight off disease and help remove wastes, toxins and damaged or cancerous cells. The five types of leukocytes are specialized for different roles in the body's defense mechanism.
- Erythrocytes contain hemoglobin, an iron-containing protein that transports oxygen.
- Erythrocytes are formed in the red bone marrow under the control of the hormone, erythropoietin. Worn and dead cells are broken down in the liver and spleen.
- Anemia is a reduction in the blood's ability to carry oxygen. There are several forms of anemia including iron-deficiency anemia, pernicious anemia, and hemolytic anemias.
- Leukemia is a cancer of the WBCs that causes the production of large numbers of ineffective cells. Infectious mononucleosis is a highly contagious viral disease of lymphocytes.
- Blood types are determined by the presence of antigens on the cell surface. One's plasma contains antibodies against the antigens of foreign blood types causing agglutination. The major blood types are ABO and Rh.
- The prevention of blood loss involves three mechanisms: blood vessel spasm to restrict the loss of blood, platelet plug formation, and clotting. Clotting is a series of chemical changes in blood proteins resulting in the production of protein fibers that form a mesh that traps red blood cells and forms the clot.
- Hemophilia is an inherited condition of the blood resulting in a failure to clot properly.

Study Tips

This chapter provides an excellent opportunity for you to practice summarizing key concepts in the form of a table or flowchart. These tools allow you to abbreviate much of the information contained in the text and will not only organize your thoughts but will also provide a study tool. As you construct the table or flowchart, you will be reworking the main concepts of the text and putting them into a new format. If making a concept map, as described in the Study Tips for Chapter 8, the

Nervous System, was valuable to you, then use it again in this chapter. Both concept maps and flowcharts allow you to show the relationships between various concepts and processes.

Review the functions of blood and then begin a flowchart or concept map of the components of blood. Begin with the word blood and then have one arrow leading to plasma and another to formed elements. Under each of those subtitles list either the components of plasma or the three main types of formed elements. As you continue through the chapter, list the role and life cycle of platelets and red blood cells. Then list the types of white blood cells and their functions. Develop a table showing the relationship between the antibodies and antigens found on red blood cells and how they affect the relationship between blood donors and recipients. Write a short explanation of the resulting coagulation that results from mistyping.

Develop a flow chart of the steps in blood clotting. Include an initial stimulus, the reactions of the muscle wall of the vessel and all the blood proteins leading to a clot. Avoid copying charts directly from the text, but try to add more detail and explanation.

Finally, review the terms at the end of the chapter and add them to your study materials. Explain your flowcharts, concept maps and tables with a study partner. Ask him or her to help you correct errors.

Review Questions

A. Complete the following paragraph about the composition of blood. Try to complete this section without referring to the text.

Whole blood is composed of plasma and ____________ ____________. Most of the dissolved substances in blood are ____________ ____________. The largest group of plasma proteins is ____________. They are responsible for the blood's ____________ abilities. The plasma proteins responsible for transporting fats, cholesterol, and fat-soluble vitamins are ____________. They also are antibodies. Fibrinogen is a plasma protein responsible for clotting. There are three main formed elements in blood. The ____________ transport ____________ and carbon dioxide. The ____________ are responsible for defending the body against disease. The ____________ play a role in ____________. All blood cells originate in the ____________ ____________ ____________ from undifferentiated cells called ____________ ____________.

B. Match the term in Column A with the function in Column B.

Column A	Column B
1. basophil	a. undifferentiated cell giving rise to all blood cells
2. eosinophil	b. term for all white blood cells
3. erythrocyte	c. gives rise to platelets
4. leukocyte	d. thrombocytes, essential for clotting
5. lymphocyte	e. gives rise to macrophages
6. megakaryocyte	f. attacks damaged or diseased cells or produces antibodies
7. monocyte	g. clear-staining WBC with multilobed nucleus
8. neutrophil	h. attacks parasites
9. platelet	i. releases histamines
10. stem cell	j. transports oxygen and carbon dioxide

C. Hemoglobin is a very interesting, globular protein molecule. It is found in red blood cells and is absolutely necessary for them to carry out the primary function of transporting oxygen. Share what else you know about the life cycle of red blood cells and hemoglobin by answering the following questions.

1. What is the primary purpose of red blood cells and how does their shape make them more efficient? ____

____

____

2. What is found inside a red blood cell? ____

____

____

3. What is the composition of hemoglobin? ____

____

____

4. The life span of a red blood cell is only about 120 days. What happens to worn out red blood cells? ____

____

____

5. What causes infant jaundice? ____

____

____

D. Complete the following table using information from Table 11-2 and Figure 11-7.

Blood Type	Antigens on Red Blood Cells	Antibodies in Plasma	Can Receive Blood From	Can Donate Blood To
A	A			A, AB
B		Anti-A		
AB			all types	
O	none			all types

E. Anti-Rh antibodies are formed when Rh negative blood comes in contact with Rh positive blood. It takes a while for the antibodies to form, but when they come in contact with Rh positive blood a second time, they will cause clotting. Complete the following sequence of events.

Cause	Primary Effect
1. During birth, the Rh- mother's blood comes in contact with Rh+ fetal blood.	______________
2. Second pregnancy of Rh- mother with Rh+ baby.	______________
3. During birth, Rh- mother's blood comes in contact with Rh- fetal blood.	______________
4. During birth, Rh+ mother's blood comes in contact with Rh- fetal blood.	______________
5. Second pregnancy of a Rh- mother. First baby was Rh-, second baby is Rh+.	______________

F. Use the following flow chart to guide the completion of the description of blood clotting. Refer to your text for further assistance.

1.

Clotting factors

2.

Prothrombin activator

Inactive blood protein

3.

Thrombin

Prothrombin

Liver

4.

Fibrin

Fibrinogen

5.

1. Injury to a blood vessel is followed immediately by ____________ spasms to close the opening and the release of ____________.

2. The ____________ produces prothrombin. Thromboplastin causes prothrombin to be converted to ____________. Thrombin then acts on ____________ another protein produced in the liver.

3. Fibrinogen is converted to long, web-like fibers called ____________. This network traps ____________ and ____________ to form a plug that stops blood flow.

4. The trapped platelets release additional ____________ causing more fibrin to be produced further covering the opening in the damaged vessel. The fibrin network draws inward pulling the opening closed.

5. After the clot has formed plasminogen is converted to ____________ by an activating factor.

Critical Thinking

Read each of the following questions carefully. Circle the most important words in the question before formulating your answer.

1. Stem cells are undifferentiated precursors of blood cells. First identify some potential uses of stem cells in the treatment of blood diseases. Next explain why umbilical cord blood is a good source of stem cells. Then identify arguments against the use of stem cells and the routine freezing of cord blood from newborns.

2. Review the regulating mechanisms for the production of red blood cells as shown in Figure 11-6. A form of erythropoietin is sometimes given to patients undergoing chemotherapy. If chemotherapy destroys cells that are rapidly dividing, why would these patients often feel tired? Why would an injection containing erythopoietin, or a similar substance, give them more energy?

3. Using your knowledge of blood types and the antibodies present on the red blood cells and the antigens present in the plasma, explain why Type AB is sometimes called the universal recipient and why Type O is referred to as the universal donor.

4. Carbon monoxide is a byproduct of combustion such as that occurring in furnaces, automobiles and factories. From your knowledge of hemoglobin and its binding capabilities with both oxygen and carbon monoxide, explain why carbon monoxide poisoning is deadly.

Practice Test

Choose the one best answer to each question that follows. As you work through these items, explain to yourself why each answer you discard is incorrect and why the answer you choose is correct.

1. What is the liquid portion of blood?
   a. erythromyocin
   b. phagocytes
   c. plasma
   d. platelets

2. Which of the following proteins is not found in blood?
   a. albumin
   b. fibrinogen
   c. keratin
   d. prothrombin

3. What cell gives rise to all other blood cells?
   a. bone cell
   b. lymphocyte
   c. megakaryocyte
   d. stem cell

4. An increase in _________ indicates the presence of parasites.
   a. basophils
   b. eosinophils
   c. monocytes
   d. neutrophils

5. Microorganisms that invade the body may be ingested by _________.
   a. antibodies
   b. red blood cells
   c. stem cells
   d. white blood cells

6. Oxygen is primarily transported by ___________.
   a. plasma
   b. platelets
   c. red blood cells
   d. white blood cells

7. What is the metal in the center of the hemoglobin molecule?
   a. calcium
   b. iron
   c. lead
   d. magnesium

8. Carbon dioxide is primarily transported by __________.
   a. erythrocytes and platelets
   b. platelets and plasma
   c. plasma and leukocytes
   d. plasma and erythrocytes

9. Why can inhaling carbon monoxide result in death?
   a. It binds more tightly to hemoglobin than does oxygen.
   b. It is a deadly poison causing damage to the central nervous system.
   c. It is a stimulant and causes a heart attack.
   d. It robs the body of iron needed to transport oxygen.

10. The yellowish color characteristic of a bruise that is healing and a jaundiced baby are both caused by the presence of ______________.
    a. bilirubin
    b. dead blood cells
    c. erythrocytes
    d. hemotoma

11. Which of the following will stimulate the production of red blood cells?
    a. albumin
    b. erythropoietin
    c. plasma proteins
    d. transfusion

12. Which of the following diseases is caused by an abnormal hemoglobin S?
    a. fetal anemia
    b. hemophilia
    c. leukemia
    d. sickle cell anemia

13. Which of the following is cancer of the white blood cells?
    a. anemia
    b. hemophilia
    c. leukemia
    d. sickle cell anemia

14. What determines the blood type of a person?
    a. antibodies in the plasma
    b. antibodies on the red blood cell surface
    c. antigens in the plasma
    d. antigens on the red blood cell surface

15. A person with blood type AB can receive blood from which of the following?
    a. A
    b. B
    c. O
    d. all of the above

16. What blood type is indicated by the following slide showing agglutination?

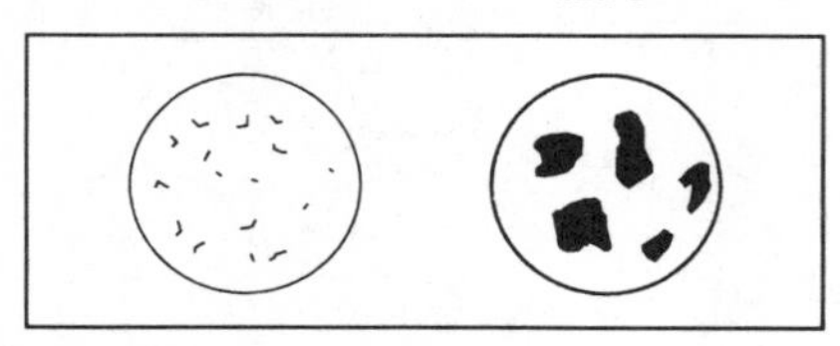

    a. type A
    b. type B
    c. type AB
    d. type O

17. When does an Rh- person make Rh+ antibodies?
    a. Rh- people have Rh+ antibodies all the time
    b. Rh- people make Rh+ antibodies at their first exposure to Rh+ blood
    c. Rh- people are incapable of making Rh+ antibodies
    d. Rh+ people are the only people who can make Rh+ antibodies

18. Assume that an Rh- mother has given birth to an Rh+ baby and was not given RhoGAM. When might hemolytic disease of the newborn occur?
    a. at the birth of the first Rh- baby
    b. during development of the next Rh+ baby
    c. all pregnancies will show hemolytic disease
    d. hemolytic disease of the newborn will never occur

19. Of the following, what is the initial stimulus for a blood clot to form?
    a. cut to the vessel
    b. trauma to the surrounding muscle
    c. presence of plasminogen
    d. presence of prothrombin

20. In which of the following is a blood clot life threatening?
    a. hemolytic disease of the newborn
    b. hemophilia patients
    c. leukemia treatment
    d. when it is an embolus

Short Answer Questions

Read each of the following questions carefully. Jot down the main points you want to include in your answer. Then write a well-organized explanation.

1. Describe the formed elements in blood. Include their structure and function.

2. A college student went to the clinic complaining of being tired. The physician took a blood sample and spun it down so that she could identify and count the cells. What type of cell was lacking? Why was the student tired? What is the most likely diagnosis?

3. Describe the process of a bone marrow transplant? What is the role of stem cells? What might be the advantage of saving one's own stem cells collected from the umbilical cord?

4. Describe the steps in clot formation. Why doesn't our blood clot all the time?

Answer Key

Review Questions

A. Whole blood is composed of plasma and formed elements. Most of the dissolved substances in blood are plasma proteins. The largest group of plasma proteins is albumins. They are responsible for the blood's water-balancing abilities. The plasma proteins responsible for transporting fats, cholesterol, and fat-soluble vitamins are globulins. They also are antibodies. Fibrinogen is a plasma protein responsible for clotting. There are three main formed elements in blood. The erythrocytes (red blood cells) transport oxygen and carbon dioxide. The leukocytes (white blood cells) are responsible for defending the body against disease. The platelets play a role in clotting. All blood cells originate in the red bone marrow from undifferentiated cells called stem cells.

B.
1. i
2. h
3. j
4. b
5. f
6. c
7. e
8. g
9. d
10. a

C.
1. Red blood cells transport oxygen to the tissues and carbon dioxide away from the tissues. The concave shape maximizes surface area and allows for the quick transfer of gases.
2. Mature red blood cells lack a nucleus or other organelles, but they contain millions of hemoglobin molecules.
3. Each hemoglobin molecule is made of four subunits. Each subunit is composed of a protein chain and a heme group. The heme group contains an iron ion that binds to the oxygen.
4. They are removed from circulation by the liver and spleen. Macrophages destroy the dying red blood cells. The liver degrades the hemoglobin to globin and heme. The protein globin is broken down into amino acids and reused. The iron from the heme is sent to the red bone marrow for recycling. The remaining heme is degraded into bilirubin and is excreted.
5. Prior to birth, babies have a different hemoglobin called fetal hemoglobin. Shortly after birth the fetal hemoglobin breaks down resulting in a lot of bilirubin. The liver is overwhelmed and cannot break it down fast enough. It accumulates under the skin giving the infant a yellowish color.

D.

Blood Type	Antigens on Red Blood Cells	Antibodies in Plasma	Can Receive Blood From	Can Donate Blood To
A	A	Anti-B	A, O	A, AB
B	B	Anti-A	B, O	B, AB
AB	A, B	none	all types	AB
O	none	Anti-A, Anti-B	O	all types

E.
1. stimulates formation of Rh+ antibodies in the mother
2. Rh+ antibodies cross placenta and destroy fetal red blood cells

3. no antibodies are formed
4. no antibodies are formed
5. no antibodies were formed during first pregnancy; antibodies may be formed during second pregnancy, but will not affect this baby.

F. 1. Injury to a blood vessel is followed immediately by muscle spasms to close the opening and the release of thromboplastin.
2. The liver produces prothrombin. Thromboplastin causes prothrombin to be converted to thrombin. Thrombin then acts on firbrinogen another protein produced in the liver.
3. Fibrinogen is converted to long, web-like fibers called fibrin. This network traps RBCs and platelets to form a plug that stops blood flow.
4. The trapped platelets release additional thromboplastin causing more fibrin to be produced further covering the opening in the damaged vessel. The fibrin network draws inward pulling the opening closed.
5. After the clot has formed plasminogen is converted to plasmin by an activating factor.

Critical Thinking

1. Stem cells hold great promise for the treatment of sickle cell anemia, beta-thalassemia, leukemia and HIV infections because they can settle in the bone marrow and produce normal cells with normal hemoglobin. Umbilical cord stem cells are a particularly abundant source of stem cells because in the fetus, stem cells continue to circulate until a few days after birth. It is difficult to find stem cells in adult bone marrow. The routine freezing of one's own umbilical cord blood would potentially ensure a supply of stem cells for everyone. The arguments against this process are few and include the fact that not enough research has been done on the viability of the stem cells and that their availability runs contrary to the natural course of disease.
2. Red blood cells are being produced at the rate of about 2 million cells per second. Their primary function is to carry oxygen to the cells where it is used by the mitochondria to make ATP, the energy currency of the cell. Without red blood cells, then there is less oxygen to the cells and less energy. The hormone erythopoietin stimulates the production of red blood cells which in turn causes more oxygen to reach the mitochondria. The resulting increase in ATP gives the patient much more energy.
3. Type AB blood plasma lacks antibodies to both Type A and Type B antigens. Therefore it will not cause clumping of donated blood that has either of those antibodies on the red blood cells. Because Type O blood has neither anti-A nor anti-B antigens, it does not stimulate clotting with any antibodies found in the blood. Therefore, it can be mixed with all other blood types and will not cause clotting.
4. Carbon monoxide binds more tightly to hemoglobin than either oxygen or carbon dioxide. It in effect monopolizes the hemoglobin so that it cannot take oxygen to the cells. If the concentration of carbon monoxide is high enough, more and more hemoglobin molecules are prevented from carrying oxygen to the cells preventing the production of ATP.

Practice Test

1. c
2. c
3. d
4. b
5. d
6. c
7. b
8. d
9. a
10. a
11. b
12. d
13. c
14. d
15. d
16. b
17. b
18. b
19. a
20. d

SHORT ANSWER QUESTIONS

1. Red blood cells, white blood cells and platelets are the three formed elements of the blood. Red blood cells are round, biconcave disks. They carry oxygen and carbon dioxide. They do not have a nucleus and are formed in the red bone marrow. White blood cells primarily fight infection. There are many types, classified by their nuclear structure and staining. They are also formed in the bone marrow. Platelets are the most numerous and smallest of the formed elements. They are important in the clotting process.
2. Most likely there was a lack of red blood cells, or there may be a lack of hemoglobin. The student was tired because there was not enough oxygen getting to the cells and the mitochondria were unable to produce enough ATP. The student most likely was anemic.
3. Acute leukemia, especially in children, is sometimes effectively treated with a bone marrow transplant. The bone marrow in the leukemia patient is destroyed by irradiation and drugs. The donor bone marrow is given to the patient intravenously. The donor bone marrow cells begin to produce healthy new white blood cells and other formed elements. When the white blood cell count reaches normal levels, the patient regains resistance to infection. All of the new white blood cells are free of cancer. Umbilical cord stem cells work exactly the same. The advantage of having one's own stem cells is not having to find a donor or risk rejection of the tissue.
4. When a blood vessel has been cut, the first response is constriction of the vessel. Next platelets form a plug by clinging to collagen. Clot formation begins when clotting factors from the injured tissue and platelets convert an inactive blood protein to prothrombin activator, which converts prothrombin to thrombin. Fibrinogen then forms long strands of fibrin which forms a web that traps blood cells and forms the clot. After the wound has healed, clots are normally dissolved by plasmin which is formed from plasminogen. The plasmin dissolves clots.

Chapter 12

The Circulatory System

Objectives

After reading the text and studying the material in this chapter, you should be able to:

- Name in order the types of vessels that carry blood as it flows from the heart through the body and back to the heart.
- Define and know how to find the pulse.
- Explain how vasoconstriction and vasodilation regulate blood flow in arteries.
- List the mechanisms responsible for the exchange of material across the capillary wall.
- Compare the structure of arteries, veins and capillaries and explain how this structure facilitates the function of each type of vessel.
- Explain how blood is returned to the heart against the force of gravity.
- Describe the structure of the heart as a double pump.
- Describe the structure and function of the heart valves.
- Name each major vessel and chamber of the heart as blood flows from the left ventricle to the left atrium.
- Describe the cardiac cycle.
- Describe control of the heartbeat.
- List three factors that determine blood pressure.
- Define hypertension and describe why it is life threatening.
- Describe how atherosclerosis develops and why it is dangerous.
- Explain the relationship between coronary artery disease, angina pectoris, and heart attack.
- List the functions of the lymphatic system and specifically of the lymph nodes.
- Compare the structure of lymphatic capillaries with vessels found in the circulatory system.

Chapter Summary

The circulatory system is composed of blood, vessels and the heart. Blood is circulated throughout the body in a branching network of vessels. Beginning at the heart, blood passes through a loop of **arteries**, **arterioles**, **capillaries**, **venules**, **veins** and back to the heart.

Arteries are thick, muscular vessels that carry blood away from the heart. They are able to withstand the high blood pressure caused by the pumping of the heart. The elasticity of the arteries maintains pressure on the blood between heartbeats to keep it flowing through the vessels. As the heart pumps blood into the arteries they expand such that one is able to feel a **pulse**. Smooth muscles in the arterial walls regulate **vasoconstriction** and **vasodilation** while precapillary sphincters regulate blood flow to the capillaries.

Capillaries have walls that are one cell thick. They form branching networks that allow for the exchange of materials between the blood and tissues. Most materials simply diffuse across the capillary cell wall into the cells that form tissues aided by the force of both blood pressure and osmotic pressure. Water diffuses especially easily and carries many other nutrients with it. Larger molecules are moved by active transport using tiny vesicles.

Veins carry blood back to the heart and also serve as a reservoir for blood volume. Blood is moved against gravity toward the heart by the contracting skeletal muscles and by pressure differences caused by the movement of the thoracic cavity during breathing. Blood is prevented from flowing backwards by valves found in the veins.

The muscular heart, contained within the **pericardium**, consists of two halves separated by a septum. Each half has two chambers, one smaller and thin-walled **atrium** and one larger, more muscular **ventricle**. **Valves** separating the atria from ventricles (**AV valves**) and the ventricles from the exit vessels (**semilunar valves**) keep blood from flowing backwards and give rise to the typical lubb-dup sounds of the heartbeat. The right side of the heart contains blood rich in carbon dioxide that flows in from the tissues and out to the lungs. The left side of the heart contains blood rich in oxygen returning from the lungs and flowing out to the tissues. The trip to and from the lungs is called the pulmonary circuit while the trip to and from the tissues is called the systemic circuit.

The **cardiac cycle** consists of the contracting of the atria followed by contracting of the ventricles followed by a rest when neither chamber is contracting. Specialized cardiac cells generate and conduct the electrical signals that control the heartbeat. The **sinoatrial (SA) node** sets the tempo and is therefore called the pacemaker. It causes contraction of the atria and sends a signal to the **atrioventricular (AV) node** which relays information to the **atrioventricular bundle** and out through the **Purkinje fibers** causing the ventricles to contract. A combination of nervous and endocrine signals control the strength and rate of contraction of the heart. An **electrocardiogram (ECG/EKG)**, a powerful diagnostic tool, is a recording of the electrical events associated with the heartbeat.

Blood pressure is its highest (**systolic**) when the ventricles contract sending blood into the arteries and at its lowest (**diastolic**) when the heart relaxes between beats. A **sphygmomanometer** measures blood pressure and can provide early identification of **hypertension**, or high blood pressure, the silent killer.

Atheroscloerosis, a narrowing of the arteries due to fatty deposits and thickening of the wall, can lead to heart attack or stroke. When this occurs in the arteries of the heart muscle, it is called **coronary artery disease**. Angiography can show coronary artery blockage, which can then be treated with **balloon angioplasty**, **atherectomy** or **coronary bypass surgery**. Heart muscle dies during a heart attack and is gradually replaced by scar tissue. During heart failure the heart becomes an inefficient pump leading to shortness of breath, fatigue and fluid accumulation. These symptoms can be effectively treated.

The lymphatic system functions to return interstitial fluid to the blood stream, to transport products of fat digestion and to defend the body against disease-causing organisms and abnormal cells. Lymph capillaries end blindly and are more permeable than blood capillaries. Interstitial fluid builds up around the cells and then enters the lymph capillaries. The fluid passes through a series of vessels and then is returned to the circulatory system. **Lymph nodes** filter lymph and contain macrophages and lymphocytes that actively defend against disease-causing organisms. Lymphoid organs include the tonsils, thymus gland, spleen, and Peyer's patches.

Key Concepts

- The cardiovascular system consists of the blood vessels and the heart.
- The heart is a muscular pump that contracts rhythmically, providing the force that drives blood through the vessels.
- Blood flows in a continuous loop from the heart to branching network of arteries and arterioles to capillaries to venules to veins and then back to the heart.
- Arteries are elastic, muscular vessels that carry blood away from the heart. Each heartbeat sends a pressure wave down the artery called the pulse.
- Arteries have smooth muscles that result in vasoconstriction and vasodilation.
- Capillaries are one cell thick and allow for the exchange of materials between the blood and the tissues by diffusion, pressure-driven movement, endocytosis or exocytosis.
- Veins return blood to the heart. The lumen is larger in a vein than an artery and the wall is less rigid.
- Veins contain valves that keep blood from flowing backward.
- The heart consists of two atria and two ventricles. The right and left halves of the heart function as separate pumps.
- The right side of the heart pumps blood through the pulmonary circuit transporting blood to and from the lungs. The left side of the heart pumps blood through the systemic circuit, which transports blood to and from the body tissues.
- The heart muscle is nourished by coronary circulation.
- The cardiac cycle is the sequence of heart muscle contraction (systole) and relaxation (diastole).
- Cardiac tissue has its own internal conduction system so that the cells contract in a coordinated manner.
- The sinoatrial node is the pacemaker, sending stimuli to the atrioventricular node down the atrioventricular bundle to the Purkinje fibers resulting in a contraction.
- The heartbeat is regulated by the autonomic nervous system and certain hormones.
- An electrocardiogram is a recording of the electrical activities of the heart.
- Blood pressure is the force blood exerts against the walls of the blood vessels. Systolic pressure is produced when the ventricles contract and diastolic pressure is produced when the ventricles relax.
- Cardiovascular disease is the single biggest killer of men and women in the US.
- High blood pressure can kill without symptoms, thus it is termed the silent killer.
- An inflammation process underlies the buildup of lipids in the artery walls called atherosclerosis. When this occurs in the coronary arteries it is called coronary artery disease and is the major cause of heart attacks.
- The lymphatic system functions in the circulatory and immune systems to return excess interstitial fluid to the bloodstream, transport products of fat digestion from the small intestine to the blood stream, and help defend against disease-causing organisms.

Study Tips

Just a quick glance through this chapter shows the value of figures. After reading the chapter, review the key concepts listed in the study guide, then begin to focus on the figures. Many times, the figures in a chapter of a science text will summarize the content quite well. Improving your "figure literacy" will enhance your ability to comprehend the sciences. When you first look at a figure, read and highlight the important words in the caption and the terms in the body of the figure used by your instructor. Next, write a description of the process or structure shown in the figure. Take time to use complete sentences and to construct a complete explanation of the figure. If you have a study partner, exchange and critique descriptions, reading for clarity and completeness. These descriptions will serve as the basis for later study.

The circulatory system consists of the blood vessels and heart. The chapter begins with a comparative description of the vessels. As you study Figure 12-2, link the structure of the vessel to its function. Your study of the heart should focus both on the anatomy of the heart and also on the electrical circuitry that causes a heartbeat. Begin by making your own drawing of the heart and labeling it. Then add arrows showing the direction of blood flow. Write on the arrows whether the blood is oxygen-rich or oxygen-poor. Next, link blood flow to and from the heart with the system of vessels. Describe the pulmonary and systemic circuits. Next focus on the heartbeat: explain electrical conduction through the heart as shown in Figure 12-13 and its recording in the ECG. Make a list or set of flash cards to review health-related problems of the cardiovascular system. Finally, be able to describe the purposes of the lymph system and link them to the circulatory system. Describe similarities and differences between the function and vessels of the two systems.

Review Questions

A. Fill in the blank.

1. The circulatory system consists of the ____________, ____________ and ____________.

2. What is the proper flow of blood through the vessels beginning at the heart and returning to the heart? Heart → ____________ → ____________ → ____________ → venule → ____________ → heart

3. The list of statements on the right gives characteristics of different types of vessels. Fill in the blanks with the proper vessel type (artery, vein, capillary) that matches the description on the right.

   a. __________________ Walls are one cell thick

   b. __________________ Contain valves

   c. __________________ Elastic to absorb pressure changes

   d. __________________ Largest total cross-sectional area

e. ____________ Thinner wall and larger lumen

f. ____________ Exchange of nutrients and gases with tissues

g. ____________ Thickest and largest vessels

B. Complete the following table, giving the characteristics of each type of vessel.

Vessel	**Relative Thickness**	**Direction of Flow**	**Primary Gases**	**Force Causing Blood Flow**
Artery				
Capillary	One cell thick	Mid point	Delivers oxygen and picks up carbon dioxide	Very little force
Vein				

C. Number the following statements in the correct order of blood flow through the body. Put that number in the first column. Then, in the next column put a ***P*** to represent pulmonary circulation and an ***S*** to represent systemic circulation.

	Order	Pulmonary/Systemic	Statement of Blood Flow
1.	______	______	Aorta
2.	______	______	Left atrium
3.	______	______	Arterioles in the thigh
4.	______	______	Pulmonary veins
5.	______	______	Femoral vein
6.	______	______	Right ventricle
7.	______	______	Right atrium
8.	______	______	Capillary beds at alveoli
9.	______	______	Capillary beds in muscle tissue
10.	______	______	Pulmonary arteries

11. ______ ______ Inferior vena cava

12. ______ ______ Left ventricle

D. Use the following terms to label the drawing of the heart shown below: aorta, aortic semilunar valve, bicuspid valve, endocardium, inferior vena cava, left atrium, left ventricle, mitral valve, myocardium, pericardium, right atrium, right ventricle, septum, superior vena cava, tricuspid valve, and trunk for pulmonary arteries.

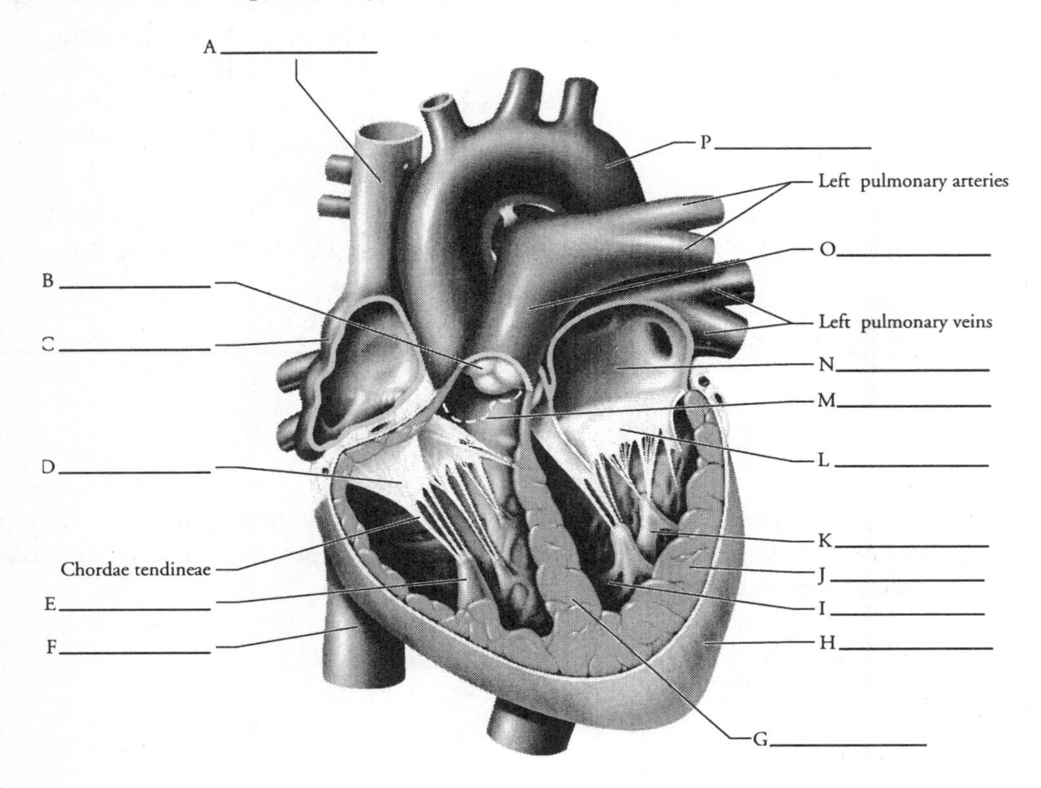

E. Put a number in Column A indicating the correct order of the events in Column B as they relate to the cardiac cycle. Then put the structure that causes that event in Column C. Each of the following structures should be used twice: AV Node, Purkinje fibers, and SA Node.

	Column A	Column B	Column C
1.	________	Initiates electrical signal	________
2.	________	Delay of signal to allow atria to fully contract	________
3.	________	Stimulate two branches in the septum	________
4.	________	Atria contract	________
5.	________	Stimulation of Purkinje cells	________
6.	________	Cause ventricles to contract	________

F. Review the electrocardiogram shown in Figure 12-14b of your text. Describe what is happening during each phase of the ECG.

1. P wave: ______________________________

2. QRS wave: ______________________________

3. T wave: ______________________________

G. Review blood pressure and the sphygmomanometer. Relate blood pressure and the sounds heard through the stethoscope to the events listed below.

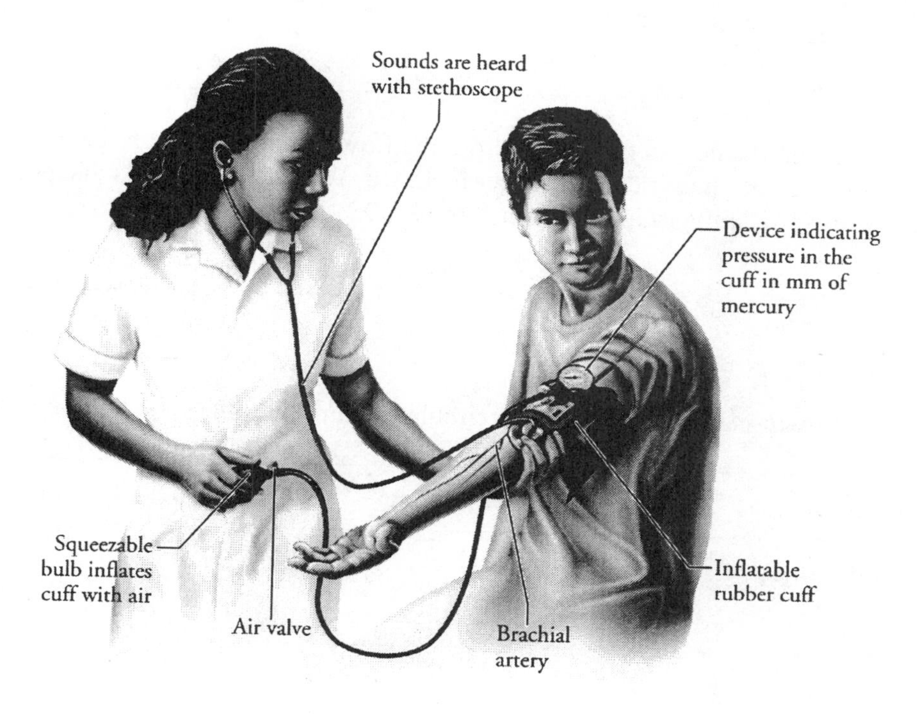

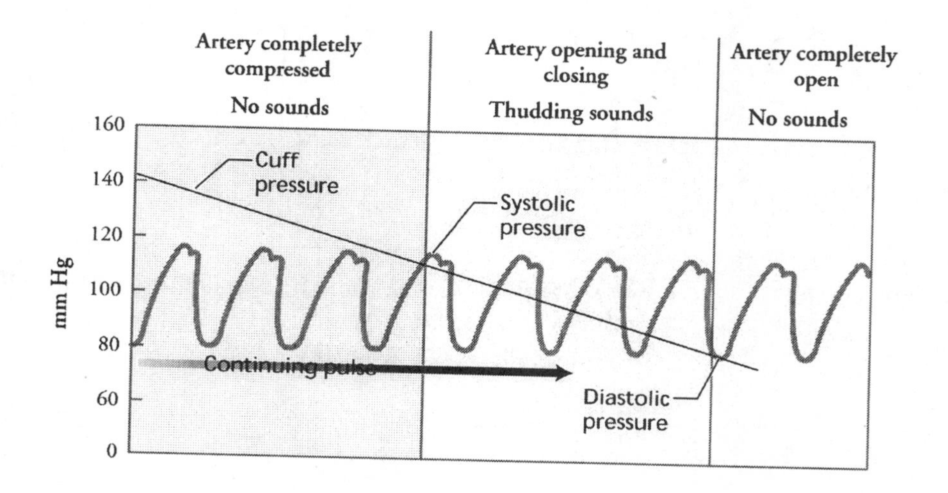

1. When the pressure in the cuff is inflated, there are no sounds heard through the stethoscope. Why? What change in the blood flow has been caused by the sphygmomanometer?

____

____

2. Next, pressure is slowly released from the sphygmomanometer cuff and soon there are sounds heard through the stethoscope. What are these sounds and how do they relate to blood pressure? How will the reading on the sphygmomanometer be recorded?

____

____

3. As the pressure in the cuff continues to drop, the flow of blood through the artery can be heard for a while and then it can no longer be heard. Why do the sounds disappear and what does the reading on sphygmomanometer represent?

____

____

H. Complete the following table comparing the circulatory and lymphatic systems.

Process/Structure	**Circulatory**	**Lymphatic**
Primary function(s)		
Fluid		
Vessels		
Force behind flow of fluids		
Structure of vessels	Closed system of vessels, vessels with varying structure but valves occur in veins	Open ended, valves in vessels

Critical Thinking

Read each of the following questions carefully. Circle the most important words in the question before formulating your answer.

1. Why is it advantageous for blood to move slowly through the capillaries? Relate this to the thickness of the capillary wall, the formation of capillary beds, and the very low pressure found in capillaries.

2. First review blood flow through the heart, then explain how the relative thickness of the walls of the chambers of the heart relate to their function.

3. Describe electrical control of the heartbeat. Then describe how an artificial pacemaker functions and how it interfaces with our natural mechanisms to correct an irregular heartbeat.

4. In the fetus, there is a hole in between the left and right atria of the heart called the foramen ovale. This provides more efficient circulation of blood for the fetus while it is in the womb. Why is this advantageous for the fetus? If this hole remains after birth, it is considered an abnormality that should be surgically corrected. Why is the hole in the septum of the adult heart a disadvantage?

5. In a small group, discuss the common cardiac risk factors and then develop a plan for improving cardiac health. What is the value of aerobic, cardiac conditioning?

Practice Test

Choose the one best answer to each question that follows. As you work through these items, explain to yourself why each answer you discard is incorrect and why the answer you choose is correct.

1. Which is the smallest blood vessel?
   a. arteriole
   b. artery
   c. capillary
   d. vein

2. The main function of valves within the circulatory system is to ____________.
   a. permit blood to circulate rapidly
   b. prevent blood from moving too rapidly
   c. prevent blood from moving in the wrong direction
   d. stop the circulation whenever necessary

3. Which of the following contains blood with the highest oxygen content?
   a. Capillaries in the tissues
   b. Pulmonary artery
   c. Pulmonary vein
   d. Superior vena cava

4. The most serious reason that aneurysms are dangerous is because they ____________.
   a. can be repaired, but it takes surgery
   b. can burst causing fatal blood loss
   c. cause distention of the artery wall
   d. cause the blood flow to be redirected

5. Which of the following most correctly traces the path of blood from the heart to a kidney and back to the heart again?
   a. Left ventricle, vena cava, renal vein, kidney, renal artery, aorta, right atrium
   b. Right atrium. aorta, renal artery, kidney, renal vein, vena cava, left atrium
   c. Left ventricle, renal artery, kidney, renal vein, vena cava, right atrium
   d. Left ventricle, aorta, renal artery, kidney, renal vein, vena cava, right atrium

6. The function of the chordae tendinae is to ____________.
   a. close the heart valves
   b. hold the heart together
   c. hold the two main heart valves in place
   d. prevent the heart valves from turning inside out under pressure

7. The heart is considered to be a double pump because ____________.
   a. the atria contract then the ventricles contract, giving two pumps
   b. there is a "back up" mechanism in the heart in case of heart attack that gives one a "second chance" – that is the double pump available to everyone
   c. the two sides of the heart pump blood that never mixes
   d. the two sides of the heart beat independently, controlled by two separate mechanisms.

8. Heart sounds result from ____________.
   a. closing of the heart valves
   b. contraction of the ventricles
   c. expansion and elastic recoil of arteries
   d. flow of blood as it enters the heart

9. Heart murmurs are usually caused by ____________.
   a. heart attacks
   b. malfunctioning of the valves
   c. normal opening and closing of the valves
   d. swishing of blood past the valves

10. What provides nutrients and oxygen to the heart muscle?
    a. blood flowing through the heart
    b. branches from the pulmonary veins
    c. capillaries from the mammary arteries
    d. cardiac circulation

11. Which of the following statements is correct for a normal heartbeat?
    a. At any given time, either the atria or ventricles are contracted.
    b. At any given time, the atria, the ventricles, or neither may be contracted.
    c. Both right chambers of the heart contract simultaneously in one large surge.
    d. Both atria and both ventricles contract simultaneously in one large surge.

12. If we had a two-chambered heart (one atrium, and one ventricle) with the vena cava entering the atrium and the aorta leaving the ventricle, which of the following would be true?
    a. no blood would reach the head and neck
    b. no oxygen would reach the cells
    c. the blood would be unable to carry food to the cells
    d. there would be no bodily circulation

13. The pacemaker of the heart is the ___________.
    a. AV node
    b. cardiac center
    c. SA node
    d. Purkinje fibers

14. In the electrocardiogram, the large QRS wave results from the ___________.
    a. large muscle mass of the ventricles
    b. many contractions of ventricular fibrillation
    c. over development of the heart muscle resulting from a heart attack
    d. quick contraction of the ventricle (QR) followed by relaxation (RS)

15. Normal blood pressure is considered to be 120/80. What does the "120" represent?
    a. The pressure in the veins
    b. The maximum pressure that can be sustained by the arteries before bursting
    c. The highest ideal arterial pressure for someone at rest
    d. The lowest pressure in the arteries that will keep someone from fainting when they stand up

16. Which of the following is the "silent killer" because it usually has no symptoms?
    a. Hemophilia
    b. High cholesterol
    c. High blood pressure
    d. Stroke

17. Why is atherosclerosis especially dangerous when found in the coronary arteries?
    a. It can restrict blood flow to the heart muscle.
    b. It can cause a heart attack.
    c. It can lead to coronary artery disease.
    d. All of the above

18. Which of the following is associated with atherosclerosis?
    a. High cholesterol diet
    b. High-fiber diets
    c. Increased exercise
    d. Low-salt diets

19. Which of the following is NOT a risk factor for cardiovascular disease?
    a. Cigarette smoking
    b. Consumption of aspirin
    c. Obesity
    d. Stress

20. Which of the following is a function of the lymphatic system?
    a. cause elephantiasis
    b. defend against disease
    c. provide a mechanism for the spread of cancer cells
    d. transport red blood cells and fibrinogen

Short Answer Questions

Read each of the following questions carefully. Review the appropriate section of the text if necessary; jot down the main points. Then formulate your answer using complete sentences in your explanation.

1. Construct a table comparing an artery, vein and capillary with regard to structure, direction of flow, pressure and the oxygenation level of the blood. Include any special features of the vessel.

2. Why is the heart called a double pump?

3. Describe the path of blood through the heart, and the pulmonary and cardiac circuits. Include the direction of gas exchange in the lungs and tissues.

3. Relate the image of an electrocardiogram with the heartbeat.

4. There are several very serious, life-threatening, diseases associated with the cardiovascular system. First name and define at least three and then describe lifestyles that will improve cardiovascular health.

5. Identify differences in the function and structure of the lymph and cardiovascular systems.

Answer Key

Review Questions

A. Fill in the blank.

1. The circulatory system consists of the blood, heart, and vessels.
2. heart → artery → arteriole → capillaries → venule → vein → heart
3. a. capillaries
   b. veins
   c. arteries
   d. capillaries
   e. veins
   f. capillaries
   g. arteries

B.

Vessel	Relative Thickness	Direction of Flow	Primary Gases	Force Causing Blood Flow
Artery	Very thick	Away from heart	Usually oxygen	Pulse, heartbeat
Capillary	One cell thick	Mid point	Delivers oxygen and picks up carbon dioxide	Very little force
Vein	Not as thick as artery, thicker than capillary	Toward the heart	Usually carbon dioxide	Skeletal muscles, breathing

C.
1. 1, S
2. 11, S
3. 2, S
4. 10, P
5. 4, S
6. 7, P
7. 6, P
8. 9, P
9. 3, S
10. 8, P
11. 5, S
12. 12, P

D.
A. superior vena cava
B. pulmonary semilunar valve
C. right atrium
D. right atrioventricular valve
E. right ventricle
F. inferior vena cava
G. septum
H. pericardium
I. endocardium

J. myocardium
K. left ventricle
L. left atrioventricular valve
M. aortic semilunar valve
N. left atrium
O. pulmonary trunk
P. aorta

E. 1. 1, SA Node
2. 3, AV Node
3. 5, Purkinje
4. 2, SA Node
5. 4, AV Node
6. 6, Purkinje

F. 1. P wave: spread of electrical signal through the artia and the resulting contraction
2. QRS wave: spread of electrical signal over the ventricles and ventricular contraction
3. T wave: ventricular repolarization/recovery

G. 1. The pressure in the sphygmomanometer has caused the blood flow to stop; therefore there is no sound. This is similar to a tourniquet.
2. As pressure in the cuff is released, blood spurts through the artery at its highest pressure. Therefore, the reading on the sphygmomanometer represents the highest pressure in the artery, systolic pressure. This is the pressure in the artery when the heart is contracting.
3. Arterial blood is forced through the sphygmomanometer cuff under pressure equal to that in the cuff. When there are no more sounds heard through the stethoscope, the pressure in the cuff has dropped below the lowest pressure in the artery. This is the diastolic pressure, the pressure in the artery when the heart is relaxed.

H.

Process/Structure	**Circulatory**	**Lymphatic**
Primary function(s)	Transport nutrients, gases and wastes throughout the body	Return excess fluid to the bloodstream, transport products of fat digestion, defend against disease
Fluid	Blood	Lymph
Vessels	Arteries, capillaries, veins	Lymphatic vessels and capillaries
Force behind flow of fluids	Force generated by heart, thoracic pressures, skeletal muscles	Thoracic pressures, skeletal muscles
Structure of vessels	Closed system of vessels, vessels with varying structure but valves occur in veins	Open ended, valves in vessels

Critical Thinking

1. Capillaries are only one cell thick to allow for diffusion. Because there are so many vessels in the capillary beds, pressure is decreased to almost nothing. Blood passes through the capillaries very

slowly allowing more time for diffusion. It is also essential that the pressure in the capillaries be very low since their thin walls provide very little strength.

2. The walls of the two atria are not very thick because blood flows by gravity from the atria to the ventricles. The right ventricle has to pump blood to the lungs, a relatively short distance. Therefore, the right ventricle is stronger than the atria but not as strong as the left ventricle. The left ventricle must pump blood throughout the body including to the head against gravity. The left ventricle is the strongest chamber of the heart.
3. The sinoatrial node serves as the heart's internal pacemaker that determines the heart rate. Electrical signals from the sinoatrial node spread through the walls of the atria causing them to contract. The signal reaches the atrioventricular node and stimulates it. The atrioventricular node relays the stimulus by means of the atrioventricular bundle. The electrical signal is conducted to the ventricles through the atrioventricular bundle and through the ventricle walls by the Purkinje fibers. A permanent pacemaker can be surgically inserted under the skin. This is the computer that controls the amplitude and frequency of the electrical charge given to the heart muscle. A wire connects the pacemaker to the apex on the heart sending a regular stimulus through the ventricles.
4. The fetus gets oxygen and removes carbon dioxide through the umbilical cord, and not through its lungs. The septum of the heart separates oxygen-rich blood from oxygen-poor blood also rich in carbon dioxide. It keeps the pulmonary and systemic circuits separate. However, since blood does not circulate to the lungs in the fetus, there is no need to separate the pulmonary and systemic circuits. In fact, it is more efficient not to.
5. The most common cardiac risk factors are genetics, too much weight, too little exercise, poor diet, and smoking. Each of these risk factors can be reduced or eliminated except the genetic factors. Get regular exercise, at least 30 minutes 6 days a week, eat a low fat/high fiber diet, do not smoke and maintain a healthy weight.

Practice Test

1. c
2. c
3. c
4. b
5. d
6. d
7. c
8. a
9. b
10. d
11. b
12. b
13. c
14. a
15. c
16. c
17. d
18. a
19. b
20. b

Short Answer Questions

1.

Vessel	Structure	Direction of Flow	Blood Quality	Pressure
Artery	Elastic, smooth muscles	Away from heart	Usually blood rich in oxygen	High pressure, pulse
Vein	Larger lumen and thinner walls than arteries, contain valves	Toward heart	Usually blood rich in carbon dioxide	Lower pressure, no pulse
Capillary	Network of vessels one cell thick	From arteriole to venules	Place where oxygen and nutrients are delivered to tissues	Due to branching, pressure drops to almost nothing

			and carbon dioxide and wastes are picked up	

2. The two sides of the heart are separated by a septum. Blood rich in oxygen never mixes with blood rich in carbon dioxide. Each side of the heart consists of one atrium and one ventricle.
3. Blood enters the heart from the body in the right atrium, goes to the right ventricle, then to the lungs via the pulmonary artery. In the lungs carbon dioxide leaves the blood entering the lungs, and oxygen is picked up from the lungs entering the blood. The blood returns to the heart via the pulmonary veins. This completes the pulmonary circuit. Blood enters the left atrium from the lungs and passes into the left ventricle. It leaves the heart by the aorta and reaches the capillary beds in the tissues where oxygen is exchanged for carbon dioxide. Blood returns to the heart entering the right atrium. This completes the systemic circuit.
3. The electrical activity of a heartbeat can be recorded using an ECG. The P wave is generated by the electrical impulses moving from the SA node across the atria. The QRS wave represents the spread of the signal through the ventricles and ventricular contraction. The T wave occurs as the ventricles recover and return to the electrical state before contraction.
4. Hypertension is high blood pressure, atherosclerosis is thickening of the arterial wall, coronary artery disease is the condition of fatty deposits forming in the coronary arteries, and heart attack results in dead heart tissue due to a lack of coronary circulation. It is important to control one's weight, exercise regularly, never smoke, limit one's dietary salt, and eat a healthy, low-fat diet.
5. The lymph system contains vessels, but no pump. In addition, the lymph system is not a closed system since the vessels are open near the cells and then open again when returning fluid to the circulatory system. Lymph vessels, similar to veins, contain valves to regulate flow. The lymph system transports interstitial fluid and soluble fats.

Chapter 13

Body Defense Mechanisms

Objectives

After reading the text and studying the material in this chapter, you should be able to:

- List the body's three lines of defense against foreign organisms, cells, or molecules.
- Describe the body's nonspecific physical and chemical surface barriers.
- Describe each of the defensive cells and their function.
- Explain how interferons and the complement system function in the body's defense mechanism.
- Give the order of events in an inflammatory response.
- Define a fever and explain its role in our immune response.
- Define antigen and MHC markers and explain their role in cell identity.
- Describe the role of various cells involved in the immune response.
- List the steps in our immune response.
- Compare an antibody-mediated immune response with a cell-mediated immune response.
- Explain immunological memory and its role in immunity.
- Describe what happens in an autoimmune disorder and give an example.
- Explain what happens when there is an allergic response.

Chapter Summary

The body's defense mechanisms target pathogens and cancerous cells. The body has three lines of defense: 1) nonspecific physical and chemical surface barriers, 2) nonspecific internal cellular and chemical defense, and 3) specific immune responses.

The skin's tough outer layer of dead cells filled with fibrous keratin makes it nearly impenetrable, waterproof, and resistant to most toxins and enzymes of invading organisms. In addition, as the dead cells flake off, they take many microbes with them. Sweat and oil glands of the skin produce chemicals that slow or prevent the growth of bacteria.

Most organisms enter the body through the mucous membranes that line the respiratory, digestive, urinary and reproductive systems. The sticky mucus of the respiratory and digestive tracts traps many microbes. The lining of the stomach produces hydrochloric acid and digesting enzymes that destroy pathogens. The acidity of urine slows bacterial growth. Saliva and tears contain an enzyme **lysozyme** that kills bacteria.

The second line of defense includes defensive cells and proteins, inflammation and fever. Defensive cells include **neutrophils** and **macrophages** that engulf pathogens, damaged tissue or dead cells by a process of phagocytosis. **Eosinophils** get close to the parasite and discharge destructive enzymes that destroy it. **Natural killer cells** search out abnormal cells, including cancerous cells, and kill them.

The body's non-specific cellular defenses use two types of defensive proteins: interferons and the complement system. Before a virally infected cell dies, it secretes small proteins called **interferons** that 1) attract macrophages and natural killer cells and 2) stimulate neighboring cells to make proteins that prevent the viruses from replicating. The **complement system** is a group of proteins that enhance both nonspecific and specific defense mechanisms by destroying pathogens, enhancing phagocytosis, and stimulating the inflammatory response.

When tissues are injured or damaged, the **inflammatory response** destroys invaders and helps repair and restore damaged tissue. The release of chemokines and cytosolic contents from injured cells stimulates mast cells to release **histamine**, which increases blood flow by dilating blood vessels and increasing the permeability of the capillaries. Fluid leaking from the capillaries causes swelling and increased blood flow causes redness and warmth. Phagocytes and other cells attracted to the area begin to clean it up. A fever is an abnormally high body temperature caused by pyrogens, chemicals that reset the brain's thermostat to a higher temperature. A moderately higher body temperature helps fight bacterial infections.

The third line of defense, the immune system, has specific targets and memory. The body's specific defenses working together in the recognition and destruction of specific pathogens are called the **immune response**. **MHC markers** are found on our own cells and mark them as belonging to us. Substances that trigger an immune response are called antigens. Each **lymphocyte** develops receptor molecules on its surface that match one particular antigen. When an antigen is detected, **B lymphocytes** and **T lymphocytes** that recognize the antigen are stimulated to divide repeatedly. Some of these cells attack and eliminate the invader while others are stored in a state of suspended animation as a form of memory of the invader and are available to attack if there is a reoccurrence.

The immune system mounts antibody-mediated and cell-mediated responses, both of which recognize and destroy the same antigens but in different ways. The **antibody-mediated immune responses** defend against antigens that are free in body fluids, including toxins or extracellular pathogens. The B-cells use antibodies to neutralize the antigen. **Cell-mediated immune responses** involve living cells and protect against cellular threats including body cells that have become infected and cancer cells.

Macrophages present antigens to **helper T-cells** to trigger an immune response. Helper T-cells activate B-cells and T-cells to destroy the specific antigen. When activated, these cells divide to form clones of cells designed to eliminate a specific antigen from the body. The effector cells produced when activated B-cells divide are called **plasma cells**. They secrete antibodies, also called **immunoglobulins** (Ig), into the bloodstream that bind to antigens. **Cytotoxic T-cells** are the effector T-cells responsible for the cell-mediated immune response. **Suppressor T-cells** turn off the immune response when the antigens no longer pose a threat.

Immunological memory allows for a more rapid response on subsequent exposure to the antigen. Although the primary response may be slow as the antibody concentration rises, the **secondary response** is strong and swift due to the large number of specific memory cells that can respond to the antigen.

In **active immunity,** the body actively participates by producing memory B-cells and T-cells following exposure to an antigen. This process can also happen through **vaccination**.

Because memory cells are produced, active immunity is relatively long lived. **Passive immunity** results when a person receives antibodies that were produced by another person or animal. This is the case when antibodies cross the placenta from mother to fetus. However, passive immunity is short lived since the recipient's body was not stimulated to produce memory cells.

Monoclonal antibodies are a group of identical antibodies that bind to one specific antigen. They are used in research, clinical diagnosis, and disease treatment because they can help diagnose certain diseases in their early stages.

Autoimmune disorders occur when the immune system attacks the body's own cells. Organ-specific autoimmune disorders are directed against a single organ and are caused by T-cells. Non-organ-specific autoimmune disorders are caused by antibodies and have effects that are widespread throughout the body. Treatment involves correcting the deficiency caused by the disorder and suppressing the immune system with drugs.

Allergies are immune responses to harmless substances called **allergens**, which cause plasma cells to release large numbers of a particular subtype of antibodies (IgE class). These antibodies bind to mast cells or basophils causing them to release histamine. The histamine causes redness, swelling, itching and other symptoms of an allergic response. Antihistamines are most effective in reducing the allergy symptoms. Allergy shots inject increasing amounts of a known allergen in an effort to desensitize the person to the offending allergens.

Key Concepts

- The body's defense system targets pathogens and cancerous cells.
- The body has three lines of defense:
  1) Physical (skin and mucous membranes) and chemical barriers that prevent entry by pathogens,
  2) Nonspecific defensive cells (phagocytes, eosinophils and natural killer cells) and proteins (interferons and complement system), inflammation and fever, and
  3) The immune system that has specific targets and memory.
- The immune system mounts antibody-mediated and cell-mediated responses.
- Substances that trigger an immune response are called antigens.
- Our own nucleated cells have MHC markers that identify them as "self."
- During maturation, B lymphocytes (B-cells) and T lymphocytes (T-cells) develop receptors on their surfaces that allow them to recognize one specific antigen and to recognize cells that belong in the body.
- When that antigen is detected they divide rapidly forming effector cells and memory cells.
- Macrophages present antigens to helper T-cells to trigger an immune response.
- Helper T-cells activate B-cells and T-cells to destroy the specific antigen.
- B-cells mount an antibody-mediated immune response against antigens free in the blood or bound to a cell surface.
- Cytotoxic T-cells mount a cell-mediated immune response to destroy antigen-bearing cells.
- Suppressor T-cells turn off the immune response
- Immunological memory allows for a more rapid secondary response on subsequent exposure to the antigen.

- Immunity can be active, when the body participates in forming memory cells against a particular antigen, or passive, when the body receives antibodies that were produced by another person or animal. Passive immunity is short lived.
- Monoclonal antibodies are used in research, clinical diagnosis, and disease tretment to identify the presence of a specific antigen.
- Autoimmune disorders occur when the immune system attacks the body's own cells.
- Allergies are immune responses to harmless substances.

Study Tips

It is critical that you conceptually understand the processes described in this chapter. Depending upon the depth of your background, much of this material may be new. First, study the chapter outline. Notice the purpose of the body's defense mechanisms and then the three lines of defense. The bulk of the chapter is spent on the third line of defense, the immune system. You will need to know the role of the various cells and chemicals involved in the immune system as well as the mechanisms of defense.

Gather some large index cards. Your challenge will be to simplify and organize the material in this chapter. You might choose to summarize the process on one side of the card and then list the special cells or chemicals on the back. Begin with the first line of defense – summarize the physical and chemical barriers. Be sure to include the information in Figure 13-3. Next summarize the second line of defense on another card and put the specialized cells and defensive proteins on the back.

Now focus on the immune response. First note how it is different from the other two lines of defense and note the definitions of key terms. Next focus on the difference between antibody-mediated and cell-mediated immune responses. Generate a card summarizing each of those mechanisms and the cells related to them.

Finally, summarize passive and active immunity and note an example for each. Then focus on autoimmune disorders as described in Table 13-6 and the allergic reaction shown in Figure 13-19. As you review the text material and check your note cards for accuracy, record the specific figures and tables that simplify the responses you are describing. With a focus on the mechanisms and the ability to define the role of various cells, you will begin to understand the body's defense mechanisms one step at a time.

Review Questions

A. List the body's three lines of defense.

1. ______________________________

2. ______________________________

3. ______________________________

B. Match each of the following examples with the general nonspecific form of defense. Each response may be used once or twice.

a. physical barrier to entry
b. chemical barrier to entry
c. internal cellular defense
d. internal chemical defense
e. inflammation
f. fever

________ 1. Interferons interfere with viral reproduction.

________ 2. Macrophages attack invading bacteria.

________ 3. Intact skin layer.

________ 4. Slightly elevated body temperature slows bacterial growth.

________ 5. Natural Killer cells attack a leukemia cell.

________ 6. Lysosomes in tears fight off bacteria.

________ 7. Complement system destroys a pathogen.

________ 8. Histamines cause swelling.

________ 9. Acidity of stomach contents and urine.

________ 10. Mucous lining of the nasal cavity.

C. Write a one-sentence description of an antibody-mediated immune response and a cell-mediated immune response. Be sure to indicate the primary difference between them.

1. Anti-body mediated immune responses ____

____

2. Cell-mediated immune responses ____

____

D. Complete the following flow chart of the steps in the immune response. Then complete the paragraph below describing the process.

Threat → ____________ → Alert → ____________ → Build specific defense →

____________ → Continued surveillance → Withdrawal of forces

The threat is posed to the body when a ____________ or molecule enters. A ____________ rushes to the scene, detects the foreign cell or molecule and ____________ it. The macrophage puts an ____________ from the pathogen on its ____________ and presents the antigen to the

___________ cell that has the correct ___________ for the antigen. The macrophage ___________ the helper T-cell. In the alarm phase, the helper T- cell activates both lines of defense to fight that specific antigen.

In an antibody-mediated defense, B-cells are activated and divide to form ___________ cells that secrete ___________ specific to the antigen. In a ___________ defense, T-cells divide to form ___________ cells that attack cells with the specific antigen. As the defense mounts, ___________ specific to the ___________ eliminate the antigen. The cytotoxic T-cells cause cells with the antigen to burst. Surveillance is continued by ___________ that are formed when helper T-cells, cytotoxic T-cells and B- cells were activated. They remain to provide a response if the antigen is detected again. The immune response ends when the ___________ has been destroyed and ___________ T- cells shut down the immune response to that antigen.

E. Match the term in Column A with the function in Column B.

Column A	Column B
1. antibody	a. secretes antibodies specific to the invader
2. antigen	b. turns off the immune response
3. B-cells	c. responsible for the memory of the immune system
4. cytotoxic T-cells	d. chemical that causes holes to form in membranes
5. helper T-cells	e. something that triggers an immune response
6. macrophage	f. responsible for cell-mediated immune responses
7. memory cells	g. proteins that bind to antigens, immunoglobulins for example
8. plasma cell	h. divides to form plasma cells and memory cells
9. porforins	i. a cell that presents the antigen
10. suppressor T-cells	j. activates B-cells and T-cells

F. Review Figure 13-12 in your text and complete the following questions describing colonal selection.

1. What is found on the surface of the B-cells? ____________ What is their function?

   ____

   ____

2. What happens after the antigens bind to the receptor sites? ____________________

   ____

   ____

3. Some of the cloned B-cells become plasma cells. What do they do? ______________

   ____

   ____

4. Other cloned B-cells become memory cells. What do they do? __________________

   ____

   ____

5. If this process described a cell-mediated response, what additional cells would be involved?

   ____

   ____

G. Immunity means that you are protected from a disease. What is the difference between passive and active immunity? Give an example of each.

1. In active immunity ____

   ____

   ____

2. In passive immunity ____

   ____

   ____

H. Identify the autoimmune disorder described below.

Description	Autoimmune Disorder
1. Inflammation of the tissues around the brain causing impaired brain functions.	1. ______________________
2. Elevated blood sugar resulting from destruction of pancreatic cells	2. ______________________
3. Severe joint pain caused by inflammation of the connective tissue around the synovial joint.	3. ______________________
4. Body attacks the myelin sheath around a nerve. When it is repaired a hardened region is formed that does not conduct nerve impulses well.	4. ______________________
5. The body's immune system attacks the heart valves and joints resulting in pain.	5. ______________________

Critical Thinking Questions

Read each of the following questions carefully. If it helps you, underline or list each component to the question. Use complete sentences in your explanation.

1. Explain why you may not want to reduce a fever. What good is it?
2. Describe how the body identifies its own cells. What is the advantage to this strategy for an immune system, that is identifying its own cells?
3. From your knowledge of viruses and the body's defense mechanisms, explain why people get so many colds.
4. Your grandmother has developed rheumatoid arthritis. Explain what caused it.

Practice Test

Choose the one best answer to each question that follows. As you work through these items, explain to yourself why each answer you discard is incorrect and why the answer you choose is correct.

1. What is our primary physical barrier against disease?
   a. bladder
   b. mucus
   c. skin
   d. stomach

2. The body has several lines of defense against foreign organisms, cells or molecules. Which of the following is the first line of defense?
   a. attack the invader with white blood cells
   b. identify and destroy the specific invader
   c. keep foreign cells or molecules out of the body
   d. provide antibodies to all cells

3. Cancerous cells routinely form within our body. What type of cell destroys them?
   a. eosinophils
   b. macrophages
   c. natural killer cells
   d. neutrophils

4. What causes the production of interferons?
   a. bacteria
   b. chemicals
   c. vaccines
   d. viruses

5. The complement system is an example of which type of nonspecific defense?
   a. barrier to entry
   b. defensive cells
   c. defensive proteins
   d. inflammation

6. How is the immune system different from the first two lines of defense?
   a. It has memory
   b. It is specific
   c. It provides a barrier
   d. Both a and b

7. What molecules are found on the surface of both T lymphocytes and B lymphocytes that identify foreign invaders?
   a. antibodies
   b. antigens
   c. receptors
   d. self markers

8. What is the function of major histocombatibility complex markers?
   a. function as antibodies
   b. identify the cell as one's own
   c. identify specific invaders
   d. produce histamines

9. Which cell type produces antibodies?
   a. helper T-cell
   b. macrophage
   c. plasma cell
   d. suppressor T-cell

10. Which cell type is associated with the cell-mediated immune response?
    a. B-cells
    b. cytotoxic T-cell
    c. helper T-cell
    d. suppressor T-cell

11. Which of the following is most important in activating both types of immune responses?
    a. B-cells
    b. helper T-cells
    c. memory helper T-cells
    d. suppressor T-cells

12. A specific antibody will bind to ____________.
    a. any antigen; antibodies are quite flexible
    b. any antigen of the same class of antigens
    c. groups of antigens with the same chemical shape
    d. only one specific antigen

13. Which cell type is the most important antigen-presenting cell (APC)?
    a. B-cell
    b. helper T-cell
    c. macrophage
    d. T-cell

14. Both cytotoxic T-cells and natural killer cells use the same method to destroy their targets. What is it?
    a. antibodies
    b. antigens
    c. immunoglobulins
    d. perforins

15. What is the importance of immunological memory?
    a. saves time
    b. saves vaccinations
    c. uses far fewer lymphocytes
    d. uses additional suppressor T cells

16. How does the secondary response different from the primary response?
    a. the primary response is more specific
    b. the primary response is much faster than the secondary response
    c. the secondary response involves additional macrophage cells
    d. the secondary response is faster than the primary response

17. Which of the following is not an autoimmune disorder?
    a. Addison's disease
    b. lupus erythematosus
    c. multiple sclerosis
    d. type 2 diabetes mellitus

18. When you receive a vaccination that includes inactivated or activated vaccines, your body is developing ____________ immunity.
    a. active
    b. aggressive
    c. passive
    d. recessive

19. When do autoimmune disorders occur?
    a. When cancerous cells are allowed to multiply unchecked.
    b. When the immune system fails to properly recognize "self" cells.
    c. When the immune system fails to produce antibodies.
    d. When the immune system has a nonspecific response to invading cells.

20. An allergic reaction is caused by ____________.
    a. autoimmune disorders
    b. environmental toxins
    c. overproduction of IgE
    d. receiving IgA from contaminated breast milk

Short Answer Questions

Read each of the following questions carefully. If it helps you, underline or list each component to the question. Formulate your answer and jot down the main points. Then use complete sentences in your explanation.

1. Compare the inflammatory response to the immune response.

2. A co-worker recently got shingles, a neurological disease caused by the chicken pox virus. The co-worker was surprised to receive a prescription for an antiviral medicine because he thought there was nothing one could do about a viral infection. If the medication is similar to natural interferons, explain how it works to your co-worker.

3. Your text claims, "Clonal selection underlies the entire immune response." First explain clonal selection and then substantiate the claim made by the text.

4. Describe an allergic reaction. Why would an allergy medicine that blocked histamines provide relief?

Answer Key

Review Questions

A. 1. Keep the foreign cells or molecules out of the body.
   2. Attack any foreign cell or molecule that enters the body.
   3. Destroy the specific type of foreign cell or molecule that enters the body.

B. 1. d
   2. c
   3. a
   4. f
   5. c
   6. b
   7. d
   8. e
   9. b
   10. a

C. 1. Anti-body mediated immune responses defend primarily against antigens that are free in body fluids, including extracellular pathogens such as bacteria or free viruses.
   2. Cell-mediated immune responses involve living cells and protect against cellular threats, including body cells that have become infected with viruses or other pathogens and cancer cells.

D. Threat → Detection → Alert → Alarm → Build specific defense → Defense → Continued surveillance → Withdrawal of forces

The threat is posed to the body when a foreign cell or molecule enters. A macrophage rushes to the scene, detects the foreign cell or molecule and engulfs it. The macrophage puts an antigen from the pathogen on its surface and presents the antigen to the helper T-cell that has the correct receptor for the antigen. The macrophage activates the helper T-cell. In the alarm phase, the helper T-cell activates both lines of defense to fight that specific antigen.

In an antibody-mediated defense, B-cells are activated and divide to form plasma cells that secrete antibodies specific to the antigen. In a cell-mediated defense, T-cells divide to form cytotoxic T-cells that attack cells with the specific antigen. As the defense mounts, antibodies specific to the antigen eliminate the antigen. The cytotoxic T-cells cause cells with the antigen to burst. Surveillance is continued by memory cells that are formed when helper T-cells, cytotoxic T-cells and B-cells were activated. They remain to provide a response if the antigen is detected again. The immune response ends when the antigen has been destroyed and suppressor T-cells shut down the immune response to that antigen.

E. 1. g
   2. e
   3. h
   4. f
   5. j
   6. I
   7. c
   8. a
   9. d
   10. b

F. 1. Receptors. They identify antigens and begin the antibody-mediated immune response.
   2. The correct cells (lymphocytes) begin to divide to produce a clone of cells all with receptors for that particular antigen.
   3. They produce antibodies specific for the antigen.
   4. They remain in the system to bring about a quick response to that antigen in the future.

5. Helper T-cells, cytotoxic T-cells, suppressor T-cells, and antigen-presenting cells (macrophages.)

G. 1. In active immunity the body participates in the immune response by producing memory B-cells and T-cells following exposure to an antigen. An example is when someone has been exposed to an infection or when he or she receives a vaccination.

2. In passive immunity the person receives antibodies that were produced by another person or animal. These antibodies are not produced by the person gaining the immunity. For example, antibodies can cross the placenta to protect the unborn fetus and later may be carried in breast milk to provide passive immunity to nursing infants.

H. 1. encephalitis

2. diabetes mellitus type 1
3. rheumatoid arthritis
4. multiple sclerosis
5. rheumatic fever

Critical Thinking Questions

1. A fever is part of the body's natural mechanism to fight infection. A mild or moderate fever helps the body fight bacterial infections by slowing the growth of bacteria. Bacterial growth is slowed because a mild fever causes the liver and spleen to remove iron from the blood. Many bacteria require iron to reproduce. Fever also increases the metabolic rate of body cells. The higher metabolic rate speeds defensive responses and the repair process.
2. Each nucleated cell in the body has molecular major histocombatibility complex markers that label it as "self." This allows the body to specifically recognize its own cells and not to have to recognize on an individual basis each of the possible invading cells. All cells that are not "self" are "foe."
3. Viruses reproduce rapidly and consequently can change their structure rapidly. Because there are so many different cold viruses, we cannot develop an effective vaccination. Also, because there are so many different viruses, our body does not develop an effective memory of the virus and must identify new antigens with each exposure.
4. Autoimmune disorders occur when the immune system attacks its own cells. This is a mistake in the system. Some of the body's defense system did not learn to distinguish between its own cells and foreign invaders. Cells attack the membranes around the joint causing swelling and scar tissue to form. The inflammation makes the joint swollen and sore.

Practice Test

1. c
2. c
3. c
4. d
5. c
6. d
7. c
8. b
9. c
10. b
11. b
12. d
13. c
14. d
15. a
16. d
17. d
18. a
19. b
20. c

Short Answer Questions

1. The inflammatory response is non specific. It includes redness, heat, swelling and pain in response to injury and invasion by a variety of organisms. The immune response is very specific. In addition it involves memory cells that provide ready antibodies should the body come in contact with those same antigens again. It involves specialized B lymphocytes and T lymphocytes.
2. A cell that has been infected with a virus cannot help itself. However, it secretes small proteins

called interferons that act to slow the viral activity. The chemical attracts macrophages and natural killer cells that destroy the infected cells immediately. It also protects cells that are not yet infected with the virus by stimulating cells to produce proteins that prevent viruses from making copies of themselves in those cells.

3. Clonal selection is the process whereby the body is able to produce huge numbers of B-cells and T-cells designed to eliminate specific antigens from the body. Most of the time, the body only has a few lymphocytes that recognize each antigen. However, when needed the lymphocytes are stimulated to produce a clone of cells, cells that are identical, to deal with the influx of antigens accompanying an infection. Without this response, there would not be enough lymphocytes to combat an infection.
4. The allergen enters the body. In response, plasma cells produce large amounts of antibodies. The antibodies attach to mast cells. When more of the same allergen invades the body, they combine with the antibodies on the mast cells and the mast cells release histamines. The histamines cause blood vessels to widen and become leaky resulting in swelling and redness. They also stimulate the release of large amounts of mucus and cause smooth muscles of the respiratory tract to contract. Antihistamines reduce the production of the histamines which in turn reduces the redness, swelling and mucus production associated with seasonal allergies.

Chapter 13A

Infectious Disease

Objectives

After reading the text and studying the material in this chapter, you should be able to:

- Define the terms infectious disease, pathogen and virulence.
- List the variety of organisms that cause disease and the related disease.
- Explain why antibiotics are effective against bacterial infections.
- Describe antibiotic resistance and its implications for society.
- Describe the structure of a typical virus and the steps in viral replication.
- List diseases caused by protozoans and how they are spread.
- Describe how fungi cause their damage to cells.
- Describe how parasitic worms cause disease.
- Explain the mechanism used by prions to destroy neural tissue.
- Define emerging and reemerging disease and describe the factors that affect their spread.
- Give examples of biological organisms and their products that may be used as biological weapons.

Chapter Summary

Infectious diseases are illnesses caused by **pathogens** or their products. The relative ability of a pathogen to cause disease is termed its **virulence**. Virulence is determined by the ease with which the pathogen enters the body, the degree of damage done to body cells, and the type of cell damaged.

Bacterial cells differ from human cells in that they are prokaryotes and have a cell wall that is sometimes encapsulated. Bacteria reproduce asexually by binary fission at a very rapid rate. Some bacteria are pathogens that damage tissues by releasing toxins into the bloodstream. Different toxins produce different responses ranging from common food poisoning to botulism, which often results in death. As you learned in the last chapter, the body produces antibodies, chemicals that inhibit the growth of microorganisms. They work by disrupting the processes that are common to bacteria but not to human cells. Unfortunately, some bacteria have become resistant to antibiotics, meaning that the drugs are no longer effective. Bacteria gain resistance to antibiotics by genetic adaptation. The genes enable the bacterium to survive exposure to an antibiotic by preventing the antibiotic from entering the cell, pumping it out of the cell, or destroying the antibiotic within the cell. Bacteria can inherit the genes, they can mutate rapidly gaining resistance, or they can get resistance genes from other bacteria. The overuse and misuse of antibiotics are largely to blame for increasing resistance.

Viruses are responsible for many human illnesses. Viruses can damage the host cell as they leave the cell after replication or when incorporated into the cell's chromosomes. A virus can replicate only when its genetic material is inside a host cell. The virus gains entry by binding to a receptor on the host cell surface. Viral genes then direct the host cell to make thousands copies of the viral DNA or RNA. After making viral proteins including coat proteins and enzymes, the viruses are assembled and put through the host cell's plasma membrane. Because viruses use host-cell structures they are very difficult to destroy.

Many other organisms act as pathogens. Protozoans cause disease by producing toxins and enzymes. Fungi often cause disease by secreting enzymes that digest cells. Most fungal cell membranes have a slightly different composition than those of human cells providing a vulnerability that allows for cure. Parasitic worms cause disease by releasing toxins, feeding off blood, or competing with the host for food. Worms can cause many very serious diseases.

Prions are infectious proteins that cause disease by causing normal proteins found on the surface of nerve cells to become misfolded and form clumps resulting in brain degeneration. Humans can become infected with prions by eating contaminated material, through tissue transplant, or through contaminated surgical instruments.

Disease is spread when a pathogen enters the body through contact, consumption, inhalation, or an animal vector. They can also be spread through indirect contact such as sneezing. An **emerging disease** is a condition with clinically distinct symptoms whose incidence has increased over the last two decades. A **reemerging disease** is a disease that has reappeared after a decline in incidence. Infectious diseases are cause for concern. Mutations are leading to the development of new organisms and organisms with antibiotic resistance. Environmental change can affect the distribution of organisms and change the region where they can live. **Epidemiology** is the study of the pattern of disease occurrence, distribution and control.

A biological weapon is an organism, or a toxin it produces, that is used intentionally to harm humans. Anthrax is caused by a bacterium that forms highly resistant, dormant spores that cause infection when they enter a person's body. There are various forms of anthrax, each with a differing degree of severity. Scientists are working on antibiotics, drugs to neutralize the toxin, and a vaccine, but anthrax remains a serious threat. Smallpox is caused by a highly contagious and deadly virus that has been eradicated worldwide by vaccination programs. However, the programs have been stopped, leaving most of the population susceptible to the virus. Botulinum toxin is a potent poison that binds to nerves and prevents muscle contraction resulting in paralysis and respiratory failure. An antitoxin is available that can prevent any free toxin from binding to nerves and causing additional harm.

Key Concepts

- An infectious disease is caused by a pathogen or its products. Sometimes they can be spread from one person to another.
- Pathogens are disease-causing organisms. A great variety of organisms can cause disease in humans.
- Although bacteria can cause disease in a number of ways, many bacteria do their damage by releasing toxins into the bloodstream or surrounding tissues.

- Some bacteria are beneficial and are normal residents of the body.
- Antibiotics kill or inhibit the growth of microorganisms. Antibiotic resistance is a growing concern.
- Viruses can damage the host cell as they leave the cell after replication or when incorporated in the cell's chromosomes. Viruses require a host cell for replication and are therefore difficult to treat and destroy.
- Protozoans cause disease by producing toxins and enzymes that prevent the host cell from functioning normally. Drugs are effective against protozoan diseases.
- Fungi cause disease by secreting enzymes that digest human cells that it is using for food.
- Parasitic worms cause disease by releasing toxins, feeding off blood, or competing with the host for food.
- Prions are infectious particles of proteins that cause disease by causing normal proteins found on nerve cells to become misfolded and form clumps.
- Humans become infected with prions by eating contaminated food, through tissue transplant or through contaminated surgical instruments.
- Disease is spread when a pathogen enters the body through contact, consumption, or an animal vector.
- Emerging and reemerging disease are cause for concern as they mutate and spread through an ever-increasing human population.
- Microorganisms and their products may be used as biological weapons.

STUDY TIPS

This is an interesting and relevant chapter that builds on and applies your knowledge of the body's defense mechanisms. The chapter is organized around the type of organism causing disease. It lends itself well to an outline. If you have time to outline the material prior to lecture, leave ample room on the side of the page to add lecture notes. Much of this material will be familiar to you; be sure to jot down your questions as you read and then ask them during lecture or discussion.

REVIEW QUESTIONS

A. Complete the table below by indicating how each organism damages the cell and giving an example of the organism or disease

Pathogen	**Mechanism of Action**	**Example Organism or Disease**
Bacteria		
	Take over the genetic and protein synthesis mechanisms of the cell, cause cell death, may cause cancer	
	Produce toxins or release enzymes that prevent normal functions of host cell	*Giardia*, malaria

Fungi		
	Release toxins, feed off blood, or compete for food with the host	

B. Fill in the blanks in the following sentences describing prions.

Prions are infectious particles of ____________. They ____________ be destroyed by heat, ultraviolet light, or most chemical agents. The general term for prion diseases is ____________ ____________ ____________, abbreviated TSEs. Prions are misfolded versions of harmless proteins found on the surface of ____________ cells. The misshapen proteins clump together and damage the ____________ ____________. Three examples of diseases caused by prions include 1. ____________ 2. ____________ and 3. ____________. Currently there is ____________ cure for any disease caused by a prion.

C. List the three most common ways that humans become infected with prions.

1. ____

2. ____

3. ____

D. Diseases spread easily. Give a common example for each means of transmission listed below.

Means of Transmission	Example
1. Direct Contact	For example, ______________________
2. Indirect Contact	For example, ______________________
3. Consumption	For example, ______________________
4. Animal Vectors	For example, ______________________

E. Complete the following question dealing with biological weapons. Then list three potential biological weapons.

1. What is the definition of a biological weapon? ______________________

____

2. List three examples of biological weapons:

    ____

    ____

    ____

Critical Thinking Questions

Read each of the following questions carefully. If it helps you, underline or list each component to the question. Use complete sentences in your explanation.

1. This chapter deals with bacteria, viruses, protozoa, parasites and fungi that are harmful. Many of these organisms are beneficial as well. What causes an organism to be classified as pathogenic? What are some nonpathogenic organisms found in or on humans?

2. Bacteria and viruses can both cause disease. Explain the differences between the two types of organisms and explain why antibiotics don't work against viruses. You may need to refer to the previous chapter on the Body's Defense Mechanisms.

3. When people become ill with a bacterial infection, they often take antibiotics. Broad spectrum antibiotics may kill a wide range of bacteria. How might this affect the normal bacteria found in the digestive tract?

4. The Center for Disease Control and Prevention (CDC) is a government agency that provides health information and statistics. It conducts research to identify the courses of diseases and to protect the public health through national and international prevention programs. Visit the web site of the CDC (http://www.cdc.gov) and investigate the epidemiology of a disease of interest to you. Click on the link to Infectious Disease Information and then choose a specific disease. Identify what the disease is, what you should know about it and what is being done to control it. Investigate diseases that are common to other parts of the world.

Practice Test

Choose the one best answer to each question that follows. As you work through these items, explain to yourself why each answer you discard is incorrect and why the answer you choose is correct.

1. An infectious disease is one that is ____________.
    a. caused by all bacteria
    b. caused by a pathogen or its products
    c. spread from person to person
    d. transferred by mosquitoes

2. What is a significant difference between human cell structure and bacteria cells?
   a. bacteria have a cell wall
   b. bacteria lack DNA
   c. human cells lack amino acids
   d. human cells reproduce both sexually and asexually

3. How do most disease-causing bacteria cause damage?
   a. causing high fevers
   b. causing lesions
   c. releasing toxins
   d. reproducing asexually

4. All bacteria are harmful.
   a. true
   b. false

5. What causes antibiotic-resistant bacteria to develop?
   a. homeopathic remedies
   b. lack of sterile conditions
   c. new antibodies
   d. overuse of antibiotics

6. Which of the following statements is true concerning viruses?
   a. all viruses are cells
   b. all viruses cause disease
   c. many do not consider a virus to be a living organism
   d. virus genetic material is only RNA

7. How do viruses gain entry to a host cell?
   a. by binding to a receptor site on the host cell
   b. by binding to an antibody site on the host cells
   c. by dissolving a piece of the host cell membrane
   d. all of the above

8. How does the virus make new genetic information and proteins?
   a. It combines with the host cell DNA.
   b. It never does reproduce; it only grows.
   c. It uses the host cell structures.
   d. Once inside the host cell it is able to reproduce sexually.

9. Most of the time, when viruses leave the host cell they cause cell death. In which case listed below does this NOT happen?
   a. antagonistic viral infections
   b. latent viral infections
   c. persistent viral infections
   d. synergistic viral infections

10. Why do most attempts to develop antiviral drugs fail?
    a. The drugs are toxic to the host cell.
    b. The drugs lead to cancer.
    c. The viruses mutate too quickly.
    d. The viruses combine with the host cell wall.

11. What is a protozoan that is commonly found in lakes and streams that can cause severe diarrhea?
    a. *E. coli*
    b. *Giardia*
    c. HIV
    d. *Streptococcus*

12. What causes the symptoms of disease causing fungi?
    a. The destruction of cell walls.
    b. The digesting of human cells for nutrients.
    c. The use of toxic antiviral medications.
    d. The rapid reproduction of the fungal cells.

13. What parasitic worm can cause lung damage and severe malnutrition?
    a. *Ascaris*
    b. *Giardia*
    c. Ringworm
    d. *Staphylococcus*

14. Which of the following is least similar to the others?
    a. Bacteria
    b. Fungi
    c. Parasitic worms
    d. Prions

15. Prions affect which tissue in the body?
    a. brain
    b. cardiac
    c. connective
    d. endocrine

16. How do animals or humans become infected with prions?
    a. contaminated food
    b. contaminated surgical instruments
    c. tissue transplant
    d. all of the above

17. Which of the following is an example of the spreading of disease by direct contact?
    a. coughing
    b. kissing
    c. sneezing
    d. sweating

18. What is the vector for Lyme disease?
    a. bee
    b. deer tick
    c. mosquito
    d. parasitic worm

19. Which of the following factors account for the rapid spread of disease?
    a. increasing population size
    b. warmer climates
    c. world travel
    d. all of the above

20. What makes the smallpox virus a potential biological weapon?
    a. It spreads through the air.
    b. It is highly contagious.
    c. Many people are not vaccinated against smallpox.
    d. all of the above

SHORT ANSWER QUESTIONS

Read each of the following questions carefully. If it helps you, underline or list each component to the question. Formulate your answer and jot down the main points. Then use complete sentences in your explanation.

1. Draw or describe the life cycle of *Plasmodium*.

2. How do different modes of transmission affect the spread and control of infectious diseases?

3. Define an emerging disease and a reemerging disease. Explain how smallpox has the potential to become a reemerging disease.

4. Anthrax is a naturally occurring soil bacterium. What makes it a biological weapon?

Answer Key

Review Questions

A.

Pathogen	Mechanism of Action	Example Organism or Disease
Bacteria	Produce toxins	*Staphylococcus*, *E. coli*
Viruses	Take over the genetic and protein synthesis mechanisms of the cell, cause cell death, may cause cancer	Flu virus, cold virus, *herpes simplex*
Protozoans	Produce toxins or release enzymes that prevent normal functions of host cell	*Giardia*, malaria
Fungi	Dissolve the host organism's cells with powerful enzymes	Histoplasmosis, athlete's foot, coccidioidomycosis
Parasitic Worms	Release toxins, feed off blood, or compete for food with the host	Schistosomiasis, trichinosis

B. Prions are infectious particles of proteins. They cannot be destroyed by heat, ultraviolet light, or most chemical agents. The general term for prion diseases is transmissible spongiform encephalolopathies, abbreviated TSEs. Prions are misfolded versions of harmless proteins found on the surface of nerve cells. The misshapen proteins clump together and damage the plasma membrane. Three examples of diseases caused by prions include 1. mad cow disease 2. Creutzfeldt-Jakob disease and 3. chronic wasting disease. Currently there is no cure for any disease caused by a prion.

C. 1. eating contaminated material
   2. tissue transplant
   3. contaminated surgical instruments

D. 1. shaking hands, hugging and kissing, being sexually intimate
   2. sneezing, coughing, touching an infected surface, sharing eating utensils or toothbrushes
   3. eating spoiled food, drinking contaminated water
   4. deer tick, mosquitoes

E. 1. A biological weapon is an organism or the toxin it produces that is used intentionally to harm humans.
   2. anthrax, smallpox, botulism toxin

Critical Thinking Questions

1. Pathogens are disease-causing organisms. They are usually transmitted from person to person and cause harm. There are a host of organisms that assist with digestion and absorption of nutrients and live in our intestinal tract including the common *E. coli* and *Streptococcus*. In addition, many fungi live on our skin.
2. Antibiotics interfere with cellular mechanisms that are present in bacteria but not in viruses. For example, many antibiotics attack the cell wall or inhibit protein synthesis common only to bacteria. Because viruses use host cell structures to carry out their metabolic functions, they are not as easy to destroy without damaging the host.

3. Wide spectrum antibiotics may kill many of the beneficial bacteria, or other microorganisms, in the intestinal tract. This may leave the natural flora unbalanced, allowing one organism to dominate as is the case when women take large doses of antibiotics and then succumb to a yeast infection. It is often recommended that one replenish their body with natural bacteria such as that found in yogurt after taking antibiotics.
4. There is a lot of very relevant information available on the CDC website that emphasizes both disease control and prevention. Students should choose a disease that has sufficient information for them to understand its spread and control.

Practice Test

1. c
2. a
3. c
4. b
5. d
6. c
7. a
8. c
9. c
10. a
11. b
12. b
13. a
14. d
15. a
16. d
17. b
18. b
19. d
20. d

Short Answer Questions

1. A drawing should replicate the information in Figure 13a-9. A description should include the following steps. Step 1: A mosquito infected with the *Plasmodium* bites a human and injects the protozoans into the person's bloodstream. Step 2: The protozoans travel to the person's liver where they mature and multiply. Step 3: The protozoans are released from the liver into the bloodstream where they enter the RBCs. Within the RBCs the protozoans grow and multiply asexually. Step 4: The infected RBCs burst, releasing parasites, gametocytes and toxins. Step 5: Another mosquito bites the infected person and ingests gametocytes, which develop into gametes. The gametes fuse and develop into new protozoans in the mosquito's gut. Step 6: The protozoans travel to the mosquito's salivary glands. The infected mosquito bites and infects another person.
2. Diseases that are spread through direct contact are more likely to infect large numbers of people in crowded cities. However, they can be controlled by hand washing and practicing safe-sex behaviors. Diseases that are spread through indirect contact are much harder to regulate. They will spread through densely populated areas quickly and there is little the recipient of the disease can to do prevent getting an airborne disease. Diseases transmitted in contaminated food or water can be controlled through proper food preparation and the boiling of contaminated water. When a disease is spread by animal vectors, the vector is usually small. Humans can take care not to come in contact with the vector by wearing appropriate clothing or netting.
3. An emerging disease is a condition with clinically distinct symptoms whose incidence has increased, particularly over the last two decades. A reemerging disease is a disease that has reappeared after a decline in incidence. Naturally occurring smallpox was eliminated by 1980 due to a well-developed vaccination program. Because the disease was eliminated, vaccinations were stopped. Should the smallpox virus reappear, many people would be unprotected and the disease would reemerge as a serious threat to human health.
4. A biological weapon is an organism or the toxin of an organism that is used intentionally to harm humans. Some of the anthrax bacteria that have been developed as biological weapons have been altered to make them easier to inhale and therefore they are more likely to cause lung anthrax.

Chapter 14

Respiratory System

Objectives

After reading the text and studying the material in this chapter, you should be able to:

- State the purpose of the respiratory system.
- Define external respiration, gas transport, and internal respiration.
- Identify and give the function of each of the organs of the respiratory system.
- Explain how to do the Heimlich maneuver and why it works.
- Explain how food and drink are prevented from entering the lungs.
- Describe the structure of the alveoli sacs and how this structure facilitates gas exchange.
- Explain how inhalation and exhalation are accomplished; include the muscles that are involved and the changes in air pressure.
- Define tidal volume, inspiratory and expiratory reserve volume, residual volume, vital capacity and total lung capacity and identify them on a graph.
- Explain how oxygen is carried in the blood and the factors that affect its release to the cells.
- Explain how carbon dioxide is carried in the blood and the role of the bicarbonate ion in buffering.
- Describe control of breathing.
- Identify various disorders of the respiratory system including their symptoms and treatment.

Chapter Summary

The function of the **respiratory system** is to provide the body with essential oxygen and dispose of carbon dioxide. Four processes play a part in respiration: **breathing (ventilating)**, **external respiration**, **gas transport** and **internal respiration**.

The **nose** cleans incoming air, warms and moistens the air, and provides for the sense of smell. The **sinuses** lighten the head and adjust air quality. The pharynx is the space behind the nose and mouth and provides a passageway for food and air. The **larynx** is an adjustable entrance to the respiratory system that controls the position of the **epiglottis** and is the source of the voice. The **trachea** is the tube that conducts air between the environment and the lungs. It divides into the bronchial tree that conducts air to each lung. The **alveoli** are the functional units of the respiratory system. These minute sacs are where oxygen diffuses from the air into the blood. Carbon dioxide produced by the cells diffuses from the blood into the alveolar air to be exhaled.

Air moves from a region of high pressure to a region of lower pressure. Pressure changes within the lungs cause breathing. When the **diaphragm** and **external intercostal** muscles contract, the volume of the thoracic cavity increases, causing the pressure in the lungs to decrease. **Inhalation** occurs when the pressure in the lungs decreases. When the same muscles relax, pressure in the lungs increases causing **expiration**. The volume of air inhaled or exhaled during a normal breath is called the **tidal volume**. The volume of air moved into and out of the lungs is an indication of health.

Although a small amount of oxygen is carried from the alveoli throughout the body dissolved in blood plasma, most is carried by the blood where it is bound to hemoglobin in a molecule called **oxyhemoglobin**. Changes in the conditions around active cells alter the behavior of hemoglobin and make oxygen delivery responsive to the needs of the cells. The carbon dioxide produced as the cells use oxygen is removed by the blood in one of three ways: dissolved in the blood, carried by hemoglobin, or as a bicarbonate ion. Most carbon dioxide is transported as bicarbonate ion that also serves as a buffer.

Breathing is controlled by respiratory centers in the brain. The basic rhythm is controlled by a **breathing center** located in the medulla. Changes in depth and rate of breathing are affected by chemoreceptors located in the medulla. Carbon dioxide is the most important chemical influencing breathing rate. However, under extreme circumstances oxygen-sensitive chemoreceptors in the aortic and carotid bodies can increase breathing.

Respiratory disorders have many causes. The **common cold** is caused by several types of viruses, some with many variants. **Influenza** is caused by only two types of viruses, but there are many variants of these two types. **Severe acute respiratory syndrome** (SARS) is caused by a coronavirus. *Streptococcus* bacteria cause **strep throat**, a soreness accompanied by swollen glands and fever. **Bronchitis** is an inflammation of the mucous membrane of the bronchi caused by viruses, bacteria or chemical irritation. The inflammation results in the production of excess mucus, which triggers a deep cough. **Emphysema** is caused by the destruction of alveoli, usually by smoking. The reduction in the surface area available for gas exchange and the increased dead air space results in shortness of breath.

Key Concepts

- The respiratory system allows for the exchange of oxygen and carbon dioxide across a moist body surface.
- External respiration is the exchange of oxygen and carbon dioxide between the gas inside the lungs and the blood; gas transport moves oxygen and carbon dioxide between the lungs and tissue; internal respiration exchanges oxygen and carbon dioxide between blood and the lungs.
- The nose warms and moistens the air and clears particles from it.
- The sinuses make the head lighter and help warm and moisten the air.
- The pharynx, or throat, is a passageway for air, food and drink from the mouth to the larynx.
- The larynx contains the epiglottis that covers the opening in the larynx and prevents food from entering the lungs. The larynx is also the source of the voice, which is generated by the vibration of the vocal cords.

- The trachea is the windpipe that conducts air between the environment and the lungs. The trachea branches into the bronchial tree, a system of air tubules.
- The alveoli are grape-like clusters of sacks that allow for the exchange of oxygen and carbon dioxide between the air and the blood.
- The contraction of the intercostals muscles and diaphragm cause an increase in the volume of the thoracic cavity. The increase in volume results in a decrease in air pressure that results in inhalation.
- The relaxation of the intercostals and diaphragm causes the volume of the lungs to decrease and the pressure to increase. This results in exhalation.
- The tidal volume is the volume of air inhaled or exhaled during normal breathing. The inspiratory and expiratory reserve volume is the additional volume of air that can be forced in or out of the lungs. The residual volume is the air that remains in the lungs after maximum exhaling. The vital capacity is the maximum amount of air that can be moved into and out of the lungs during forceful breathing. The total lung capacity is the total volume of air contained in the lungs after the deepest possible breath.
- Blood transports oxygen to the cells and carbon dioxide away from the cells.
- Most oxygen is carried by hemoglobin in the form of oxyhemoglobin.
- Most carbon dioxide is transported as bicarbonate ion, although some is carried by hemoglobin and some is dissolved in the blood.
- The rhythm of breathing is controlled by the breathing center in the medulla.
- Changes in the depth and rate of breathing are affected by chemoreceptors in the aortic and carotid bodies and the medulla. Carbon dioxide controls the rate of breathing.
- Respiratory disorders include the common cold, influenza, SARS, strep throat, bronchitis, and emphasema.

Study Tips

The authors have organized this chapter for you with their bold-faced headers. The early part of the chapter deals with the anatomy of the respiratory system, then the mechanism of breathing is discussed. Next your attention is focused on the transport of gases and blood chemistry. The chapter ends with common respiratory disorders — their causes and symptoms.

A few key figures summarize the material and can serve as a means of review. After reading the chapter, review the key concepts listed in the study guide, then begin to focus on the figures. As you read the text, refer to Figure 14-2. Highlight important terms and add notes from your lecture. Figure 14-9 summarizes the mechanisms of breathing. Be sure you can explain that figure to a study partner. Figure 14-11 summarizes the movement of gases in internal and external respiration and provides a resource to help you explain the purpose of the respiratory system.

Note cards are an excellent way to memorize material. Put each respiratory illness on a note card. On the back of the card give a description of the cause and symptoms of the illness. Although you can use Figure 14-2 to study the organs of the respiratory system, some students prefer to write each organ on a note card with the function on the back.

Review Questions

A. Fill in the blank.

1. The function of the respiratory system is to provide the body with ____________ and to dispose of __________ ____________.

2. The following four processes play a part in the exchange of gases:

   a. ____________: moving ____________ rich air into the lungs and ____________ ____________ laden air away from the lungs.

   b. ____________ respiration: moving ____________ from the lungs to the blood and carbon dioxide from the ____________ to the ____________.

   c. Gas exchange: ____________ of oxygen from the lungs to the ____________ and ____________ ____________ from the cells to the ____________.

   d. Internal respiration: moving ____________ from the blood to the ____________ and ____________ ____________ from the cells to the ____________.

B. Match the organ/structure in Column A with the function in Column B.

Column A	Column B
1. alveoli	a. connects larynx with bronchi
2. bronchi	b. filters, warms and moistens air; sense of smell
3. bronchioles	c. produces voice
4. epiglottis	d. microscopic chambers for gas exchange
5. larynx	e. cavities in skull to lighten head
6. lungs	f. two branches of trachea
7. nasal cavity	g. connects mouth and nasal cavity to larynx
8. pharynx	h. narrow passageways to connect bronchi to alveoli

9. sinuses

10. trachea

i. covers larynx during swallowing

j. structures that contain alveoli

C. Label the following organs of the respiratory system: alveoli, bronchi, bronchioles, diaphragm, epiglottis, intercostal muscles, larynx, lungs, nasal cavity, pharynx, sinuses and trachea.

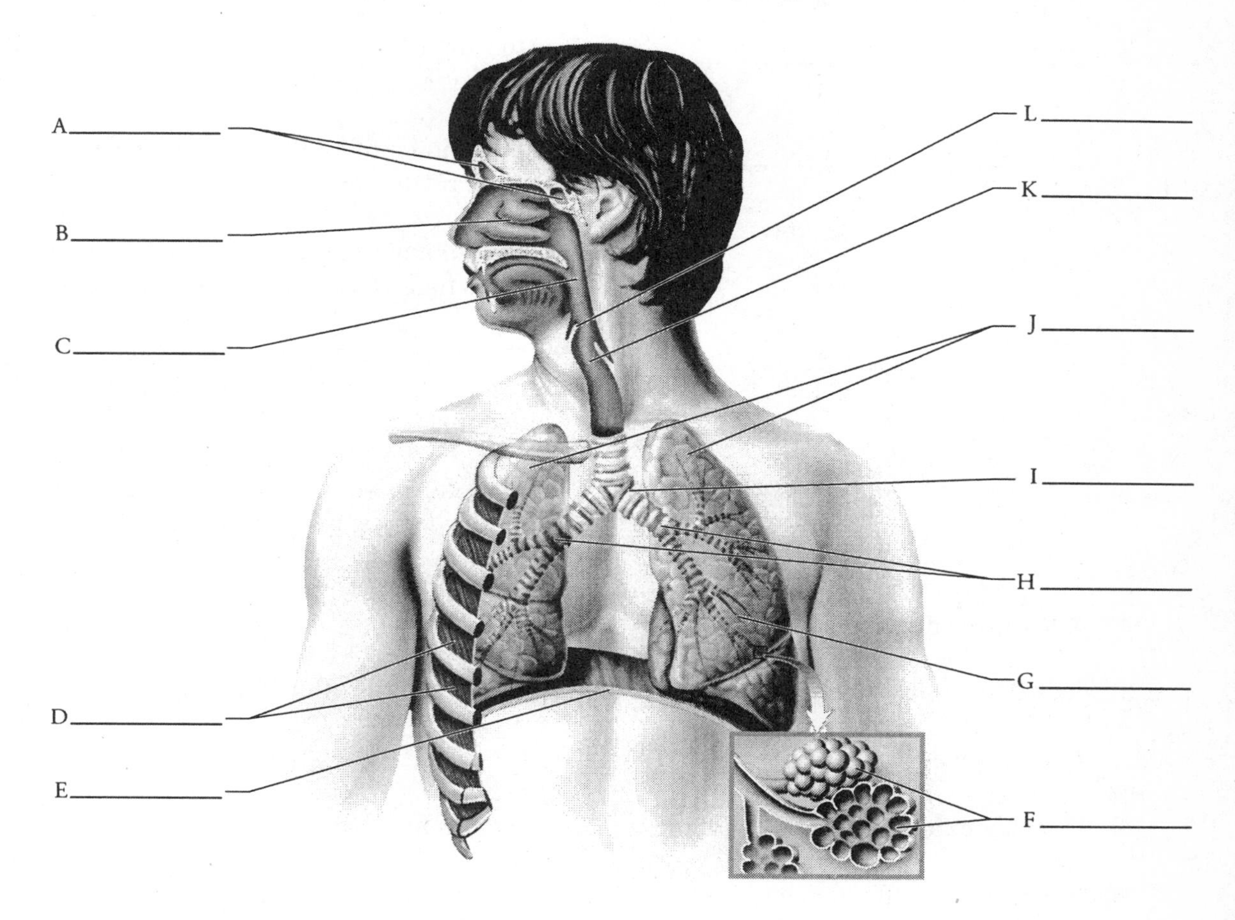

D. Complete the following table describing what happens during inhalation and exhalation.

Organ/structure/action	Inhalation	Exhalation
Diaphragm	Contracts, moves down	
Intercostal muscles		Relaxes, moves in
Thoracic volume	Increases	
Air pressure in lungs		
Movement of air	In	Out

E. Match the following terms with the definitions below: expiratory reserve volume, inspiratory reserve volume, residual volume, tidal volume, total lung capacity, vital capacity. Next answer the questions that follow.

1. ______________________________ is the amount of air inhaled or exhaled during normal, quiet breathing

2. ______________________________ is the amount of air that can be expelled from the lungs after the tidal volume

3. ______________________________ is the amount of air that can be inhaled after the normal tidal volume

4. ______________________________ is the amount of air that remains in the lungs after as much air as possible has been exhaled

5. ______________________________ is the maximum amount of air that can be moved in and out of the lungs during forced breathing

6. ______________________________ is the total amount of air contained in the lungs after the deepest possible breath

7. What is the normal tidal volume? ____________

8. Jack had a tidal volume of 500ml and an inspiratory reserve of 2800ml. What is his inspiratory capacity? ____________

9. If Jack's expiratory reserve volume is 900ml, what is his vital capacity? ____________

10. If Jack's residual volume is 1200ml, what is his total lung capacity? ____________

F. Write an explanation for how oxygen and carbon dioxide are carried in the blood in response to the questions below:

1. Explain how most of the oxygen is transported by the blood. ________________________

____

2. Explain how the remaining 1.5% is transported. ________________________________

____

3. Explain how most of the carbon dioxide is carried in the blood. Include the relevant chemical reaction. ____

____

____

____

4. Explain how carbon dioxide is transported by hemoglobin. ______________________

____

5. Explain how the remaining 7-10% of carbon dioxide is transported. ________________

____

G. At each pause in the scenario below indicate specifically what area is controlling breathing.

As Karma sat in the meeting, her breathing was shallow and slow. (It was under control of the ____________.) Finally, she could stand it no longer; the discussion was way off topic and she gave a heavy, noisy sigh. (It was under control of the ____________ ____________.) Soon the meeting ended and she bolted out of the room and ran to the student recreation center. When she arrived she was breathing faster and harder than normal. (The increased rate and depth of breathing were controlled by the ____________ in the medulla.) She met a friend and changed to go swimming. Her friend challenged her to an underwater swim. The person swimming farther would be treated to ice cream. Karma knew to exhale several times before starting her swim. This was to lower the carbon dioxide levels in her blood. She was off to a good start, but as she approached the other end of the pool she had a really strong urge to inhale. (The desire to inhale was not due to too much carbon dioxide, but rather to ____________ as perceived by the chemoreceptors in the ____________.) Karma swam well and enjoyed ice cream later that evening.

Critical Thinking Questions

Read each of the following questions carefully. If it helps you, underline or list each component to the question. Use complete sentences in your explanation.

1. A group of students were watching a weekend game on television and snacking on sodas and nachos. They were talking loudly and laughing when one of the students began choking and coughing. What mechanism is in place to prevent this and what was going on that caused this person to start choking?

2. Sometimes when the Olympic games are held at high altitudes where there is less oxygen in the air, teams go early to train and build their red blood cell count. Why would this be an important component to their performance?

3. Sinus headaches are common in many parts of the country. Explain what causes the head to ache?

4. The alveoli are often compared to balloons. However, when a balloon is blown up and then the air is let out, it collapses. Sometimes the moisture inside the balloon makes it easier to blow up the second time and if left for a day or so, it will dry out and be very difficult to inflate. Explain why the alveoli don't collapse, stick together and dry out.

5. Diffusion is the *random* movement of molecules from a greater concentration to a lesser concentration. Our life depends upon the exchange of gases between the air, our blood and our tissues. Explain how this can occur through *random* diffusion.

Practice Test

Choose the one best answer to each question that follows. As you work through these items, explain to yourself why each answer you discard is incorrect and why the answer you choose is correct.

1. What is the function of the respiratory system?
   a. To provide the body with essential nutrients.
   b. To provide the body with oxygen and dispose of carbon dioxide.
   c. To provide the body with the enzymes and hormones needed to function.
   d. All of the above.

2. Which of the following is the correct pattern of airflow during inhalation?
   a. larynx, pharynx, trachea, bronchi, bronchioles
   b. nasal cavity, larynx, pharynx, trachea, bronchioles, bronchi
   c. nasal cavity, pharynx, larynx, trachea, bronchi, bronchioles
   d. pharynx, trachea, bronchi, lungs

3. If someone had an upper-respiratory infection, where might it be located?
   a. bronchioles
   b. larynx
   c. lungs
   d. sinuses

4. After working outdoors in the garden, you come indoors and blow your nose to find dirt in the mucous. What function of the respiratory system is your nose performing?
   a. adding moisture to the air
   b. causing an allergic reaction
   c. filtering the air
   d. warming the air

5. What connects the throat with the middle ear so that air pressure can be equalized on the eardrum?
   a. Eustachian tubes
   b. larynx passageways
   c. pharynx
   d. tracheolas

6. What prevents food from going down the trachea?
   a. epiglottis
   b. esophagus
   c. glottis
   d. tongue

7. What is it called when the vocal cords are swollen and thick due to inflammation?
   a. cough and cold
   b. laryngitis
   c. pharyngitis
   d. sinusitis

8. The large muscle involved in breathing that separates the thoracic and abdominal cavities is the __________.
   a. breathing muscle
   b. bronchiolar muscles
   c. diaphragm
   d. intercostal muscles

9. Contraction of the intercostal muscles and diaphragm cause the thoracic cavity to ____________ and the air pressure in the lungs to ____________.
   a. contract, decrease
   b. contract, increase
   c. expand, decrease
   d. expand, increase

10. The amount of air that leaves the lungs under normal conditions is called the __________.
   a. residual volume
   b. tidal volume
   c. total lung capacity
   d. vital capacity

11. Which of the following best describes the vital capacity?
    a. expiratory reserve plus residual volume
    b. inspiratory reserve plus expiratory reserve plus tidal volume
    c. tidal capacity plus inspiratory capacity
    d. total lung capacity

12. Nearly all of the oxygen is carried in the blood as ________________.
    a. bicarbonate
    b. erythroglobin
    c. gaseous oxygen
    d. oxyhemoglobin

13. Which gas is carried mostly by the plasma?
    a. carbon dioxide
    b. oxygen
    c. both carbon dioxide and oxygen
    d. neither carbon dioxide nor oxygen

14. Hemoglobin can carry up to four oxygen molecules. Most of the time it delivers only one oxygen to the cells. What can cause this to change?
    a. greater concentration of carbaminohemoglobin
    b. greater concentration of oxygen in the blood
    c. greater demand for oxygen by the cells
    d. greater demand for carbon dioxide by the mitochondria

15. The breathing center ________________.
    a. is located in the chest
    b. is located in the medulla
    c. is stimulated by oxygen levels
    d. sends nerve impulses directly to the alveoli

16. The rate of breathing is primarily controlled by ________________.
    a. concentration of carbon dioxide in the blood
    b. concentration of oxygen in the blood
    c. demand for carbon dioxide by the tissues
    d. demand for oxygen by the tissues

17. The aortic and carotid bodies respond to changes in the concentration of ______.
    a. carbon dioxide
    b. carbonic acid
    c. hemoglobin
    d. oxygen

18. Colds are most commonly spread by contact with ___________.
    a. coughs and sneezes
    b. hands and other contaminated objects
    c. nasal mucus
    d. phlegm

19. Which of the following is a respiratory disorder that results in permanent damage to the alveoli?
   a. common cold
   b. emphysema
   c. influenza
   d. SARS

20. Which of the following CANNOT be caused by a virus?
   a. bronchitis
   b. cold
   c. influenza
   d. strep throat

Short Answer Questions

Read each of the following questions carefully. If it helps you, underline or list each component to the question. Formulate your answer and jot down the main points. Then use complete sentences in your explanation.

1. Trace a molecule of oxygen from the air outside your lungs to its entrance into the capillaries.

2. Compare internal and external respiration in terms of where they are located and the gases that are exchanged.

3. First describe the role of the bronchial tree in respiration. Then explain what causes asthma and why those conditions result in difficult breathing. Finally explain how asthma treatments work.

4. Explain the reaction shown below including where it occurs. Then explain the role of carbonic anhydrase.

$$CO_2 + H_2O \rightarrow H_2CO_3 \rightarrow HCO_3 + H^+$$

Answer Key

Review Questions

A. 1. The function of the respiratory system is to provide the body with oxygen and to dispose of carbon dioxide.

2. The following four processes play a part in the exchange of gases:
   a. Breathing/ventilating: moving oxygen rich air into the lungs and carbon dioxide laden air away from the lungs.
   b. External respiration: moving oxygen from the lungs to the blood and carbon dioxide from the blood to the lungs.
   c. Gas exchange: transport of oxygen from the lungs to the cells and carbon dioxide from the cells to the blood.
   d. Internal respiration: moving oxygen from the blood to the cells and carbon dioxide from the cells to the blood.

B. 1. d
2. f
3. h
4. i
5. c
6. j
7. b
8. g
9. e
10. a

C. A. sinuses
B. nasal cavity
C. pharynx
D. intercostal muscles
E. diaphragm
F. alveoli
G. bronchioles
H. bronchi
I. trachea
J. lungs
K. larynx
L. epiglottis

D.

Organ/structure/action	**Inhalation**	**Exhalation**
Diaphragm	Contracts, moves down	Relaxes, moves up
Intercostal muscles	Contract, move out	Relaxes, moves in
Thoracic volume	Increases	Decreases
Air pressure in lungs	Decreases	Increases
Movement of air	In	Out

E.
1. tidal volume
2. expiratory reserve volume
3. inspiratory reserve volumen
4. residual volumen
5. vital capacity
6. total lung capacity
7. 500ml
8. 500ml + 2899ml = 3300ml
9. 500ml + 2800ml + 900ml = 4200ml
10. 500ml + 2800ml + 900ml + 1200ml = 5400ml

F.
1. 98.5% of the oxygen is carried bound to hemoglobin as oxyhemoglobin (HbO_2).
2. The remaining 1.5% of the oxygen is dissolved in the plasma.
3. 70% of the carbon dioxide is transported as bicarbonate ion dissolved in the plasma. Carbon dioxide (CO_2) reacts with water (H_2O) in the red blood cells and plasma and forms carbonic acid (H_2CO_3). Carbonic acid dissociates to form hydrogen ions (H^+) and bicarbonate ions (HCO_3^-).
4. About 20% of the carbon dioxide is combined with hemoglobin to form carbaminohemoglobin.
5. Between 7 and 10% of the carbon dioxide is transported dissolved in the blood as molecular carbon dioxide.

G. As Karma sat in the meeting, her breathing was shallow and slow. (It was under control of the breathing center located in the medulla.) Finally, she could stand it no longer;the discussion was way off topic and she gave a heavy, noisy sigh. (It was under control of the cerebral cortex.) Soon the meeting ended and she bolted out of the room and ran to the student recreation center. When she arrived she was breathing faster and harder than normal. (The increased rate and depth of breathing were controlled by the chemoreceptors in the medulla.) She met a friend and changed to go swimming. Her friend challenged her to an underwater swim. The person swimming farther would be treated to ice cream. Karma knew to exhale several times before starting her swim. This was to lower the carbon dioxide levels in her blood. She was off to a good start, but as she approached the other end of the pool she had a really strong urge to inhale. (The desire to inhale was not due to too much carbon dioxide, but rather to a lack of oxygen as perceived by the chemoreceptors in the aortic and carotid bodies.) Karma swam well and enjoyed ice cream later that evening.

Critical Thinking Questions

1. Under normal conditions, the epiglottis keeps food and drink from entering the lower respiratory system during swallowing. That is why you cannot breathe or talk while swallowing. However, during times of excitement when people are talking and laughing, both of which require air, while they are drinking, the epiglottis becomes confused and drink goes down the windpipe (trachea).
2. At high altitudes there is less oxygen in the air. The body responds by making more red blood cells so that more oxygen can reach the cells. Oxygen is critical for the efficient conversion of glucose to ATP production.
3. The sinuses are connected to the nasal cavities and under the best of conditions the mucous from the sinuses can drain out the nose. However, when the mucous membranes of the sinuses become inflamed, the swelling can block the connection between the nasal cavities and the sinuses. The fluid remains in the sinuses and the pressure causes pain, usually in the front of the head. Decongestants that reduce the swelling in the tubes connecting the sinuses with the nasal cavity usually relieve the condition.
4. The alveoli are kept open by the presence of a surfactant. This prevents the water molecules from being attracted to one another and prevents the collapsing of the air chambers. The upper respiratory system warms and moistens the air before it gets to the alveoli.
5. Carbon dioxide and oxygen cross the membranes of the respiratory and circulatory systems by diffusion. Oxygen concentration is highest in the outside air, which is in the alveoli. It diffuses

across the alveolar membrane, through the capillary wall and into the red blood cell. At the tissues it diffuses from the red blood cell across the plasma membrane, through the capillary wall and into the cells. Carbon dioxide follows the same path in reverse, although most of the carbon dioxide is carried in the plasma and not in the red blood cell. The process of diffusion works because there is always a driving force in the concentration gradient, opposing pressures are absent and there are only two layers of cells to cross.

Practice Test

1. b
2. c
3. d
4. c
5. a
6. a
7. b
8. c
9. c
10. b
11. b
12. d
13. a
14. c
15. b
16. a
17. d
18. b
19. b
20. d

Short Answer Questions

1. nasal cavity → pharynx → epiglottis → larynx → trachea → bronchi → bronchioles → alveoli
2. External respiration is the exchange of gases between the outside air and the blood. It occurs at the alveoli where oxygen diffuses from the air inside the lungs into the blood and carbon dioxide diffuses from the blood across the alveolar membrane into the air inside the lungs. Internal respiration occurs at the interface between the tissues and the capillaries. Here oxygen diffuses from the blood into the tissues and carbon dioxide diffuses from the tissues into the blood.
3. The bronchial tree is the system of air tubules that leads from the trachea to the alveoli. Asthma is a chronic condition characterized by recurring attacks of wheezing and difficulty in breathing. The difficult breathing is caused by spasms of the bronchial muscles and worsened by persistent inflammation of the airways. Asthma is usually triggered by allergies. Inhalants improve the condition by relaxing the bronchial muscles or by reducing the inflammation of the air tubules usually with steroids.
4. $CO_2 + H_2O \rightarrow H_2CO_3 \rightarrow HCO_3 + H^+$
   This is the reaction that occurs when carbon dioxide reacts with water in the red blood cells and plasma to form carbonic acid. The carbonic acid quickly dissociates to form hydrogen ions and bicarbonate ions which serve as a buffer for the blood. When this reaction occurs in the red blood cells, there is an enzyme present, carbonic anhydrase, that makes it occur much faster.

Chapter 14a

Smoking and Disease

Objectives

After reading the text and studying the material in this chapter, you should be able to:

- Provide statistical evidence supporting the claim that smoking is the leading cause of death in the United States.
- Identify the predominant poisons and cancer-causing chemicals in smoke and describe their effects.
- Explain the addictive effects of nicotine.
- Describe the effect of carbon monoxide on the tissues of the smoker and a fetus.
- Describe the progressive lung diseases that result from smoking.
- Describe the development of lung cancer.
- Explain how smoking leads to cardiovascular disease.
- Describe the increased health risks of women who smoke and the associated risks to the fetus of smokers.
- Compare the effects of smoke pollution on the nonsmoker and smoker.
- Describe and provide data on the benefits of quitting.

Chapter Summary

Smoking is the greatest single preventable cause of death in the United States. The economic loss, shared by all of society, due to missed workdays, premature death and health care costs caused by cigarette smoking is great.

Cigarette smoke contains poisons and cancer-causing substances that damage every tissue it touches. The mouth, tongue, throat, esophagus, air passageways, lungs and stomach are obvious tissues in line for damage. But the metabolized byproducts of smoke can cause damage to the liver, bladder, pancreas and kidneys. Of more than 4000 chemicals in smoke, at least 50 substances have been shown to cause cancer, plus poisons such as hydrogen cyanide, carbon monoxide and creosols. However, the most dangerous substances in smoke are nicotine, carbon monoxide and tar.

Nicotine is the most addictive of these reaching the brain in less than 7 seconds with feelings of pleasure and relaxation. Surprisingly, nicotine is a stimulant that raises the heart rate and blood pressure doing unseen damage to the circulatory system. Nerve cells become hyperactive when nicotine is removed causing symptoms of withdrawal that last for several weeks.

The amount of carbon monoxide in cigarette smoke is 1600 ppm exceeding the industry standard of 10 ppm. Carbon monoxide binds to hemoglobin reducing its oxygen-carrying capacity and further straining the heart. Tar is a collection of thousands of substances in smoke that can cause cancer and reduce the elasticity of the lung tissue.

Smoking causes lung disease that is gradual and progressive. Smoke damages the ability of the cilia and macrophages to sweep debris from the air passageways. The nicotine and sulfur dioxide in the smoke paralyze cilia and the cyanide kills ciliated cells. Smoke stimulates the mucus secreting cells in the linings of the respiratory passageways plugging the smaller bronchiole tubes and resulting in cough and congestion. Smokers' lungs are chronically inflamed but the work of the macrophages is hampered by the paralyzing effects of the chemicals in smoke. In time, most smokers will develop chronic bronchitis and repeated bacterial infections. Emphysema is often the next stage in the progressive damage to the lungs. The walls of the alveoli are destroyed reducing the surface area for gas exchange and leaving the smoker breathless. The inability to clear the lungs increases the concentration of poisons deep in the lungs.

Smoking causes cancer by stimulating cell division. Some of the carcinogens in smoke change the structure of the genetic material and other cause enzyme changes that allow cells to become cancerous. Other components of smoke supplement the action of carcinogens or promote cell division after tumor growth has begun.

Smoking stresses both the heart and blood vessels resulting in about 25% of all fatal heart attacks. Nicotine constricts blood vessels raising blood pressure. Smoking increases atherosclerosis by decreasing the levels of good cholesterol (HDL) and by promoting cholesterol deposits in the vessels. The cholesterol deposits cause inflammation and higher blood pressure. Smoking also increases the chances of forming blood clots in the vessels by increasing the number of platelets and fibrinogen. Finally it can initiate coronary artery spasms that may lead to heart attack.

Smoking causes other health problems of varying severity. It impairs the body's ability to make collagen and thus slows the healing of skin and bones, causes cataracts, and urinary incontinence.

The hazards of smoking may be greater for women than men. Women are more likely to get lung cancer because more women carry the *K-ras* mutation that helps the lung tumor to grow. Another gene that causes lung cancer growth is located on the X chromosome and is activated by nicotine. Women who smoke are three times more likely to develop cervical cancer and have higher rates of osteoporosis. The combination of smoking and contraceptives leads to a higher rate of strokes, heart attacks, and blood clots. Women who smoke are less fertile and more likely to miscarry than those who don't. The carbon monoxide in smoke reduces the oxygen available to the fetus causing its heart rate to increase and blood pressure to rise.

Smokers affect the environment of those who do not smoke. The smoke leaving the burning end of a cigarette contains the same cancer-causing chemicals as inhaled smoke, but in higher concentrations because it is not filtered through the tobacco. Long-term exposure to secondary tobacco smoke affects the nonsmoker in the same ways as direct inhalation affects the smoker.

Filters and low-tar, low-nicotine cigarettes do not provide a safe guard against cancer and other health risks. It is very difficult to quit smoking but the benefits are enormous. Blood pressure and heart rate drop, as does the risk for heart attack, and the life expectancy increases for those who quit smoking.

Key Concepts

- Smoking is the leading cause of death in the United States.
- Smoking costs society in lost workdays, premature death and increased health-care costs.
- Smoke damages every tissue it touches and the organs that metabolize the poisons.
- Smoke contains poisons including hydrogen cyanide, carbon monoxide, and cresols, in addition to radioactive substances.
- Tar, nicotine and carbon monoxide are the three most dangerous substances in smoke.
- Nicotine is a stimulant that increases blood pressure, heart rate and the chance for abnormal clotting. It makes nerve cells hyperactive which is the addictive characteristic that causes the withdrawal symptoms.
- Carbon monoxide restricts the amount of oxygen that can be carried by hemoglobin reducing the amount of oxygen to the tissues.
- Tar contains cancer-causing substances and chemicals that destroy the elasticity of the lung.
- Smoking causes lung disease by paralyzing the cilia and macrophages and killing ciliated cells. Mucus secretion is stimulated but cannot be cleared from the lungs leading to coughs, infections, and chronic bronchitis.
- Emphysema results when the alveoli are destroyed decreasing the oxygen-exchange ability of the lungs and stressing the cardiovascular system.
- Smoking causes cancer by changing the structure of DNA, by acting as a co-carcinogen, and by promoting tumor growth.
- The nicotine in smoke raises blood pressure and heart rate. Smoking also increases the chances of atherosclerosis, can lead to blood clots and can cause arterial spasms that can cause a heart attack.
- Smoking also reduces the body's ability to make collagen, causes cataracts and urinary incontinence.
- The risks of cancer are increased in women because more women carry the *K-ras* mutation and a gene on the X chromosome that increase tumor growth.
- Women also experience greater risks because of complications with contraceptive medication, increased heart attacks and blood clots, decreased fertility and increased incidence of miscarriage.
- Smoke pollution affects nonsmokers in the same way as it does smokers.
- Filtered, low-tar, low-nicotine cigarettes are not a safe substitute for standard cigarettes.
- Major benefits to quitting smoking include a longer life, reduced chances of heart attack and lung cancer.

Study Tips

The claims that smoking negatively affects one's health are supported with data throughout the chapter. It is important that you take time to read and understand the graphs and tables that are included in the text material. First identify the labels of the rows and columns of a table or the axes of a graph. What changes are being described as you move along the line of a graph or the columns of a table? Take time to carefully read the caption and then try to summarize what is shown in the table or figure.

The material in this chapter is very interrelated. This is a good time to develop a concept map to summarize the material and serve as a tool for review. Concept mapping was introduced in the study guide in Chapter 8. If you want to review an example and learn more about the technique visit the web site, http://cmap.coginst.uwf.edu/info/printer.html, The Theory Underlying Concept Maps and How to Construct Them. Center your concept map around the chapter title, Smoking and Disease. One branch might include the chemicals contained in smoke and another the diseases. Connect the chemicals to the diseases. Use one branch of the concept map to describe the special risks among women and connect them to the diseases and chemicals. Finally include the risks smoke brings to the nonsmoker and the benefits of quitting. As you review your concept map add the statistics that are provided in the chapter.

Review Questions

A. What are three costs shared by society that increase among smokers?

1. ______________________________

2. ______________________________

3. ______________________________

B. List the tissues or organs that are damaged by smoking.

1. ____________________	6. ____________________
2. ____________________	7. ____________________
3. ____________________	8. ____________________
4. ____________________	9. ____________________
5. ____________________	10. ____________________

C. Match the chemical in Column A with its description in Column B.

Column A	Column B
1. carbon monoxide	a. poisonous gas used to kill people and animals
2. hydrogen cyanide	b. addictive drug, acts as a stimulant
3. nicotine	c. replaces oxygen in hemoglobin
4. radioactive substances	d. brown sticky substance that can destroy elasticity of lungs
5. tar	e. unstable form of a chemical, can cause mutations

D. Use your knowledge from this chapter and your knowledge of the circulatory and respiratory systems to complete the following paragraph on the harmful effects of carbon monoxide.

Carbon monoxide is a poison found in car exhaust and ____________. The amount of carbon monoxide in smoke is ___________ (higher, lower, the same) than that found in industry or urban air. Carbon monoxide displaces ___________ in the hemoglobin molecule. Unlike oxygen or carbon dioxide, it remains attached to the hemoglobin. Thus, the more one smokes, the more the oxygen level to the tissues ___________. This reaction is similar in a fetus that gets its oxygen from the mother. If the mother smokes, the oxygen available to the fetus is __________. This can result in ______ development and _______ birth weight.

E. What causes "smoker's cough"? ____

____

____

F. Explain the progression of lung cancer development as shown in the following diagrams.

1. ____

____

____

2. ____

____

____

3. ____

____

____

4. ____

____

____

G. Complete the following paragraph dealing with women and the health risks of smoking by circling the correct word.

Women who smoke are (more, less) likely to develop lung cancer than men who smoke. Women are more than (3, 10) times more likely to carry the genetic mutation *K-ras* that promotes tumor growth than men. Another gene that promotes tumor growth is found on the (X, Y) chromosome. Women who smoke are (more, less) likely to develop cervical cancer and osteoporosis than women who do not smoke. The death rate due to strokes, heart attacks, and blood clots in the legs is (a little, much) higher in women who smoke and take the pill. Women who do not smoke are (no more, twice as, four times as) likely as women who smoke to conceive. A woman smoker is (twice, five times) as likely to miscarry as a nonsmoker. Nicotine affects the fetus by (constricting, dilating) the vessels in the umbilical cord and (increasing, decreasing) the amount of oxygen in the hemoglobin. The incidence of stillbirth among women who smoke is about (the same, two times, four times) that of nonsmokers.

H. Just the facts, please. Add the appropriate statistic to each statement below.

1. Each cigarette a person smokes reduces their life expectancy by _______ minutes.
2. Annually, nearly ____________ Americans will die prematurely from smoking.
3. Smoking-related illnesses cost the United States about ________________ each year.
4. Of the ____________ substances found in smoke, at least _____ have been shown to cause cancer.
5. The average heart rate is about 72 beats a minute. The nicotine from a cigarette can raise that ________ beats each minute.
6. There are approximately _______ times more deaths from smoking than from alcohol each year in the United States.
7. The amount of carbon monoxide in cigarette smoke is __________ ppm, which greatly exceeds the _________ ppm considered dangerous in industry.

8. Carbon monoxide remains in the bloodstream for up to _____ hours after smoking a cigarette.
9. Smoking is responsible for _______ of all cancer deaths.
10. _____ of all people diagnosed with lung cancer die within _____ years.
11. Smokers are _________ times more likely to develop lung cancer than nonsmokers.
12. As estimated by the American Heart Association, about _________ of all fatal heart attacks are caused by cigarette smoke.
13. Smoking is responsible for _____ of all cases of cataracts.
14. After an hour in a smoke-filled room, a nonsmoker may inhale an amount of cancer-causing nitrosamines equivalent to smoking _____ filter cigarettes.
15. ________________ Americans quit smoking each year.

CRITICAL THINKING QUESTIONS

Read each of the following questions carefully. If it helps you, underline or list each component to the question. Use complete sentences in your explanation.

1. What is the warning on each pack of cigarettes from the Surgeon General? Interview some friends who are smokers. Does the warning have any impact on them?

2. Cigarette smoking and cigarette smoke cause harm, even death, to oneself and to others. So do alcohol and firearms which are far more regulated. Use the Internet to identify the number of deaths per year due to smoking, alcohol, and firearms. The CDC Fact Sheet on Actual Causes of Death in the United States, 2000 is a good place to start: http://www.cdc.gov/nccdphp/factsheets/death_causes2000.htm
   What is the relationship between the regulations and the number of deaths?

3. Write a one or two sentence summary of Figure 14a-1 showing the causes of death and of Figure 14a-8 showing a comparison of lung cancer deaths among lifetime smokers and those who quit. Now combine these data and write a one paragraph letter to a friend stating reasons why he or she should quit smoking.

4. What would you say to someone who argues that they aren't doing damage to their body because they are smoking low-tar and low-nicotine cigarettes? Use data in your response.

5. Jenny is a waitress in a smoke-filled tavern. She is concerned that although she doesn't smoke that the "second hand" smoke is hurting her health. Are her concerns founded? Provide solid information regarding the risks of smoke to non-smokers.

Practice Test

Choose the one best answer to each question that follows. As you work through these items, explain to yourself why each answer you discard is incorrect and why the answer you choose is correct.

1. Which of the following organs are affected by cigarette smoke?
   a. esophagus
   b. lungs
   c. throat
   d. all of the above

2. How can smoke cause damage to the bladder, pancreas and kidneys when it doesn't come in contact with them?
   a. byproducts of the breakdown of smoke cause damage
   b. nicotine circulates through the blood
   c. urine causes cancer of the bladder
   d. it doesn't cause damage to these organs

3. Which of the following is the most addictive?
   a. carbon monoxide
   b. cresol
   c. nicotine
   d. tar

4. Although nicotine gives a feeling of relaxation, it is actually a stimulant that affects the brain.
   a. true
   b. false

5. The number of deaths caused by smoking in the United States each year is ____ times the number of deaths due to motor vehicle accidents.
   a. 2
   b. 5
   c. 10
   d. 50

6. The amount of carbon monoxide in cigarette smoke is __________ that which is considered dangerous in an industrial or urban situation.
   a. much less than
   b. about the same as
   c. much more than
   d. There are no data on this fact.

7. Smoking is not the single major cause of lung cancer.
   a. True
   b. False

8. What is it about carbon monoxide that makes it so dangerous to the smoker, the passive smoker and the fetus of a smoker?
   a. It leads to diabetes.
   b. It prevents hemoglobin from carrying oxygen.
   c. It replaces carbon dioxide in the plasma.
   d. It uses needed glucose for metabolism

9. A research study followed smokers for 40 years. About what percentage of the group who smoked 25 or more cigarettes a day were still alive at age 70?
   a. 10%
   b. 25%
   c. 50%
   d. 75%

10. Smoking causes which of the following changes in respiratory tissue?
    a. Decreased cilia action.
    b. Increased production of mucus.
    c. Paralysis of macrophages.
    d. All of the above

11. Damage to the lungs from smoking results in inflammation and congestion. What disease is characterized by a deep cough that brings up mucus?
    a. chronic bronchitis
    b. emphysema
    c. pneumonia
    d. SARS

12. Which of the following is characterized by damaged alveoli and reduced elasticity in the lung tissue?
    a. chronic bronchitis
    b. emphysema
    c. influenza
    d. tuberculosis

13. Which of the following statements is true regarding the relationship between DNA and the chemicals in smoke?
    a. Cigarette smoke does not affect DNA.
    b. Cigarette smoke results in an increase in the amount of DNA in the cells it touches.
    c. Cigarette smoke results in mutations to the DNA.
    d. Cigarette smoke results in mutations to the RNA.

14. Most smokers diagnosed with cancer die within _______ years.
    a. 1-2
    b. 5
    c. 10
    d. 15-20

15. Where do the cellular changes occur that lead to lung cancer?
    a. basal cells
    b. basement membrane
    c. ciliated columnar cells
    d. epithelial cells of the throat

16. Smoking is linked to cardiovascular disease. Which of the following might be a cause of cardiovascular disease?
    a. elevated levels of HDL
    b. higher blood pressure
    c. increased red blood cell count
    d. vasodilatation caused by nicotine

17. What could be the explanation behind the statement that smoking causes wrinkles?
    a. smoking causes cataracts
    b. smoke impairs the body's ability to make collagen
    c. smokers neglect their personal care
    d. smokers squint to keep the smoke from burning their eyes

18. Which of the following is true regarding smoking and women?
    a. Rate of lung cancer among women is increasing.
    b. Smoking can lead to miscarriage.
    c. Women who smoke are more likely to develop cervical cancer than nonsmokers.
    d. All of these statements are true.

19. Filtered cigarettes and the brands advertising a reduced tar and nicotine level are safe to smoke.
    a. True
    b. False

20. Which of the following statements is true regarding quitting smoking?
    a. Few smokers really want to quit.
    b. Former smokers have a dramatically reduced chance of dying from lung cancer compared to those who continue to smoke.
    c. Former smokers have a lower mortality rate from lung cancer than those who have n never smoked.
    d. Smokers who quit have a reduced social life.

Short Answer Questions

Read each of the following questions carefully. If it helps you, underline or list each component to the question. Formulate your answer and jot down the main points. Then use complete sentences in your explanation.

1. Describe the effects of smoke on the cilia and cells found in the air passageways. What are the resulting symptoms expressed by the smoker?

2. Why are cancer of the bladder, pancreas and kidney common to smokers? The smoke doesn't reach these organs.

3. Describe the role of a co-carcinogen.

4. Describe the effects of a smoking on the cardiovascular system.

5. What is it about cigarette smoking or cigarette smoke that makes it such a hard habit to break?

ANSWER KEY

REVIEW QUESTIONS

A. 1. missed work
   2. increased health-care
   3. premature death and related costs

B. mouth, tongue, throat, esophagus, air passageway, lungs, stomach, liver, pancreas, bladder, kidneys, cervix

C. 1. c
   2. a
   3. b
   4. e
   5. d

D. Carbon monoxide is a poison found in car exhaust and cigarette smoke. The amount of carbon monoxide in smoke is higher than that found in industry or urban air. Carbon monoxide displaces oxygen in the hemoglobin molecule. Unlike oxygen or carbon dioxide, it remains attached to the hemoglobin. Thus, the more one smokes, the more the oxygen level to the tissues decreases. This reaction is similar in a fetus that gets its oxygen from the mother. If the mother smokes, the oxygen available to the fetus is reduced. This can result in poor development and low birth weight.

E. Smoking stimulates the secretion of mucus in the airways. It also paralyzes the cilia and kills new ciliated cells. This results in an inflammation of the bronchiole tubes. Smokers cough to remove particles and mucus from the airways.

F. 1. The number of layers of basal cells increases.
   2. Basal cells replace ciliated cells.
   3. The nuclei of basal cells change and basal cells become disorganized.
   4. Cancer cells break through the basement membrane.

G. Women who smoke are more likely to develop lung cancer than men who smoke. Women are more than 3 times more likely to carry the genetic mutation *K-ras* that promotes tumor growth than men. Another gene that promotes tumor growth is found on the X chromosome. Women who smoke are more likely to develop cervical cancer and osteoporosis than women who do not smoke. The death rate due to strokes, heart attacks, and blood clots in the legs is much higher in women who smoke and take the pill. Women who do not smoke are twice as likely as women who smoke to conceive. A woman smoker is twice as likely to miscarry as a nonsmoker. Nicotine affects the fetus by constricting the vessels in the umbilical cord and decreasing the amount of oxygen in the hemoglobin. The incidence of stillbirth among women who smoke is about twice that of nonsmokers.

H. 1. Each cigarette a person smokes reduces their life expectancy by 5-7 minutes.
   2. Annually, nearly 500,000 Americans will die prematurely from smoking.
   3. Smoking-related illnesses cost the United States about $150 billion each year.
   4. Of the 4000 substances found in smoke, at least 50 have been shown to cause cancer.
   5. The average heart rate is about 72 beats a minute. The nicotine from a cigarette can raise that 33 beats each minute.
   6. There are approximately 430/81 = 5.3 times more deaths from smoking than from alcohol each year in the United States.
   7. The amount of carbon monoxide in cigarette smoke is 1600 ppm, which greatly exceeds the 10 ppm considered dangerous in industry.
   8. Carbon monoxide remains in the bloodstream for up to 6 hours after smoking a cigarette.
   9. Smoking is responsible for 30% of all cancer deaths.
   10. 90% of all people diagnosed with lung cancer die within 5 years.
   11. Smokers are 15-25 times more likely to develop lung cancer than nonsmokers.

12. As estimated by the American Heart Association, about 25% of all fatal heart attacks are caused by cigarette smoke.
13. Smoking is responsible for 20% of all cases of cataracts.
14. After hour in a smoke-filled room, a nonsmoker may inhale an amount of cancer-causing nitrosamines equivalent to smoking 15 filter cigarettes.
15. 30 million Americans quit smoking each year.

Critical Thinking Questions

1. The current warning states, "Surgeon General's Warning: Smoking Causes Lung Cancer, Heart Disease, Emphysema, And May Complicate Pregnancy." It is difficult to predict the effect it has on smokers. Many young people will report that they know about the warning but plan to quit soon, others admit to enjoying smoking and disregard the warning.
2. Smoking causes more deaths per year in the United States than any other source. Specifically tobacco related deaths totaled 435,000, deaths from alcohol consumption totaled 85,000 and deaths from firearms totaled only 29,000. Someone 18 years or older can purchase tobacco, but one must be 21 to purchase alcohol, and 18 to purchase a firearm (many times firearm training is also required.) It appears as though there is not as much regulation for the number one cause of death.
3. Figure 14a-1 shows that by far most deaths are caused by smoking. Figure 14a-8 shows that there are immediate benefits to quitting and the sooner someone quits the better it is. The letter should indicate that smoking is the number one cause of deaths in the United States and that all of these deaths are preventable. It should encourage the smoker to quit as soon as possible because the rewards compound with every year he or she does not smoke.
4. Low-tar and low-nicotine cigarettes do reduce the levels of poisons entering the body. However, smokers of these cigarettes still have 6.5 times the risk of developing lung cancer of that or a nonsmoker. Because of the addictive qualities of nicotine, those who smoke low-tar/low-nicotine cigarettes often smoke more.
5. The smoke that enters the environment either from the burning end of a cigarette or exhaled from a smoker contain dangerous poisons. The smoke that leaves the burning end of the cigarette is actually more dangerous than smoke inhaled through tobacco and a filter; therefore, presenting a greater threat to the nonsmoker. After an hour in a smoke-filled room, the blood levels of carbon monoxide and nicotine match those of someone who has smoked a cigarette.

Practice Test

1. d
2. a
3. c
4. a
5. c
6. c
7. b
8. b
9. c
10. d
11. a
12. b
13. c
14. b
15. a
16. b
17. b
18. d
19. b
20. b

Short Answer Questions

1. Smoke paralyzes the cilia and macrophages found in the air passageways. This prevents the cilia from naturally clearing the mucus and particles from the bronchioles. The smoker coughs reflexively to remove these materials.
2. The substances in smoke must be broken down by the body. This happens in the liver but the by-products of these metabolites can cause cancer in other excretory and auxiliary organs including the bladder, kidneys and pancreas.

3. Co-carcinogens are chemicals that enhance the action of other carcinogens or promote cell division. This allows cancer to grow faster once the tumor has started.
4. Nicotine is a vasoconstrictor, meaning that it causes the small blood vessels to constrict which raises blood pressure. It also makes the heart beat faster at the same time that carbon monoxide reduces the amount of oxygen reaching the cells. Smoking can lead to increased incidents of atherosclerosis. It decreases the levels of the protective cholesterol (HDL) leaving more bad (LDL) cholesterol in the blood. The increased blood pressure stresses arteries and can lead to inflammation. Nicotine causes spasms of the arteries, which can lead to heart attack. Finally nicotine causes an increase in the number of platelets and the amount of fibrinogen in the plasma. This can lead to spontaneous clotting resulting in a stroke or heart attack.
5. Nicotine, one of the primary chemicals in cigarette smoke, is addictive much like cocaine or opium. It can provide a pleasant, calming feeling that reduces anxiety and agitation. However, it also stimulates the cells of the nervous system and cells become hyperactive when it is withdrawn.

Chapter 15

The Digestive System

Objectives

After reading the text and studying the material in this chapter, you should be able to:

- Describe the passage of food through the gastrointestinal tract from the mouth to the anus.
- Explain the function of each organ and accessory organ of the digestive system.
- Describe the role of the teeth, tongue and salivary glands in the preparation of food for swallowing.
- Explain the structure of a tooth and the factors affecting decay and gum disease.
- Describe swallowing and the role of the palate and epiglottis.
- Describe peristalsis as a mechanism for moving food through the digestive system.
- Explain how the structure of the stomach lining protects it from high acid concentrations.
- Describe the chemical breakdown of different types of foods (carbohydrates, proteins, and fats) as they pass through the digestive tract.
- Contrast the chemical and mechanical breakdown of food in the stomach.
- Describe the digestive activities of the small intestine including the secretion of intestinal and pancreatic enzymes and bile.
- Describe the structure of the villi and how they function in the absorption of food molecules.
- Explain the function of the liver as it regulates blood sugar levels.
- Describe the role of bacteria in the large intestine.
- Compare neural and hormonal control of digestion.

Chapter Summary

The digestive system is arranged as a series of organs along a tube called the **gastrointestinal (GI) tract**. Each of these organs helps to breakdown the food you eat into molecules small enough to be absorbed into the bloodstream and utilized by the body. The digestive organs (mouth, esophagus, stomach, small intestine, and large intestine) are aided by several accessory organs (salivary glands, pancreas, and liver) as they complete these processes.

Digestion begins in the **mouth** where food enters the digestive process. Here it is prepared for swallowing and monitored to determine if it is good, spoiled, or even poisonous. The teeth bite, tear, and crush the food into smaller pieces, increasing the amount of surface area available for the action of digestive enzymes. Teeth are alive and are surrounded by living tissue that physically supports the teeth and help maintains their health. Three pairs of **salivary glands** release secretions, called **saliva**, into the mouth. The saliva moistens food, dissolves the chemicals in the food, and contains the enzyme, **salivary amylase**, which begins digestion of carbohydrates.

The tongue helps form food into a **bolus**, a soft mass of food, suitable for swallowing. The bolus passes through the **pharynx**, the place where the nasal and oral cavities join, and then it passes the **epiglottis** on the way down the **esophagus**. During swallowing the epiglottis covers the trachea, or windpipe, preventing choking. The esophagus is merely a tube that transports food from the mouth to the stomach; no digestive processes occur here. Food is pushed through our digestive system by a series of muscular contractions called **peristalsis**.

The **stomach** is a major organ of the digestive system responsible for the storage of food, turning food into a soupy mixture called **chyme**, and the addition of digestive enzymes and acids that begin chemical digestion of proteins. Two **sphincter muscles** control the movement of food in and out of the stomach. Food enters the stomach from the esophagus and is released into the small intestine. While in the stomach, **hydrochloric acid (HCl)** kills bacteria, begins the breakdown of protein, and activates the digestive enzyme, **pepsin**, which also acts to split proteins. The stomach itself is protected in several ways from being digested by this caustic combination of chemicals.

The **small intestine** is perhaps the most important digestive organ. It is here where food is both digested into its smallest chemical components and where it is absorbed into the blood. The complex chemicals of carbohydrates, proteins, fats and nucleic acids are broken down into their simplest forms in the small intestine by enzymes of the **pancreas** and the small intestine aided by **bile** from the gall bladder. The pancreas is an **accessory organ** of the digestive system that produces enzymes and acid-neutralizing agents. Together intestinal and pancreatic enzymes break the biological polymers of foods into their monomers: carbohydrates into monosaccharides, proteins into amino acids, and triglycerides (fats) into fatty acids and glycerol. **Bile**, produced in the liver and concentrated in the **gallbladder**, aids in the digestion and absorption of fats.

Once food is broken down into its smallest components, it is ready to be absorbed into the bloodstream for transport to the cells. The small intestine has several physical features that allow it to efficiently move small chemical molecules from the inside of the small intestine to the wall of capillaries. The lining of the small intestine is pleated and has numerous finger-like projections called **villi** to increase surface area. In addition, each villus is covered with **microvilli,** giving the small intestine a velvety appearance called the **brush border**. Each villus contains a network of capillaries and a lymph vessel that carry away the products of digestion; a process called absorption.

The network of vessels in the villi contains a high concentration of digestive products. They are taken directly to the **liver**, an important auxiliary organ of the digestive system. The liver monitors blood glucose levels, packages fat molecules with proteins to aid transport in the blood, and removes poisonous substances.

Undigested and indigestible materials that have not been absorbed by the small intestine move to the **large intestine**. The large intestine absorbs the water, ions, and vitamins found in these materials. The remaining waste matter is stored for controlled elimination from the body. Eventually the **feces**, now in a consistency suitable for elimination, pass into the **rectum** and leave the body through the **anal canal**.

The gastrointestinal tract is long and complex; involving many organs, enzymes, and special structures. The controlled movement of partially digested materials and the secretion of enzymes at the right time are critical to healthy and comfortable digestion. Processes that are quick or anticipatory are controlled by the nervous system, while processes that take more time and are able to move at a slower pace, are controlled by hormones. **Gastrin**, released by the stomach lining, and **vasoactive intestinal peptide (VIP)**, **secretin**, and **cholecystokinin**, released by the lining of the small intestine, all cause the release of digestive enzymes. In addition, the digestive process is partially controlled by physical sensations such as the stretching of the stomach or the detection of food particles in the organs.

Key Concepts

- The gastrointestinal tract consists of the mouth, pharynx, esophagus, stomach, small intestine, and large intestine. Auxiliary organs include the salivary glands, liver, gallbladder, and pancreas.
- The mouth prepares food for swallowing and monitors food quality.
- Teeth mechanically break food into smaller fragments for swallowing.
- In the mouth, salivary amylase begins to digest complex carbohydrates into shorter chains of sugars.
- Food moves into the pharynx, then into the larynx, past the epiglottis and down the esophagus to the stomach.
- Food moves throughout the gastrointestinal tract by peristalsis.
- The stomach functions to store food, digest proteins and kill ingested microorganisms.
- The small intestine is the major site of digestion and absorption of nutrients.
- Intestinal enzymes, pancreatic enzymes and bile break proteins, polysaccharides, polypeptides and triglycerides (fats) into their component parts.
- The pleated lining of the small intestine and the villi increase surface area of the small intestine to aid absorption of nutrients.
- The liver produces bile and keeps blood glucose levels steady.
- The gallbladder stores, concentrates, and releases bile into the small intestine where it emulsifies fats.
- The large intestine absorbs water and vitamins that remain in the indigestible food, stores feces, and eliminates waste from the body.
- Both neural and hormonal mechanisms regulate the release of digestive secretions.

Study Tips

As you study the digestive system, first get a grasp of the organs and their relative order in the digestive process. Then focus on the function of each organ and any special structures that allow it to do its job more efficiently. Solidify your understanding by describing the path of a piece of food from the mouth to elimination at the anus. Reread the textbook material. Then study each figure and read the caption. Focus especially on the figures as you study the organs, their function, location and structure.

Chemical digestion and mechanical breakdown of food are at the heart of the digestive process. Concentrate on each organ and each accessory organ, one at a time and review the digestive processes that occur there. Use Tables 15-1 and 15-2 to answer the questions: What enzymes are present? What are they breaking down? and What is the final product?

Lastly, review Tables 15-4 and 15-5 to develop an understanding of the control mechanisms for digestion. Answer these questions in your own mind: What processes are controlled by the nervous system? What processes are controlled by hormones? and How do the physical stimuli of food molecules and the stretching of organs stimulate and control the process of digestion?

Review Questions

A. Use the following terms to label the figure below: anus, esophagus, gallbladder, large intestine, liver, mouth, pharynx, pancreas, salivary glands, small intestine, stomach, and rectum.

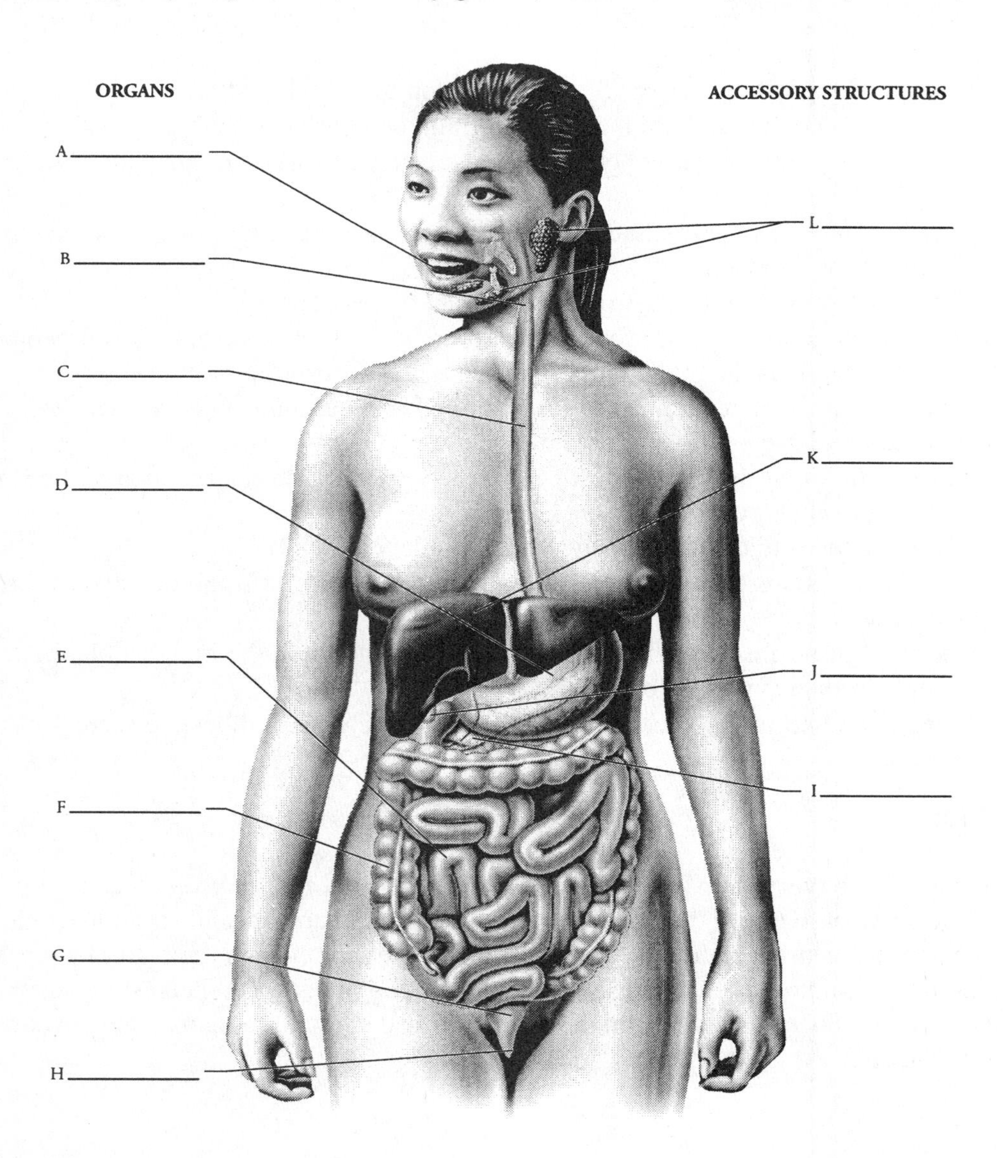

B. Match the organ or structure in Column A with the function in Column B.

Column A	Column B
1. anus	a. mechanically break food into smaller pieces
2. epiglottis	b. digestion is completed and nutrients are absorbed
3. esophagus	c. terminal outlet of the digestive tract
4. gallbladder	d. stores partially digested food and begins protein digestion
5. large intestine	e. first part of the digestive tract to receive food
6. liver	f. secrete saliva and begin carbohydrate digestion
7. mouth	g. tube transporting food from the mouth to stomach
8. pharynx	h. stores bile and releases it into the small intestine
9. pancreas	i. produces bile and regulates blood sugar
10. salivary glands	j. covers the opening to the airway during swallowing
11. small intestine	k. control the passing of material from one organ to another
12. sphincter muscles	l. absorbs water and forms and expels feces
13. stomach	m. produces digestive enzymes that enter the small intestine
14. teeth	n. increases surface area in the small intestine
15. villi	o. area where the oral and nasal cavities come together

C. Label the following structures of a villus found in the small intestine: absorbing cell, capillary, lacteal, microvilli, and mucus secreting cell.

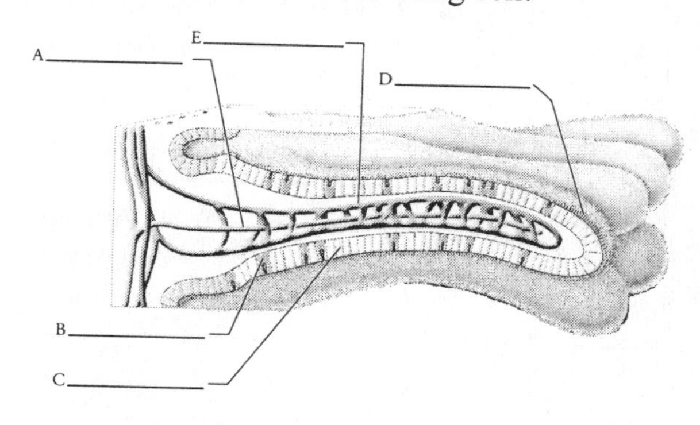

D. Fill in the blank.

1. In the center of each tooth is the ____________ which contains blood vessels and nerves. It is surrounded by a hard, bone-like substance, called ____________.
2. The crown of the tooth is covered with a hard, non-living material called the ____________.
3. Below the gum line is the ____________ of the tooth, covered by a living and sensitive connective tissue called ____________.
4. Blood vessels and nerves enter the root and reach the pulp through the ____________ ____________.
5. Tooth decay is caused by ____________ produced by ____________ living in the mouth.

E. Put these 5 statements in the correct order.

_____ Bolus is formed.

_____ Bolus is pushed into the esophagus and moves by peristalsis to the stomach.

_____ Epiglottis covers the windpipe (trachea).

_____ The mouth, including the teeth and salivary glands, prepare food for swallowing.

_____ Soft palate rises to prevent food from going into the nasal cavity.

F. Complete the table below using your knowledge of how materials cross the lining of the small intestine.

	Carbohydrates (polysaccharides)		**Proteins**	**Fats**
Digestive Product (monomer)	Glucose	Fructose		
Mechanism for entering epithelial cells of villi			Active Transport	

Mechanism for leaving the epithelial cells		Diffusion		
Transport Vessel			Blood	
Final destination	Liver			

G. Fill in the blanks with the name of a food type (carbohydrate, protein, or fat), the enzyme acting on it or the location of the action.

1. Food enters the mouth where the teeth mechanically break it down. ____________, secreted by the ____________ ____________ acts on ____________ here in the mouth.

2. The ____________ produces ____________ ____________ and pepsin which breaks down ____________.

3. The small intestine is a major site of chemical digestion. The small intestine secretes the following three enzymes that act on disaccharides: ____________, ____________, and ____________. In addition, it secretes ____________ which acts on ____________.

4. The pancreas secretes ____________ and ____________ which both act on proteins and polypeptides. It also secretes ____________ that acts on only polypeptides. The enzyme ____________ breaks down fats and ____________ work on nucleic acids.

5. Bile is produced in the liver and concentrated in the ____________. It enters the small intestine via the ____________ ____________ ____________ and emulsifies ____________. Since bile is rich in ____________ it sometimes leads to the formation of ____________.

H. Complete the following series of statements describing the cause and effect of various digestive stimuli. Circle the appropriate word to indicate if the response is neural or hormonal.

1. The scent of food causes ________________________. This is a (neural or hormonal) response.

2. While chewing is still taking place, the stomach is stimulated to secrete ____________ ____________. This is a (neural or hormonal) response.

3. Stretching of the stomach and the presence of partially digested proteins cause ____________. This is a (neural or hormonal) response.

4. The most important stimulus for the release of intestinal enzymes is ____________. This is triggered by a (neural or hormonal) response.

5. Acid chyme is also responsible for the release of three chemicals, ____________, ____________, and cholecystokinin that cause the release of digestive enzymes, bile, and sodium bicarbonate. This is a (neural or hormonal) response.

Critical Thinking Questions

Read each of the following questions carefully. If it helps you, underline or list each component to the question. Use complete sentences in your explanation.

1. What is the primary purpose of digestion and why is it necessary to humans?

2. It is common for young people to eat a cheeseburger. Trace the cheeseburger through your digestive tract, include each organ it passes through. Describe where the bread (a carbohydrate), the cheese and meat (proteins), and fats are digested and what enzymes are involved.

3. Think about the order that food is digested. What foods are digested first, second, and last? Now think about the function of each of these foods. Why is it efficient for the body to digest foods in the order that it does when one considers the function of the nutrients?

4. The movement of food through the digestive system is very controlled as is the secretion of the enzymes. Why is it necessary to control the passage of food through the gastrointestinal tract? Why is it advantageous to have several control mechanisms?

PRACTICE TEST

Choose the one best answer to each question that follows. As you work through these items, explain to yourself why each answer you discard is incorrect and why the answer you choose is correct.

1. Which of the following gives the correct order for food as it passes through the organs of the digestive system?
   a. esophagus, epiglottis, stomach, liver, small intestine, large intestine, anus
   b. mouth, epiglottis, liver, stomach, pancreas, small intestine, large intestine, anus
   c. mouth, esophagus, stomach, gallbladder, pancreas, large intestine, small intestine, anus
   d. mouth, pharynx, esophagus, stomach, small intestine, large intestine, anus

2. Which of the following tooth parts are not living?
   a. cementum
   b. enamel
   c. pulp
   d. root

3. Which of the following describes how to keep teeth and gums healthy?
   a. avoid sticky foods
   b. brush your teeth twice a day and floss regularly
   c. choose a diet low in sugar
   d. use fluoridated toothpaste or get fluoride treatments from your dentist
   e. all of the above are good dental practices.

4. The tongue functions to ____________.
   a. add moisture and enzymes to the food
   b. grind and crush the food
   c. position food for swallowing
   d. protect the opening of the pharynx during swallowing

5. If the epiglottis does not function correctly, what might happen?
   a. acid reflux disease will damage the esophagus
   b. one might choke
   c. peristalsis will stop
   d. swallowing will be difficult or impossible

6. Which of the following best describes the role of the esophagus in digestion?
   a. Connects the mouth to the stomach and has no function in chemical digestion
   b. Digestive enzymes are secreted as food passes from the esophagus to the stomach
   c. Mucous is secreted to protect the esophagus from the stomach enzymes
   d. Serves a minor role in the chemical digestion of fats

7. The primary function of the stomach is to ____________.
   a. Break down fats and proteins for further digestion
   b. Provide HCl to be used by the small intestine
   c. Store, digest, and absorb food nutrients
   d. Store and digest food

8. Although the stomach contents are very acidic, the stomach usually does not cause damage to itself. Why?
   a. Mucus forms a thick protective coating for the stomach
   b. Food and water dilute the gastric juices
   c. The release of gastric juices is controlled to avoid too high a concentration
   d. The stomach lining is quick to repair itself
   e. All of the above are correct

9. To which combination of organs in the digestive system is chemical digestion of proteins confined?
   a. Mouth and large intestine
   b. Mouth, stomach, small intestine
   c. Stomach, esophagus, and small intestine
   d. Stomach, small intestine

10. The fingerlike projections along the surface of the small intestine area called ____________.
    a. Capillaries
    b. Circular folds
    c. Lacteal
    d. Villi

11. The primary function of the small intestine is to ____________.
    a. digest and absorb nutrients
    b. kill bacteria in the food
    c. maintain constant levels of nutrients in the blood
    d. pass indigestible waste from the body

12. The pancreas is an important auxiliary organ of the digestive system because it ____________.
    a. aids in water balance as food moves through the digestive system
    b. provides insulin to the small intestine
    c. provides several digestive enzymes to the small intestine
    d. secretes bile

13. What is the function of bile?
    a. aids in the digestion of fats
    b. causes several serious diseases of the liver
    c. helps dissolve gallstones
    d. is first produced in the gallbladder

14. Villi, microvilli and the folds of the small intestine all function to ____________.
    a. increase surface area
    b. move chemicals across the membrane
    c. prevent the intestine from digesting itself
    d. secrete important enzymes

15. Fatty acids, glycerol, cholesterol and phospholipids cross the intestinal lining and enter the ____________.
    a. circulatory system
    b. liver
    c. lymphatic system
    d. mucosa

16. Which of the following is NOT a function of the liver?
    a. aid in digestion by producing bile
    b. convert byproducts of protein digestion to urea
    c. digest food as it passes through the liver on the way to the large intestine
    d. monitor glucose levels in the blood

17. What is the primary role of the large intestine in digestion?
    a. absorb ions, water and vitamins to maintain homeostasis
    b. add the enzyme lactase to the digestive process
    c. complete the breakdown of indigestible food
    d. produce essential bacteria for one's health

18. If the large intestine becomes irritated and peristalsis increases, ____________ may result.
    a. appendicitis
    b. constipation
    c. diarrhea
    d. hernias

19. What is primary the role of the bacteria found in the large intestine?
    a. absorb critical vitamins
    b. cause disease
    c. digest food and produce vitamins
    d. provide a natural source of protein

20. Digestive processes that need to respond quickly are under the control of the ____________.
    a. digestive enzymes
    b. digestive hormones
    c. endocrine system
    d. nervous system

Short Answer Questions

Read each of the following questions carefully. If it helps you, underline or list each component to the question. Formulate your answer and jot down the main points. Then use complete sentences in your explanation.

1. Describe the four basic layers of the structures of the digestive system. Include the function of each layer.

2. What are the problems associated with a receding gum line?

3. Why is the pain from heartburn sometimes confused with that from a heart attack?

4. The liver, gallbladder and pancreas are considered accessory organs of the digestive system. Why aren't they primary organs and what is their role?

5. When people become ill due to a virus or bacterial infection, they sometimes get diarrhea. What is the connection between the infectious organisms and diarrhea?

Answer Key

Review Questions

A. A. mouth
   B. pharynx
   C. esophagus
   D. stomach
   E. small intestine
   F. large intestine
   G. rectum
   H. anus
   I. pancreas
   J. gallbladder
   K. liver
   L. salivary glands

B. 1. c
   2. j
   3. g
   4. h
   5. l
   6. I
   7. e
   8. o
   9. m
   10. f
   11. b
   12. k
   13. d
   14. a
   15. n

C. A. lacteal
   B. mucus secreting cell
   C. absorbing cell
   D. microvilli
   E. capillary

D. Fill in the blank.
   1. In the center of each tooth is the pulp which contains blood vessels and nerves. It is surrounded by a hard, bone-like substance, called dentin.
   2. The crown of the tooth is covered with a hard, non-living material called the enamel.
   3. Below the gum line is the root of the tooth, covered by a living and sensitive connective tissue called cementum.
   4. Blood vessels and nerves enter the root and reach the pulp through the root canal.
   5. Tooth decay is caused by acid produced by bacteria living in the mouth.

E. 1. The mouth, including the teeth and salivary glands, prepares food for swallowing.
   2. Bolus is formed.
   3. Soft palate rises to prevent food from going into the nasal cavity.
   4. Epiglottis covers the windpipe (trachea).
   5. Bolus is pushed into the esophagus and moves by peristalsis to the stomach.

F.

	Carbohydrates (polysaccharides)		Proteins	Fats
Digestive Product (monomer)	Glucose	Fructose	Amino acids	Fatty acids, glycerol, phospholipids, cholesterol
Mechanism for entering epithelial cells of villi	Active transport	Facilitated diffusion	Active transport	Diffusion
Mechanism for leaving the epithelial cells	Diffusion	Diffusion	Diffusion	Exocytosis
Transport vessel	Blood	Blood	Blood	Lactea
Final destination	Liver	Liver	Liver	Thoracic duct

G. 1. Food enters the mouth where the teeth mechanically break it down. Amylase, secreted by the salivary glands acts on polysaccharides/carbohydrates here in the mouth.
2. The stomach produces hydrochloric acid and pepsin which breaks down protein.
3. The small intestine is a major site of chemical digestion. The small intestine secretes the following three enzymes that act on disaccharides: maltase, sucrase, and lactase. In addition, it secretes aminopeptidase which acts on peptides.
4. The pancreas secretes trypsin and chymotrypsin which both act on proteins and polypeptides. It also secretes carboxypeptidase that acts on only polypeptides. The enzyme lipase breaks down fats and nucleases work on nucleic acids.
5. Bile is produced in the liver and concentrated in the gallbladder. It enters the small intestine via the common bile duct and emulsifies fats. Since bile is rich in cholesterol it sometimes leads to the formation of gallstones.

H. 1. The scent of food causes saliva to be released. This is a neural response.
2. While chewing is still taking place, the stomach is stimulated to secrete gastric juices. This is a neural response.
3. Stretching of the stomach and the presence of partially digested proteins cause the stomach to release gastrin. This is a hormonal response.
4. The most important stimulus for the release of intestinal enzymes is acidic chyme and bile in the small intestine. This is triggered by a neural response.
5. Acid chyme is also responsible for the release of three chemicals, vasoactive intestinal peptide, secretin, and cholecystokinin that cause the release of digestive enzymes, bile, and sodium bicarbonate. This is a hormonal response.

Critical Thinking Questions

1. Digestion takes complex molecules and breaks them down into their chemical subunits small enough to be absorbed into the bloodstream and delivered to body cells. They may then be used to provide energy for daily activities or they may provide materials for growth and repair. Humans, like all animals, are incapable of making their own food. Therefore, they must obtain it, eat it, and break it into the building blocks they need to make the other necessary molecules.
2. The cheeseburger enters the mouth where it is chewed and where salivary enzymes begin to digest carbohydrates. The food is swallowed passing the pharynx and entering the esophagus. The food now enters the stomach where the churning of the stomach mixes the food with gastric juice and protein digestion by HCl and pepsin begins. Next the food enters the small intestine where

carbohydrate, protein and fat digestion are completed by enzymes of the small intestine aided by pancreatic enzymes and bile. Absorption of most nutrients occurs in the small intestine. Finally, the digested food enters the large intestine where water and vitamins are absorbed and feces is formed and expelled.

3. Carbohydrates are used for quick energy and are digested first. Proteins are used for the development of new tissues but may also be used for energy and are digested next. Lipids are used for stored energy and are not immediately accessible. They are digested last.
4. The release of enzymes must be timed with the arrival of food. Also, the length of time that food spends in an organ determines the amount of time that an enzyme has to digest the food. The nervous is a quick response system that starts the process quickly while the endocrine system secretes hormones that have a slower, longer lasting effect.

Practice Test

1. d
2. b
3. e
4. c
5. b
6. a
7. d
8. e
9. d
10. d
11. a
12. c
13. a
14. a
15. c
16. c
17. a
18. c
19. c
20. d

Short Answer Questions

1. The innermost layer is the mucosa that helps lubricate and protect the digestive tube. Next is the submucosa that is made of connective tissue containing a blood supply and nerves that coordinate contractions. The next layers are the muscularis which occur in different directions to maximize the churning and mixing actions of the various organs. The outermost layer is the serosa a connective tissue that secretes a fluid that reduces friction with contacting surfaces.
2. The gums adhere tightly to the enamel of the tooth protecting it from contact with food and bacteria. A receding gum line increases the sensitivity of the tooth to temperature. If a pocket forms between the gum and the tooth, it can trap plaque that can destroy the bone and soft tissues around the tooth leading to loss of the tooth.
3. The pain from heartburn is caused when the stomach juices back up into the esophagus. The pain is located in the upper abdomen or behind the breastbone where pain from a heart attack may begin.
4. Food does not pass through any of these organs and they do not share the same structure as those of the digestive system;, therefore, they are not considered primary digestive organs. They are necessary because they produce or concentrate materials needed to aid the digestive process.
5. Colon contractions, or peristalsis, can be stimulated by the toxins produced by microorganisms. This causes the material to pass through the large intestine too quickly for all of the water to be absorbed. The result is a very liquid feces.

Chapter 15A

Nutrition and Weight Control

Objectives

After reading the textbook and studying the material in this chapter, you should be able to:

- Describe how food is used by the body.
- State the function of fats, carbohydrates and proteins in the diet, the recommended daily calorie intake from each and the calories gained per gram of each.
- Describe any health risks associated with the consumption of various foods.
- Differentiate between the different types of cholesterol and their effects on the body.
- Explain the glycemic load and how it can be used to increase the amount of complex carbohydrates in the diet.
- Explain the value of dietary fiber and give examples of good sources of fiber.
- Explain the importance of essential amino acids and how to ensure they are in the diet in the correct proportions.
- Differentiate between minerals and vitamins and explain their role in cellular functioning.
- Explain how the body uses energy and what happens to excess food calories.
- Define Body Mass Index (BMI) and explain how it can be used to identify a desirable weight.
- List the dietary guidelines for Americans to promote health and explain how you might apply them to your own life.
- Describe the characteristics of successful weight-loss programs.
- Describe obesity, anorexia nervosa and bulimia explain how they are serious health risks.

Chapter Summary

Food provides fuel for cellular activities, metabolic regulators, and building blocks for cell division, maintenance, and repair. Water is necessary for a proper cellular environment and cellular reactions. A nutrient is a substance in food that provides energy or plays a structural or functional role to promote normal growth, maintenance or repair.

Lipids include fats, oils and cholesterol. **Saturated fats** are solid at room temperature and come from animal sources, whereas **unsaturated fats** are liquid at room temperature and come from plant sources. Fat serves as a ready source of energy, insulation, and cushion for vital organs. On a cellular level, fats are components of the cell membrane, used to construct myelin sheaths, and absorb fat-soluble vitamins. Cholesterol is the structural basis for the steroid hormones.

A high-fat diet is related to obesity and to certain cancers, high blood pressure, and increased risk of diabetes. Another health risk of a high-fat diet is developing atherosclerosis. To minimize this risk, adults should have blood cholesterol levels under 200 mg/dl[2]. Cholesterol moves through the blood or lymph as a lipoprotein. Low-density lipoproteins (LDLs) are bad because they deposit cholesterol in the artery walls. In contrast, high-density lipoproteins (HDLs) help the body to eliminate cholesterol. Only 20-30% of the calories in the diet should come from fats, and most of those should be monounsaturated and polyunsaturated fats.

Carbohydrates provide fuel for our bodies and are found in our diet as sugars and starches. Dietary fiber is a carbohydrate in plant foods that humans cannot digest. It is important in our diets because water-soluble fiber lowers LDL and is good for the heart and blood vessels. It also absorbs water and gives bulk to the feces.

Nutritionists recommend that 45-65% of the calories in our diet come from carbohydrates, including both simple carbohydrates (sugars) and complex carbohydrates (starches and fiber). The glycemic load is a measure of how quickly a serving of food is converted to blood sugar. A healthful diet should contain more foods with a low glycemic load to reduce the risk of heart disease and diabetes.

Proteins are chains of amino acids that are digested and delivered to the cells. They are the structural components of every cell, regulate body processes, are antibodies, and can be used for energy. **Complete dietary proteins** contain all 9 of the **essential amino acids** that your body cannot synthesize, while **incomplete proteins** are lacking one or more of the essential amino acids. **Complementary proteins** are combinations of proteins that together supply enough of all the essential amino acids.

Nutritionists recommend that protein represent 10-35% of the dietary calories. It is usually recommended to eat plant and animal sources of protein and restricting the amount of red meat.

Vitamins are organic compounds that are needed only in minute quantities. Most function as coenzymes. Vitamins are either water-soluble or fat-soluble. Except for vitamin D, our cells cannot make vitamins and must obtain them from food. Adequate daily supplies of some vitamins are required to prevent birth defects, heart disease, cancer, macular degeneration and atherosclerosis. Minerals are inorganic substances that are essential to a healthy diet.

Water is perhaps the most essential nutrient. Water transports materials through our blood and lymph, provides a medium for chemical reactions, is a lubricant, can serve as a protective cushion, and plays an important role in the regulation of body temperature.

Body energy balance depends on the calories gained in food and the calories used. Carbohydrates and proteins provide 4 calories per gram and fats provide 9 calories per gram. Food energy not used for the body's activities is stored as fat or glycogen. The body requires energy for maintenance of basic body functions (**basal metabolic rate** (**BMR**)), physical activity, and processing of the food that is eaten. Obesity is body weight 20% or more above the body weight standard. The Body Mass Index (BMI) evaluates your weight in relation to your height. The most important reason to maintain a healthy weight is to avoid cardiovascular disease, diabetes, and degenerative joint disease.

Dietary guidelines for Americans include fitness, choosing healthy foods according to the Food Guide Pyramid, and avoiding foods that lead to disease. Successful weight loss programs usually include a reduction in the number of calories consumed, an increase in energy expenditure, and behavior modification. The number of calories required daily depends upon your activity level and age. The easiest way to reduce calorie intake is to cut back on fatty foods and to increase the amount of fiber in the food.

Although obesity is dangerous, weight loss can be dangerous also as in the case of anorexia nervosa or bulimia. The change in eating habits associated with these disorders is thought to be the result of physiological, social and psychological factors. Both eating disorders can be fatal.

Key Concepts

- The Food Guide Pyramid helps us plan a balanced diet to improve health and reduce the risk of serious chronic diseases.
- Food provides fuel needed for cellular activities, metabolic regulators, and building blocks needed for cell division, maintenance and repair.
- Water is needed for the proper cellular environment and for certain cellular reactions.
- A nutrient is a substance in food that provides energy or plays a structural or functional role to promote normal growth, maintenance, or repair.
- Lipids include fats, oils, and cholesterol. They are used for energy, insulation, protection of vital organs, and the transport of fat-soluble vitamins.
- Dietary fat should make up no more than 20-35% of daily calorie intake. Monounsaturated and polyunsaturated fats lower total blood cholesterol and low-density proteins in the blood.
- Carbohydrates include simple carbohydrates (sugars) and complex carbohydrates (starches). They provide fuel for the body.
- Dietary fiber is the indigestible part of plants. It adds bulk to the feces and can lower low-density lipid levels.
- Carbohydrates should make up 45-65% of daily calorie intake with an emphasis on whole grains, vegetables and fruits.
- The glycemic load is a measure of how quickly a serving of food is converted to blood sugar and how high the spike in blood sugar is. A healthful diet contains more foods with low glycemic loads.
- Proteins are chains of amino acids that are digested to provide the raw materials to replace and repair cells, regulate body processes, form antibodies, help maintain water balance, and serve as a source of energy.
- Dietary protein should make up 10-35% of our daily calorie intake. Of the 20 amino acids used by the body, we are able to synthesize all but the 9 essential amino acids that must be eaten.
- Vitamins are organic compounds that are needed only in minute quantities but are essential for health and growth. They function as coenzymes and help fend off disease.
- Minerals are inorganic substances that are essential to a wide range of life processes.
- Water functions as a solvent, provides a medium for chemical reactions, is a lubricant, forms a protective cushion around vital organs, and helps regulate body temperature. It is essential to consume enough water daily.
- Body energy balance depends on calories consumed versus calories used. The body uses energy to maintain itself (basal metabolic rate) and for physical activity.
- Obesity is a serious health risk because it leads to cardiovascular disease, raises cholesterol levels, and increases the risk of high blood pressure, diabetes, kidney disease and degenerative joint diseases.
- Dietary guidelines recommend that we aim for a healthy weight, participate in daily physical activity, and make wise food choices. Successful weight-loss programs usually involve reducing calorie intake, increase energy expenditure and behavior modification.
- Anorexia nervosa and bulimia are eating disorders that can be fatal.

STUDY TIPS

The information found in this chapter can benefit you the rest of your life. Rather than simply learning the material as part of this class, focus on applying it to yourself as a means to permanent health improvement. As you read about the dietary and exercise recommendations reflect on your own habits and commit to positive change.

The first part of this chapter applies the basic biochemistry covered in Chapter 2. If you are unsure of the structure of the biological macromolecules or their monomers, review that information first. As you read and attend lectures, arrange your notes by food type. You can distill them into a table that would include a column for lipids, carbohydrates, and proteins with rows listing the monomer, examples, use to the body, recommended dietary intake, and associated problems. Then focus on the value of water, minerals and vitamins to the body.

The second part of the chapter applies this nutritional biochemistry to a healthy body weight. Organize this part of your notes around the recommendations for good nutrition, the need for exercise, and how to maintain a healthy weight. Make notes to yourself regarding possible changes in your lifestyle that might improve your health. Finally include in your notes a section that defines obesity, anorexia nervosa and bulimia and explains the health risks of each.

REVIEW QUESTIONS

A. What are the three major ways that the body uses the food you eat?

1. ______________________________

2. ______________________________

3. ______________________________

B. Complete the following table.

Dietary Macromolecule	**Lipids**	**Carbohydrates**	**Proteins**
Monomer	Triglyceride (3 fatty acids and a glycerol)		
Function			Structural component of all cells, needed for repair and replacement of cells, regulate body processes, form antibodies

% Daily Calorie Recommendation			
Calories/gram			4 calories/gm
Health Risks/Concerns		Weight gain and associated cardiovascular problems	The fat in animal proteins can lead to cardiovascular problems
Good foods to eat		Complex carbohydrates, fiber, and food with low glycemic load	
Foods to avoid	Saturated and trans fats in red meat, butter, cheese, and whole milk		

C. Sometimes all types of blood cholesterol are lumped together. Total blood cholesterol is the sum of high-density lipoproteins (HDLs), which is referred to as "good" cholesterol, low-density lipoproteins (LDLs) and the very low-density lipoproteins (VLDLs). The LDLs are usually referred to as "bad" cholesterol. Write a sentence or two explaining why HDLs are good and LDLs are bad for our health. Then include ways to increase the good cholesterol and decrease the bad.

1. High-density lipoproteins are good for our health because they ______________________________

____

____

2. Ways to increase HDLs include ______________________________________

____

____

3. Low-density lipoproteins are bad for our health because they ______________________

____

____

4. Ways to decrease LDLs include ______________________________________

____

____

D. Water is our most essential nutrient. List at least 5 ways it benefits the body.

1. ____

2. ____

3. ____

4. ____

5. ____

E. Use Figure 15A-9 to find your Body Mass Index. Put a check next to the correct weight evaluation.

Height = ___________ inches   Weight = ___________ lbs   BMI = ___________

Underweight ________   Healthy ________   Overweight ________   Obese ________

F. Mark each of the following statements as true (T) or false (F). If the statement is false, correct it so that it is true.

1. _______ Our body cannot synthesize the 9 essential amino acids.

2. _______ Compared to other foods such as kidney beans or most fruits and vegetables, whole wheat bread is high in fiber.

3. _______ Olive and canola oils are good choices of healthy oils.

4. _______ Adding brown sugar and honey to your cereal is a good way to increase your intake of complex carbohydrates.

5. _______ Vitamins are organic compounds needed in small amounts.

6. _______ Tuna canned in water is an excellent choice of low-fat protein.

7. _______ Vitamin C is a fat-soluble vitamin that remains in your system for a long time.

8. _______ Salt is an essential mineral and should be added in cooking and at the table.

9. _______ Vegetables are an excellent source of vitamins, minerals, water, fiber and carbohydrates.

10. _______ Since about half of our water intake comes in the fruits and vegetables we eat, the remaining water can come from coffee and soft drinks.

G. Use your knowledge from this chapter to complete the following paragraph on maintaining a healthy weight.

A healthy BMI is in the range of ____________. The BMI evaluates body weight relative to ____________. The number of Americans dying from weight-related illnesses is second only to deaths caused by ____________. It is not surprising that the Dietary Guidelines for Americans recommend a base of healthy foods and then sensible choices of foods that should be limited, but it also recommends ______________________ (more than one word). To be successful, a weight loss program must have these three characteristics (more than one word): 1) ______________________, 2) ______________________, and 3) ______________________. Fad diets are almost always ____________ in the long term. Obesity is an eating disorder associated with ____________ ____________. Anorexia nervosa and bulimia are similar in that they result in a deficit of ____________. However, they are different in that an anorexic may exercise excessively and eat ____________ calories while a bulimic will eat ____________ calories and then purge the body of the calories. Both of these disorders if untreated may become ____________. All three eating disorders are treatable with professional help.

Critical Thinking

Read each of the following questions carefully. Circle the most important words in the question before formulating your answer.

1. Your text presents the current Food Pyramid. As noted near the end of the chapter, that will be updated in 2005. Use the Internet to find a suggested New Food Pyramid. How do they compare? What are the most striking differences? This Harvard School of Public Health web site provides a discussion of food pyramids http://www.hsph.harvard.edu/nutritionsource/pyramids.html.

2. Keep a log of what you eat for a representative 24 hours (not a time when you are too busy to eat well or a time when you are eating out often.) Keep track of the calories of fats, carbohydrates and proteins. Figure the percentage of each and compare your intake to the recommended percentages.

3. Your text presented the BMI as one way to determine if you are obese or overweight. Some people disagree with this measure. They say that it does not take into account weight due to large muscle mass. Use Figure 15A-9 in your text to test a few examples of athletes you know. What is their BMI? Is it in the healthy range? What are your thoughts on the validity of this measure of "healthy weight"?

4. If you and your friends decided to order pizza late one night while watching a movie, you might consume two pieces of pizza and a large soft drink. That could easily total 1000 calories just before bed. Use Table 15A-5 to determine what type of exercise and how much would be required to burn those extra calories.

5. Interview a professional counselor or dietician and ask about current fad diets. Are the diets effective? Are they safe? Are they popular among college students? Ask also about the most common eating disorders they see in college students. Is it obesity, anorexia or bulimia? What do they believe is the cause of the eating disorder? Do most students recover? What weight control plan do they recommend?

Practice Test

Choose the one best answer to each question that follows. As you work through these items, explain to yourself why each answer you discard is incorrect and why the answer you choose is correct. Do not refer to your text or notes as you take this practice test.

1. One purpose of the Food Guide Pyramid is to ____________.
   a. assist with planning an effective exercise program
   b. help us plan a healthy diet
   c. provide a calorie guide
   d. provide guidance for vegetarians on protein choices

2. The foods at the very top of the pyramid ____________.
   a. are high in nutritional value
   b. are of little nutritional value
   c. are of the highest quality
   d. should make up 3-4 servings a day

3. What does it mean when a food is described as *essential*?
   a. it can be synthesized by the liver
   b. it is required as an amino acid
   c. it must be eaten as part of our diet and cannot be synthesized
   d. it forms the base of the Food Guide Pyramid

4. Which of the following is NOT a function of lipids in the body?
   a. cushions vital organs
   b. insulation
   c. provides quick energy
   d. source of stored energy

5. What is the clearest risk of a high fat diet?
   a. developing anorexia
   b. developing atherosclerosis
   c. maintaining a layer of "baby fat" for insulation
   d. promoting the build up of HDLs

6. Choose the best type of fat to eat from the list below.
   a. animal fat
   b. saturated fat
   c. trans fat
   d. unsaturated fat

7. What is the primary function of carbohydrates in our diet?
   a. cushion internal organs and eyes
   b. insulate against heat and cold
   c. source of glucose
   d. source of minerals and vitamins

8. What is the value of dietary fiber?
   a. helps prevent constipation, hemorrhoids and diverticulosis
   b. lowers LDLs
   c. helps remove cancer-causing chemicals from the body
   d. all of the above

9. What is the recommended percent of daily calories that should consist of carbohydrates?
   a. 10-30%
   b. 20-35%
   c. 45-65%
   d. 50-75%

10. Which of the following is the best choice of dietary fiber?
    a. broccoli
    b. cooked carrots
    c. kidney beans
    d. whole wheat bread

11. Proteins are components of which of the following?
    a. antibodies
    b. cell structure and membranes
    c. hormones
    d. all of the above

12. What is a good choice of food that is high in protein but low in fat?
    a. cheddar cheese
    b. oatmeal
    c. tuna in water
    d. whole milk

13. Essential amino acids must be consumed together in the diet. Nonessential amino acids ____________.
    a. are stored in body fat until needed.
    b. are seldom used
    c. can be synthesized in the body.
    d. can be taken as supplements in a multiple vitamin

14. Why are vitamins needed in only very small amounts?
    a. They are only needed when we are children.
    b. They are very high in calories.
    c. They are very expensive.
    d. They can be used over and over as coenzymes.

15. Tony read that sodium was essential to health and that it was found in table salt so he salted his food at the table. Is this a good way to get dietary sodium?
    a. Yes
    b. No

16. What does the recommended 8 glasses of water a day mean?
    a. 8 8-ounce glasses of plain water
    b. at least 8 glasses of plain water plus fruits and vegetables
    c. water in fruits and vegetables plus about 4 glasses of plain water
    d. water in fruit, meat, coffee, tea, and plain water

17. The minimum energy needed to keep an awake, resting body alive is called the ____________?
    a. basal metabolic index (BMI)
    b. basal metabolic rate (BMR)
    c. resting metabolic rate (RMR)
    d. stable metabolic rate (SMR)

18. What is a primary reason that people who do not change their calorie intake begin to gain weight at about age 35?
    a. BMR rises slightly until menopause
    b. metabolic rate continues to increase until age 50
    c. muscle mass and metabolic rate decrease with age
    d. muscle mass decreases but metabolic rate increases with age

19. What is the best way to successfully and safely lose weight?
    a. change your behaviors
    b. decrease the amount of calories consumed
    c. increase the level of exercise daily
    d. combine all of the above

20. Anorexia nervosa, bulimia and obesity are all considered to be ____________.
    a. eating disorders
    b. eating disorders characterized by bingeing
    c. eating disorders characterized by too much exercise
    d. neurotic disorders

SHORT ANSWER QUESTIONS

Read each of the following questions carefully. Jot down the main points you want to include in your answer. Then write a well-organized explanation.

1. Why do vegetarians have to be more concerned with the choice of protein in their diet than non-vegetarians?

2. The nutrition label from a box of crackers indicates that a serving size of 16 crackers contains 4g of fat, 21 g of carbohydrates and 2g of proteins. Calculate the calories from fats, carbohydrates, and protein. Then calculate the total number of calories in one serving of crackers.

3. If you calculate your BMR you will be surprised at how few calories that is. Why would you require more calories than the calculated BMR?

4. Jeff was taking large doses of vitamins in an effort to be very healthy. After taking biochemistry class, he restricted the amount of Vitamins A and E. What could be dangerous about very high levels of these vitamins?

5. If someone was fed only glucose for energy and water for hydration, they would waste away and die although they had plenty of calories. Why?

ANSWER KEY

REVIEW QUESTIONS

A. 1. Fuel for cellular activities.
   2. Building blocks for cell division, maintenance and repair.
   3. As metabolic regulators.

B.

Dietary Macromolecule	Lipids	Carbohydrates	Proteins
Monomer	Triglyceride (3 fatty acids and a glycerol)	Glucose	Amino acids
Function	Stored energy, insulation, protect vital organs, transport and absorption of fat-soluble vitamins	Fuel for cellular processes	Structural component of all cells, needed for repair and replacement of cells, regulate body processes, form antibodies
% Daily Calorie Recommendation	No more than 20-35%	45-65%	10-35%
Calories/gram	9 calories/gm	4 calories/gm	4 calories/gm
Health Risks/Concerns	Atherosclerosis, high blood cholesterol	Weight gain and associated cardiovascular problems	The fat in animal proteins can lead to cardiovascular problems
Good foods to eat	Nuts and fish, corn, safflower, olive, canola and peanut oils	Complex carbohydrates, fiber, and food with low glycemic load	Plant proteins, fish, and non-red meats
Foods to avoid	Saturated and trans fats in red meat, butter, cheese, and whole milk	Simple carbohydrates, sugars, and foods with high glycemic load	Red meat

C. 1. High-density lipoproteins are good for our health because they help eliminate bad cholesterol from the body and are protective against heart disease.
   2. Ways to increase HDLs include weight loss, exercise, and eating monounsaturated found in olive, canola and peanut oils and nuts, and polyunsaturated fats such as omega-3 fatty acids found in fish oils, and omega-6 fatty acids found in corn and safflower oils.
   3. Low-density lipoproteins are bad for our health because they they deposit cholesterol in the artery walls causing increased blood pressure and atherosclerosis
   4. Ways to decrease LDLs include reduce saturated and trans fats in the diet, lose weight and engage in exercise.

D. 1. transports materials in the blood and lymph
   2. medium for chemical reactions
   3. lubricant
   4. protective cushion
   5. regulation of body temperature

E. BMI = weight (lbs) x 700/height (in)^2 Healthy = 18.5-25

F. 1. T
   2. F, low in fiber
   3. T
   4. F, increase simple sugars
   5. T
   6. T
   7. F, water soluble vitamin that can be washed out of your system
   8. F, do not add salt to cooking or served food
   9. T
   10. F, the remaining water should not come from carbonated, caffeinated, or alcoholic drinks

G. A healthy BMI is in the range of 18.5-25. The BMI evaluates body weight relative to height. The number of Americans dying from weight-related illnesses is second only to deaths caused by tobacco. It is not surprising that the Dietary Guidelines for Americans recommend a base of healthy foods and then sensible choices of foods that should be limited, but it also recommends maintaining a healthy weight with daily physical activity. To be successful, a weight loss program must have these three characteristics: 1) reduction in number of calories, 2) increased calorie use, and 3) behavior modification. Fad diets are almost always ineffective in the long term. Obesity is an eating disorder associated with over eating. Anorexia nervosa and bulimia are similar in that they result in a deficit of calories. However, they are different in that an anorexic may exercise excessively and eat few calories while a bulimic will eat many calories and then purge the body of the calories. Both of these disorders if untreated may become deadly. All three eating disorders are treatable with professional help.

Critical Thinking Questions

1. Flaws include treating all fats and all carbohydrates the same. Not including a recommendation for exercise.
2. Students should develop a table that lists all foods eaten, grams fat, grams protein and grams carbohydrate. Then multiply the grams fat by 9 calories and the grams of carbohydrates and proteins by 4 calories to determine the calorie intake of each food type. Find the percentages and compare to the recommendations.
3. A typical college football lineman might be 6'3" tall and weigh 275 pounds. This person would have a BMI in the range of obese although his extra weight is probably due to a very large muscle mass. As with all recommendations of ideal weight, the BMI must be considered on an individual basis. Someone within the healthy weight range may have a lot of fat and little muscle while a very muscular person may have a BMI in the obese range and not be obese at all.
4. It is surprising, but burning 1000 calories will take a lot of effort. A combination of exercises over a few days is the most likely response. Jogging burns the most calories per hour, but few people can jog for an entire hour.
5. Most of the current fad diets are low in carbohydrates, but high in fat causing cholesterol levels to rise in some people. The best diet is a well balanced diet that matches your level of activity. Obesity is the most common eating disorder on all college campuses. Obesity is on the rise because people are eating too much food, food that is high in calories, and not engaging in enough exercise. Although genetics plays a role in the likelihood of eating disorders, culture seems to have the greatest effect.

Practice Test

1. b
2. b
3. c
4. c
5. b
6. d

7. c
8. d
9. c
10. c
11. d
12. c
13. c
14. d
15. b
16. c
17. b
18. c
19. d
20. a

Short Answer Questions

1. The pool of amino acids available for protein synthesis must always contain sufficient amounts of all the essential amino acids. Animal proteins are generally complete, but plant proteins are generally incomplete proteins are low in one or more of the essential amino acids. Incomplete proteins from two or more plant sources can be combined so that, after digestion, the pool of amino acids available for protein synthesis will contain ample amounts of all the essential amino acids. Combinations such as these are called complementary proteins. A vegetarian must make careful food choice to be sure to consume complementary proteins.
2. Multiply the grams of fat by 9 calories and the grams of proteins and carbohydrates by 4 calories. Then add them together. ((4g fat x 9 calories/gm) + (21 g carbohydrates x 4 calories/gm) + (2g proteins x 4 calories/gm) = 128 calories per serving.)
3. Total body energy demands include those needed to stay alive represented by the BMR and the calories expended in exercise. In addition, gender, muscle mass, activity, age, hormonal activity, and many other factors affect a person's energy requirements.
4. Excess water-soluble vitamins are usually excreted in the urine. In contrast, excess fat-soluble vitamins are stored in fat and can accumulate in the body causing serious problems and poisoning. Vitamins E and A are both fat-soluble.
5. The structural components of the body are made of proteins and proteins are made of amino acids. The body cannot synthesize the essential amino acids and therefore will not have the raw materials for tissue growth and repair.

Chapter 16

The Urinary System

Objectives

After reading the text and studying the material in this chapter, you should be able to:

- List the organs and systems that eliminate waste.
- Identify and give the function of each of the organs of the urinary system.
- Describe the structure of the kidney.
- Trace in detail the flow of filtrate through the nephron.
- Describe the processes of glomerular filtration, tubular reabsorption, and tubular secretion.
- Explain the role of the kidney in the maintenance of pH balance, water retention, red blood cell production and activation of vitamin D.
- Describe hormonal regulation of kidney function and name the hormones that are involved.
- Differentiate between acute and chronic renal failure and describe the various types of dialysis available to patients in renal failure.
- Describe the process of a kidney transplant and the problems that often occur.
- Explain the voluntary and involuntary components of urination and the causes of urinary incontinence.
- Identify various urinary tract infections and common causes.

Chapter Summary

Many organs from several systems dispose of the wastes from cellular reactions. Lungs and skin eliminate heat, water, and carbon dioxide. Skin excretes salts and urea. Organs of the digestive tract eliminate solid wastes and many other substances.

The function of the **urinary system** is to regulate the volume, pH, pressure and composition of the blood. The **kidneys** are the primary organs of the urinary system. The kidneys are located between the back abdominal wall and the parietal peritoneum supported by connective tissue and protected by fat. They are responsible for excreting nitrogen-containing wastes, water, carbon dioxide (as HCO_3^-), inorganic salts and hydrogen ions as **urine**. Urine from the kidneys travels down the **ureters** to the **urinary bladder** a muscular organ that temporarily stores urine until it is excreted from the body through the **urethra**.

Each kidney has three regions: the **renal cortex**, the **renal medulla** and the **renal pelvis**. The nephron is the functional unit of the kidney responsible for the formation of urine. Each kidney contains 1-2 million

nephrons with about 80% located entirely in the renal cortex and 20% extending into the renal medulla. Each nephron consists of the **renal corpuscle** where fluid is filtered consisting of the **glomerulus**, the network of capillaries, and **Bowman's capsule** which surrounds it. The renal tubule is the site of reabsorption and secretion and consists of three sections: the **proximal convoluted tubule**, the **loop of Henle**, and the **distal convoluted tubule**.

The nephron performs three functions: glomerular filtration, tubular reabsorption and tubular secretion. **Glomerular filtration** occurs as blood pressure forces water, ions and other small molecules in the blood through the pores in the glomerulus and into Bowman's capsule. The concentration of this filtrate is very close to that of the blood. **Tubular reabsorption** is the process that removes useful materials from the filtrate as it passes through the proximal convoluted tubule with numerous microvilli to increase surface area. About 99% of the filtrate, including water, essential ions, and nutrients, is returned to the blood. Tubular secretion, occurring along the proximal and distal convoluted tubules, removes wastes and excess ions from the body by actively transporting substances that escaped glomerular filtration into the renal tubule. After passing through numerous nephrons, the blood has been cleansed and the urine contains all of the materials not reabsorbed.

The kidneys help regulate the pH of the blood by reabsorbing bicarbonate ions and returning them to the blood, thus restoring the carbonic acid buffer system, and by removing excess hydrogen ions from the blood. The nephrons that extend into the renal medulla are responsible for conserving water by regulating the concentration of the filtrate as it passes through the tubules. Along the descending limb, water leaves the filtrate by osmosis, creating a greater concentration of solutes including NaCl in the filtrate. Just past the loop of Henle, large amounts of NaCl are pumped out of the filtrate into the interstitial fluid of the renal medulla. As the filtrate moves through the collecting ducts in the renal cortex, water moves into the interstitial tissue by osmosis concentrating urea in the filtrate. As the filtrate moves through the collecting duct in the renal medulla, urea moves into the tissue fluid. This causes even more water to leave the filtrate and results in a more concentrated urine.

Three hormones play important roles in adjusting kidney function. **Antidiuretic hormone (ADH)** regulates the amount of water reabsorbed by the distal convoluted tubules and collecting ducts of the nephrons. The hypothalamus responds to changes in the concentration of water in the blood by inversely altering secretion of ADH. The release of a series of enzymes causes changes in proteins that ultimately increase blood pressure, stimulate the thirst center and increase release of ADH. **Aldosterone** increases reabsorption of sodium by the distal convoluted tubules and collecting ducts in response to blood pressure monitored by the **juxtaglomerular apparatus**. **Atrial natriuretic peptide (ANP)** is a hormone released from the heart in response to increased blood volume and pressure relieving both of these conditions with increased urine output and inhibiting ADH and rennin.

The kidneys release **erythropoietin**, a hormone that stimulates the production of red blood cells. They also transform vitamin D from our diet into calcitriol, which promotes absorption and use of calcium and phosphorus by the body.

Renal failure can be acute or chronic but in either case results in irreversible decline in the rate of glomerular filtration. This can lead to acidosis, anemia, edema, hypertension, and accumulation of nitrogenous wastes in the blood, resulting in death within a few days. **Hemodialysis** uses artificial devices to cleanse the blood. **Continuous ambulatory peritoneal dialysis (CAPD)** uses the peritoneum, a selectively permeable membrane, as the dialyzing membrane. Kidney transplants provide the recipient with a healthy kidney.

Urination is the process by which the urinary bladder is emptied. Urine passes down the ureters by peristalsis, when 200ml or more of urine stimulates the stretch receptors of the bladder, they in turn cause the **internal urethral sphincter** to let urine flow from the bladder into the urethra. Voluntary relaxation of the **external urethral sphincter** allows for urination. Lack of voluntary control over urination is called **urinary incontinence**. Incontinence may be caused by injury to nerves, muscle spasms, or infection. **Urinary retention** is the failure to expel urine from the bladder.

Microorganisms can enter the urethra from the rectum or as STDs and cause **urinary tract infections (UTIs)**. These include **urethritis** (infection of the urethra), **cystitis** (infection of the bladder) and **pyelonephritis** (infection of the kidneys), each of which is serious and should be treated immediately.

Key Concepts

- The skin and organs from the circulatory, respiratory, digestive and urinary systems eliminate waste.
- The main function of the urinary system is to regulate the volume, pressure and composition of the blood.
- The kidneys filter wastes and excess materials from the blood, assist in the regulation of blood pH, and maintain fluid balance by regulating the volume and composition of blood and urine.
- Nephrons are the functional units of the kidneys. They carry out glomerular filtration, tubular reabsorption and tubular secretion.
- Glomerular filtration occurs as blood pressure forces water and dissolved substances from the blood in the glomerulus to the inside of Bowman's capsule.
- Tubular reabsorption is the process that removes useful materials from the filtrate and returns them to the blood.
- Tubular secretion removes wastes, ions and large molecules from the blood wastes that escaped glomerular filtration.
- The kidneys help maintain pH balance through the reabsorption of bicarbonate ions, and by removing excess hydrogen ions from the blood.
- The kidneys help conserve water through the production of concentrated urine. Urine concentration is largely determined by the concentration of solutes in the interstitial fluid from the cortex to the medulla of the kidneys.
- Three hormones adjust kidney function.
    - ✓ Antidiuretic hormone (ADH) regulates the amount of water reabsorbed by the distal convoluted tubules and collecting ducts of the nephrons.
    - ✓ Aldosterone increases reabsorption of sodium by the distal convoluted tubules and collecting ducts.
    - ✓ Atrial natriuretic peptide (ANP) decreases water and solute reabsorption by the kidneys by either increasing the permeability of the glomerular filter or by dilating afferent arterioles.
- The kidneys release erythropoietin that stimulates the production of red blood cells and transform vitamin D into its active form, calcitriol.

- Hemodialysis, the use of artificial devices to cleanse the blood, and transplant surgery help during renal failure.
- Urination involves both voluntary and involuntary actions. The lack of voluntary control over urination is called urinary incontinence.
- Microorganisms can enter the urinary system and cause urinary tract infections (UTIs).

Study Tips

This chapter, more than the other chapters describing organ systems, focuses on the function of one organ – the kidney. Furthermore, the emphasis is placed on the mechanism of the nephron and how it maintains homeostasis.

First, glance through the entire chapter reading the boldfaced headings and looking at the figures. Notice that an overview of excretion is presented, then a description of the urinary system in general, followed by a very detailed discussion of the kidney and nephron. The final sections of the chapter discuss some familiar health issues related to the urinary system.

After reading the chapter, review the key concepts listed in the study guide, then begin to focus on the figures for a review of the general anatomy of the urinary system (Figure 16-3), the structure of the kidney (Figure 16-4), and the structure of the nephron (Figure 16-5). Once you have the structure of the system and organs in mind, focus on the function of the nephron. Study Figure 16-8 and then explain the processes to yourself and to a study partner. Once you understand what happens in the nephron, dig deeper. Figure 16-9 explains the mechanism of urine formation and water conservation. Follow the steps and explain the movement of water, ions and urea. Relate the function of the nephron to a dialysis machine. Finally, make a list of the health problems associated with the urinary system, what causes them and possible treatments.

A study partner is very valuable when the material requires an understanding of physiological mechanisms and the relationship between structure and function. Spend time together explaining the various mechanisms shown in the text.

Review Questions

A. Complete the following table indicating the waste and the organ that eliminates it. Try to complete the table on your own, then use Figure 16-2 if you need help.

Organ	Waste
Skin	
	Water, heat, CO_2
	Food waste, metabolic toxins
Kidneys	

B. Match the following organs with their functions: Bowman's capsule, distal tubule, glomerulus, kidneys, loop of Henle, nephrons, proximal tubule, urethra, ureters, and urinary bladder.

1. ______________	transport urine from kidneys to urinary bladder
2. ______________	secretes drugs, H^+, K^+, NH_4^+
3. ______________	capillary bed where filtration occurs
4. ______________	release erythropoietin, activate vitamin D
5. ______________	reabsorbs water, glucose, amino acids, some urea, Na^+, Cl^-, and HCO_3^-
6. ______________	stores urine
7. ______________	functional unit of the kidneys
8. ______________	reabsorbs K^+, Na^+, Cl^-
9. ______________	cuplike structure surrounding the glomerulus
10. ______________	transports urine from the bladder to the outside of the body

C. Label the following structures of the kidney and nephron: ascending limb of loop of Henle, Bowman's capsule, collecting duct, descending limb of the loop of Henle, distal tubule, loop of Henle, nephrons, proximal tubule, renal columns, renal cortex, renal medulla, renal pelvis, and renal pyramid.

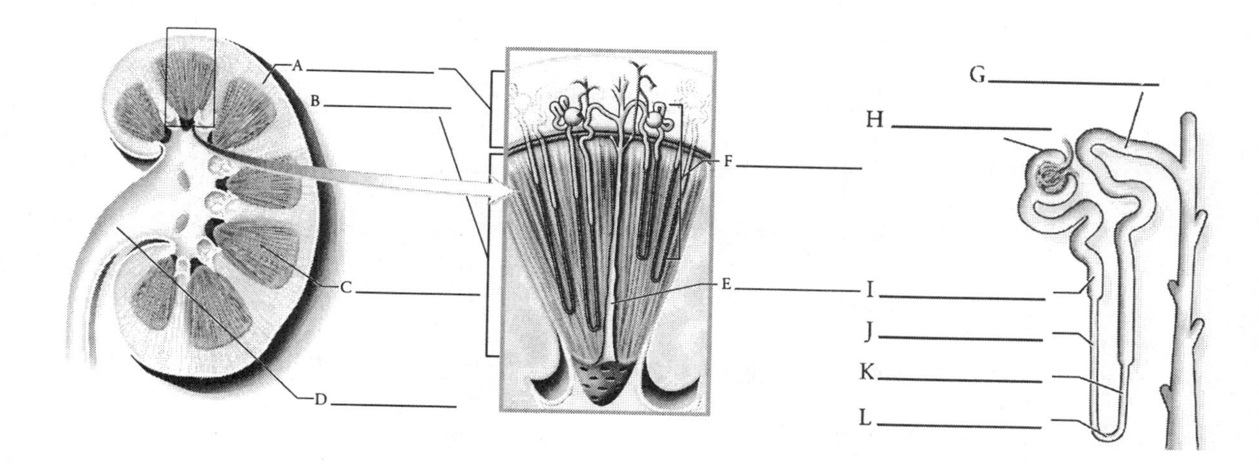

D. Write a sentence describing each of the three functions of the nephron.

1. Glomerular filtration: ____

____

2. Tubular reabsorption: ____

____

3. Tubular secretion: ____

____

E. Complete the following sentences describing the steps that conserve water and concentrate urine and the action of the controlling hormones.

1. In the descending limb and the loop of Henle water ____________ making the urine ____________ concentrated.

2. In the ascending limb of the loop of Henle NaCl is actively pumped out of the tubule fluid whereas water cannot leave the tubules, therefore the urine becomes ____________ concentrated.

3. In the collecting duct, water ____________ making the urine ____________ concentrated.

4. Antidiuretic hormone is manufactured in the ____________ and released from the ____________ ____________. Decreases in the concentration of water in the blood stimulate ____________ secretion of ADH and a more ____________ urine.

5. Aldosterone is released by the ____________ ____________. Aldosterone ____________ reabsorption of sodium by the distal convoluted tubules and collecting ducts which results in ____________ urine.

6. Atrial natriuretic peptide is released from the ____________ (more than one word) stimulated by ____________ blood volume and pressure. Atrial natriuretic peptide ____________ water and solute reabsorption of the kidneys resulting in ____________ volume of urine.

F. Match the treatment or disorder in Column A with its description in Column B.

Column A	Column B
1. acute renal failure	a. infection of the kidneys
2. chronic renal failure	b. hard crystals that can block the flow of urine
3. continuous ambulatory peritoneal dialysis	c. abrupt, complete or nearly complete irreversible kidney damage
4. cystitis	d. surgical implantation of another kidney
5. hemodialysis	e. infection of the urethra
6. kidney stones	f. infection of the urinary bladder
7. kidney transplant	g. progressive decline in kidney function over several years
8. pyelonephritis	h. lack of voluntary control over urination
9. urethritis	i. dialysis fluid flows into the abdominal cavity
10. urinary incontinence	j. use of an artificial kidney machine to cleanse blood

Critical Thinking Questions

Read each of the following questions carefully. If it helps you, underline or list each component to the question. Use complete sentences in your explanation.

1. Explain what materials enter and leave the nephron tubules as urine is formed, water is conserved and urine is concentrated. Recall your knowledge of diffusion, osmosis, and active transport (or review the information in Chapter 3 if necessary).

2. The kidney functions to help maintain acid-base balance in the body. It does this by altering both the hydrogen ion and bicarbonate concentrations. Explain this process using the equation below:

$$CO_2 + H_2O \leftrightarrow H_2CO_3 \leftrightarrow H^+ + HCO_3^-$$

3. When two things are *juxtaposed* that means that they are side by side. How does this definition apply to the juxtaglomerular apparatus? What happens because the afferent arteriole is juxtaposed to the distal convoluted tubule?

4. Hemodialysis is a treatment that uses artificial devices to cleanse the blood. Sometimes this is done using a kidney machine whereby blood flows through tubing made of a selectively permeable membrane surrounded by a dialysis solution. Or it may be done using continuous

ambulatory peritoneal dialysis whereby the dialysis solution enters a person's abdomen and the peritoneum is the membrane. What would happen if the dialysis solution was not changed? What are the advantages and disadvantages to each method?

5. Blood donation is very common and is considered a volunteer activity since it is a true donation and not a sale. However, for-profit companies will purchase plasma and some people use it as a source of income. Most people have two functioning and healthy kidneys. When called upon to donate one for a relative, most people are willing, yet there remains a severe shortage of organs for transplant. Discuss with your classmates or friends what you think the effect would be if someone could sell a kidney to another person in need. How many of the people participating in the discussion would be willing to do this? Use the Internet to answer questions concerning cost and the long waiting list for donor.

PRACTICE TEST

Choose the one best answer to each question that follows. As you work through these items, explain to yourself why each answer you discard is incorrect and why the answer you choose is correct.

1. What is the source of the waste products that are excreted by the urinary system?
   a. excess food
   b. metabolic processes
   c. pollutants in the water and food
   d. undigested fiber

2. What holds the kidneys in place?
   a. connective tissue
   b. epithelial tissue
   c. muscles
   d. tendons

3. What is the function of the urinary bladder?
   a. aid defecation
   b. contain bacteria to fight infection
   c. filter the blood
   d. store urine

4. The ureter connects which part of the kidney to the bladder?
   a. renal cortex
   b. renal medulla
   c. renal pelvis
   d. urethra

5. What is the tuft of capillaries that brings blood to the kidney for filtration?
   a. Bowman's capsule
   b. Bowman's space
   c. glomerulus
   d. proximal arteriole

6. Most nephrons have short loops of Henle and are located almost entirely in the ___________.
   a. renal cortex
   b. renal medulla
   c. renal pelvis
   d. ureters

7. What best describes the process whereby blood pressure forces water and other dissolved substances from the blood into Bowman's capsule?
   a. glomerular filtration
   b. nephritic excretion
   c. tubular reabsorption
   d. tubular secretion

8. What best describes the process that concentrates the urine as it passes the proximal convoluted tubule?
   a. glomerular filtration
   b. nephritic excretion
   c. tubular reabsorption
   d. tubular secretion

9. During which of the following processes would dyes or prescription drugs enter the urine?
   a. glomerular filtration
   b. nephritic excretion
   c. tubular reabsorption
   d. tubular secretion

10. The kidneys help maintain pH balance by ___________.
    a. adding soda lime to the filtrate
    b. dissolving the calcium and sodium crystals in the urine
    c. removing bicarbonate ions from the blood
    d. removing hydrogen ions from the blood

11. Which of the following mechanisms are responsible for water conservation in the tubules as they pass through the renal medulla?
    a. decreasing concentration of solutes in the interstitial fluid
    b. decreasing concentration of plasma proteins in the interstitial fluid
    c. increasing concentration of solutes in the interstitial fluid
    d. decreasing concentration of plasma proteins in the interstitial fluid

12. Antidiuretic hormone (ADH) ___________.
    a. causes the blood to be more dilute
    b. changes in the permeability of the nephron tubules
    c. causes filtration rates to increase
    d. causes the secreting of more sodium ions in the interstitial fluid

13. What structure in the nephron is responsible for monitoring blood pressure?
    a. Bowman's capsule
    b. glomerulus
    c. juxtaglomerular apparatus
    d. proximal convoluted tubule

14. Which is the correct sequence of events leading to the production and release of aldosterone?
    a. angiotensinogen is converted to renin which is converted to aldosterone
    b. renin is converted directly to aldosterone
    c. renin is converted to angiotensin I which is converted to angiotensin II which stimulates the adrenal gland to release aldosterone
    d. renin converts angiotensinogen to angiotensin I which is converted to angiotensin II which stimulates the adrenal gland to release aldosterone

15. How are the kidneys related to red blood cell production?
    a. all blood cells are produced in the kidneys
    b. erythropoietin is released by the kidneys
    c. precursor red blood cells develop in the renal medulla
    d. the kidneys are not related to red blood cell production

16. Acute renal failure is very serious and can lead to death within a few days.
    a. True
    b. False

17. What is used as the dialysis membrane in CAPD?
    a. dialysis tubing
    b. filters in the kidney machine
    c. implanted kidney tubules
    d. peritoneum

18. Which of the following would be symptoms of serious kidney problems?
    a. acidosis and build up of nitrogenous wastes in the blood
    b. anemia and edema
    c. hypertension and water retention
    d. all of the above

19. Which of the following is true concerning kidney transplants?
    a. removal of the donor kidney is very risky, often fatal
    b. there is a shortage of kidneys for transplant
    c. there is a shortage of people needing kidneys
    d. there is nearly a 100% success rate using artificial kidneys for transplant

20. Why are women more susceptible to urinary tract infections than men?
    a. the urethra is shorter in women and the opening is closer to the anus
    b. the urethra is shorter in men but farther from the anus
    c. the urethra combines with the reproductive system in men
    d. the urethra combines with the vagina in women

Short Answer Questions

Read each of the following questions carefully. If it helps you, underline or list each component to the question. Formulate your answer and jot down the main points. Then use complete sentences in your explanation.

1. Explain the interconnectedness of the systems involved in the elimination of waste from the body.

2. Identify the path that urine takes from the first drop that diffuses from the glomerulus until it leaves the body through the urethra.

3. Describe two roles of the kidney that are not directly connected to the formation of urine.

4. Describe voluntary and involuntary control of urination.

Answer Key

Review Questions

A.

Organ	Waste
Skin	Heat, water, urea, HCO_3^-, salt
Lungs	Water, heat, CO_2
Digestive organs	Food waste, metabolic toxins
Kidneys	Nitrogenous wastes, water, salts, excess ions

B.
1. ureters
2. distal tubule
3. glomerulus
4. kidneys
5. proximal tubule
6. urinary bladder
7. nephrons
8. loop of Henle
9. Bowman's capsule
10. urethra

C.
A. renal cortex
B. renal medulla
C. renal pyramid
D. renal pelvis
E. collecting duct
F. nephron
G. distal tubule
H. Bowman's capsule
I. proximal tubule
J. descending limb loop of Henle
K. ascending limb loop of Henle
L. loop of Henle

D.
1. Glomerular filtration filters the blood as pressure forces water and dissolved substances into Bowman's capsule.
2. Tubular reabsorption removes useful materials from the filtrate and returns them to the blood.
3. Tubular secretion removes wastes and excess ions from the blood as well as drugs.

E.
1. In the descending limb and the loop of Henle water <u>leaves</u> making the urine <u>more</u> concentrated.
2. In the ascending limb of the loop of Henle NaCl is actively pumped out of the tubule fluid whereas water cannot leave the tubules, therefore the urine becomes <u>less</u> concentrated.
3. In the collecting duct, water <u>leaves</u> making the urine <u>more</u> concentrated.
4. Antidiuretic hormone is manufactured in the <u>hypothalamus</u> and released from the <u>posterior pituitary</u>. Decreases in the concentration of water in the blood stimulate <u>increased</u> secretion of ADH and a more <u>concentrated</u> urine.
5. Aldosterone is released by the <u>adrenal</u> <u>cortex</u>. Aldosterone <u>increases</u> reabsorption of sodium by the distal convoluted tubules and collecting ducts which results in <u>concentrated</u> urine.

6. Atrial natriuretic peptide is released from the (more than one word) right atrium of the heart stimulated by increased blood volume and pressure. Atrial natriuretic peptide decreases water and solute reabsorption of the kidneys resulting in increased volume of urine.

F.
1. c
2. g
3. i
4. f
5. j
6. b
7. d
8. a
9. e
10. h

Critical Thinking Questions

1. Due to both blood pressure and diffusion, small molecules move from the blood to the filtrate in Bowman's capsule. As materials move down the proximal tubule, water and ions move back into the tissues out of the kidney tubules due to osmosis and diffusion. This makes the urine more concentrated. As materials move up the ascending limb, NaCl leaves the urine and moves into the tissues making the urine less concentrated. Tubular secretion occurs in the proximal and distal convoluted tubules where wastes and excess ions are removed from the blood. As urine passes through the collecting duct, water moves into the tissue due to osmosis concentrating the urine.
2. The kidneys help maintain pH balance by the reabsorption of bicarbonate ions. This means that bicarbonate is reabsorbed and returned to the blood were it can serve as a buffer. In addition, the kidneys remove excess hydrogen ions from the blood.
3. The juxtaglomerular apparatus is a group of cells located where the distal convoluted tubule comes next to the afferent arteriole. This apparatus allows for monitoring of blood pressure. When the blood pressure in the afferent arteriole drops, so does the glomerular filtration rate. The drop in filtration rate, reduces the volume of filtrate within the nephrons. In response, the juxtaglomerular cells in the afferent arteriole secrete renin. This ultimately causes increased blood volume and pressure resulting in an increased volume of filtrate.
4. The dialysis solution is made to draw waste materials from the blood. As time passes the dialysis solution becomes more and more similar to the blood. If the solution were not changed, it would cease to be effective and would no longer cleanse the blood. The kidney machine is inconvenient and time consuming, but the environment is controlled. CAPD allows people to remain at home, but often results in more incidents of infection.
5. Discussion will, of course, vary but might include the black market that might develop, the ethics of making a profit from your own organs that you received free of charge, the sacrifice of those incapable of making their own decisions, sale of organs from minors, etc.

Practice Test

1. b
2. a
3. d
4. c
5. c
6. a
7. a
8. c
9. d
10. d
11. c
12. b
13. c
14. d
15. b
16. a
17. d
18. d
19. b
20. a

Short Answer Questions

1. The lungs and skin eliminate heat, water, and carbon dioxide (as the gas CO_2 from the lungs and as bicarbonate ions from the skin). Skin also excretes salts and urea. Organs of the gastrointestinal tract eliminate solid wastes and many other substances. The urinary system eliminates nitrogenous wastes, water, salts and excess ions. The liver processes metabolic products and toxins while the circulatory system transports the materials.
2. Glomerulus → Bowman's capsule → proximal convoluted tubule → descending limb → loop of Henle → ascending limb → distal convoluted tubule → collecting duct → renal pelvis → ureter → urinary bladder → urethra
3. The kidneys release erythropoietin that stimulates the production of red blood cells in the red bone marrow. The kidneys also transform vitamin D into its active form, calcitriol which promotes absorption and use of calcium and phosphorus by the body.
4. Urine moves through the ureters and into the bladder by involuntary smooth muscle contractions. When 200 ml of urine has accumulated in the bladder, stretch receptors in the bladder wall cause the smooth muscles in the internal urethral sphincter to relax. This lets urine flow involuntarily from the bladder into the urethra. Voluntary urination occurs when the external urethral sphincter is relaxed.

Chapter 17

Reproductive Systems

Objectives

After reading the text and studying the material in this chapter, you should be able to:

- Name and describe the function of each organ of the male and female reproductive systems.
- Describe sperm development and the formation of semen.
- Explain the role of testosterone in the male reproductive process and how it is regulated.
- Describe the sequence of events in the development of the follicle and release of the egg. Explain the hormonal control of this cycle.
- Describe the uterine cycle and the hormones that control it.
- Explain the coordination and interplay of the ovarian and uterine cycles.
- Define menopause and describe the resulting physiological effects.
- Explain the cause and severity of health problems with the female reproductive system.
- Name the stages involved in the human sexual response and the physiological changes that accompany them.
- Name each method of birth control presented in the chapter, explain how it works, why it prevents pregnancy and its relative effectiveness.

Chapter Summary

The function of the reproductive system is to produce the cells that fuse to form a new individual, called **gametes**, and sex hormones. Each gamete contains one-half the number of chromosomes of the adult.

The reproductive strategy of the male is to produce millions of sperm and deliver them to the female reproductive system. The male **gonads** are the **testes** that produce **sperm** and **testosterone**. The testes are held outside the body cavity in the **scrotum**, which allows for temperature regulation. Sperm are produced at the rate of over 100 million per day in the **seminiferous tubules**. Sperm mature in the **epididymis**. During ejaculation, they leave the testes through the **vas deferens,** passing out the **urethra** as they leave the body. **Semen** neutralizes the passageways of the sperm, provides nutrition for them and cause uterine contractions that take them closer to the egg. The fluid of the semen is produced by the **prostate gland**, the paired **seminal vesicles** and the paired **bulbourethral glands**. The penis transfers sperm to the female during ejaculation following an erection.

Spermatogenesis, the development of sperm, occurs in the wall of the seminiferous tubule. Undifferentiated cells divide, reducing the number of chromosomes by one-half, and change in structure to become mature sperm.

Testosterone production is regulated by a negative feedback loop beginning when the hypothalamus releases **gonadotropin-releasing hormone** (**GnRH**). This hormone stimulates the anterior pituitary gland to secrete **luteinizing hormone** (**LH**), which stimulates the production of testosterone by the interstitial cells of the testes. Finally, the rising testosterone level inhibits the release of GnRH to complete the loop.

Sperm production is also controlled by a negative feedback loop. It begins with **follicle-stimulating hormone (FSH)** produced by the anterior pituitary. FSH stimulates the seminiferous tubules to become more sensitive to testosterone, which then increases sperm count. As the seminiferous tubules produce more sperm they also increase production of inhibin, which inhibits the production of FSH, and possibly GnRH. As inhibin levels fall, FSH production increases.

The female gonads are the **ovaries.** They produce **eggs** through a process called oogenesis and the hormones, **estrogen** and **progesterone**. The female also nourishes and protects the developing offspring. The oviducts transport the immature egg from the ovaries to the **uterus** and are most commonly the site of fertilization. The wall of the uterus consists both of a muscular layer and the **endometrium**. The **cervix** forms the end of the uterus that extends into the **vagina.** The vagina both receives the penis during intercourse and serves as the birth canal during delivery. The **vulva** and **clitoris** are female **external genitalia**. In women, the breasts produce milk that nourishes the newborn.

The **ovarian cycle** involves egg production and ovulation. Immature eggs are formed when the woman is a developing fetus and remain dormant until puberty. The immature eggs, called primary oocytes, are surrounded by a layer of granulose cells within primordial follicles. Follicle maturation takes about 10-14 days and usually occurs once a month. At **ovulation**, the mature follicle ruptures releasing the oocyte. The **corpus luteum** forms from the remaining follicle cells and secretes estrogen and progesterone. If pregnancy does not occur, the corpus luteum degerates eliminating hormonal support for the thickened endometrium, which is sloughed off taking the unfertilized ooctye with it. If the sperm reach the egg at the right time, there will be **fertilization** creating a new cell called a **zygote**.

The interplay of hormones coordinates the ovarian and uterine (menstrual) cycles. The first day of menstrual flow is considered day 1 of the **uterine cycle**. At this time, estrogen and progesterone levels are the lowest allowing for the production of FSH by the anterior pituitary. This stimulates development of the follicle in the ovary resulting in an increase of estrogen, which causes the endometrium to thicken. When the egg and follicle are nearly mature, estrogen levels rises rapidly causing a release of LH. This causes the egg to undergo cell division and triggers ovulation. The egg bursts out of the ovary and the corpus luteum continues to develop producing estrogen and progesterone. The rising levels of estrogen and progesterone inhibit the secretion of FSH and LH, preventing the development of a new follicle. If fertilization does not occur, the corpus luteum degenerates resulting in falling levels of estrogen and progesterone. Menstruation follows and the drop in estrogen and progesterone levels cause FSH and LH levels to climb, beginning the cycle again. However, if fertilization occurs, the embryo produces **human chorionic gonadotropin** (**HCG**) that maintains the corpus luteum keeping estrogen and progesterone levels high preventing menstruation.

A woman's fertility gradually declines until ovulation and menstruation completely stop, an event called **menopause**. This results in a drop in estrogen and progesterone levels that have many physiological effects ranging from poor temperature regulation to failure to absorb calcium and increased risk of atherosclerosis.

Problems associated with the female reproductive system vary in severity. Premenstrual syndrome (PMS) may be due to a decrease in the levels of progesterone and includes depression, irritability, fatigue and headaches. High levels of prostaglandins produced by the endometrial cells can cause menstrual cramps. Endometriosis is the condition when tissue from the lining of the uterus is found outside the uterine cavity. Vaginitis is an inflammation of the vagina and pelvic inflammatory disease (PID) is an infection of the pelvic organs, both caused by a variety of microorganisms.

In both men and women, sexual arousal and sexual intercourse involve vasocongestion, where certain tissues fill with blood, and myotonia, where certain muscles show sustained or rhythmic contractions. The four stages of sexual arousal and intercourse are excitement, plateau, orgasm, and resolution.

Birth control prevents pregnancy and in some cases can also reduce the risk of spreading sexually transmitted infections. Abstinence involves refraining from intercourse and most reliably avoids both pregnancy and the spread of STDs. Surgical sterilization involves cutting and sealing gamete transport tubes permanently preventing fertilization. Hormonal contraception works because synthetic female hormones inhibit FSH and LH production preventing the maturing and release of an egg. Intrauterine devices prevent the union of sperm and egg and/or implantation. Barrier methods including the diaphragm, cervical cap, contraceptive sponge and condoms prevent fertilization. Spermicidal preparations kill sperm and therefore prevent fertilization. They are often used in conjunction with more mechanical methods. Fertility awareness is the avoidance of intercourse when fertilization is likely to occur. Morning-after pills are high doses of hormones taken after unprotected intercourse; the mechanism of action is unknown.

Key Concepts

- The gonads (testes and ovaries) produce gametes (sperm and eggs) as well as sex hormones. The male sex hormone is testosterone and the female sex hormones are estrogen and progesterone.
- The male reproductive strategy is to produce millions of sperm and deliver them to the female. On the other hand, the female usually produces only one mature egg per month and is responsible for nourishing and protecting the developing embryo and later fetus.
- The male reproductive system consists of the testes, held outside the body in the scrotum, a series of ducts (the epididymis, vas deferens and urethra), accessory glands (the prostate gland, seminal vesicles and bulbourethral glands), and the penis.
- The development of sperm, spermatogenesis, occurs continually in the seminiferous tubules of the testis. They are stored and mature in the epididymis and leave the body through the vas deferens and urethra during ejaculation.
- Semen is produced by the accessory glands neutralizing the passageway of the sperm, providing them with an energy source and causing uterine contractions.
- Male reproductive processes are regulated by a negative feedback loop involving FSH and LH from the anterior pituitary gland, GnRH from the hypothalamus and testosterone and inhibin from the testis.

- The female reproductive system consists of the ovaries that produce eggs and the female hormones, estrogen and progesterone, the oviducts, uterus, which supports implantation of and the development of the embryo, vagina and external genitalia.
- In women, the mammary glands produce milk.
- The ovarian cycle involves egg production that began during fetal development of the woman and ovulation that usually releases one egg each month from puberty until menopause.
- Hormones that regulate the ovarian and uterine cycles involve a negative feedback loop of FSH, estrogen, LH and progesterone. If fertilization occurs, the young embryo produces HCG which can be detected by pregnancy tests and which maintains the corpus luteum.
- At menopause, both the uterine and ovarian cycles stop, resulting in a drop in progesterone and estrogen levels.
- Possible problems with the female reproductive system include premenstrual syndrome, menstrual cramps, endometriosis, vaginitis and pelvic inflammatory disease.
- The human sexual response is the sequence of events occurring during sexual intercourse, including excitement (period of increased arousal), plateau (continued arousal), orgasm (climax and ejaculation) and resolution (the return to normal functioning).
- Birth control methods may be behavioral, including abstinence and natural family planning (rhythm method); chemical including hormonal contraception, spermicides, and the morning-after pill; mechanical including all barrier methods; and surgical sterilization.
- Only abstinence and the latex condom prevent the transmission of STDs.

Study Tips

This chapter presents the male and female reproductive systems, the development of egg and sperm, the human sexual response, methods of contraception and health problems associated with these systems. Some of the material will be familiar to you and some, especially the role and interaction of hormonal control, may be new. As you read the text material, take time to study each figure and table. Use them to help you identify the male and female reproductive organs, describe their function and explain their role in the reproductive process.

Study the stages of sperm development and then explain Figure 17-3 to a study partner. Sperm production and testosterone levels are controlled by a negative feedback loop of hormones. A negative feedback loop is the process whereby a drop in one hormone causes an increase in another. As you read the text material describing these processes, sketch the relationships among the hormones and compare your drawing to Figure 17-5. Make corrections and then use your drawing as a study tool. Do the same thing for the ovarian and uterine cycles. First be able to explain the development of a follicle and the release of an egg to a study partner. Next describe the build up and breakdown of the endometrial lining and relate this to the ovarian cycle and preparation for fertilization of the egg. Explain the rise and fall of the hormones involved in these processes as shown in Figure 17-9 to a study partner.

Make a list of the potential health problems associated with both the male and female reproductive systems giving a definition of the terms, causes and treatments. As you study birth control, focus on *how* each method prevents pregnancy in relation to the anatomy and physiology of the reproductive systems. Highlight important information in Table 17-6.

Review Questions

A. Use the following terms to label the drawing of the male reproductive system shown below: bulbourethral gland, epididymis, erectile tissue of penis, glans penis, penis, prostate gland, seminal vesicle, testis, urethra, and vas deferens.

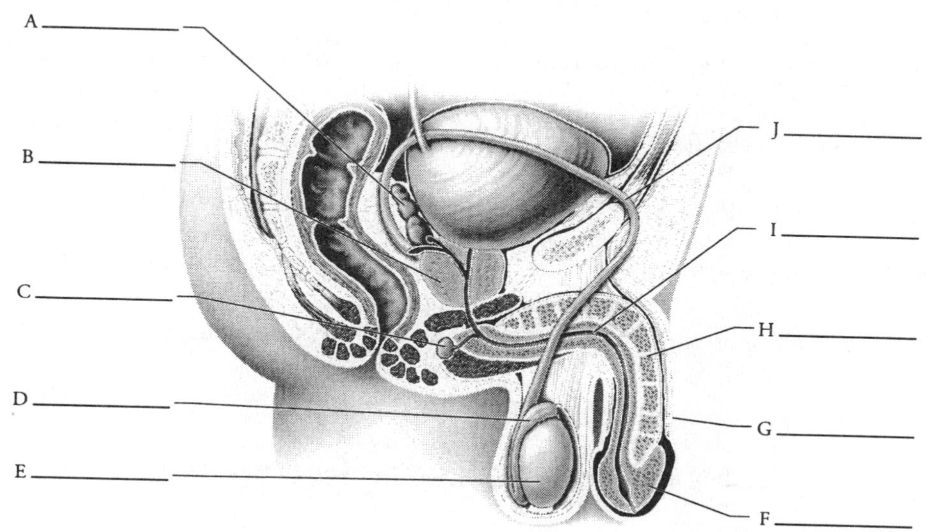

B. Study the text material and then match the organ or structure in Column A with the function in Column B. If you need help, refer to Figure 17-1 and Table 17-1.

Column A	Column B
1. bulbourethral gland	a. produces testosterone and sperm
2. epididymis	b. delivers sperm to female
3. erectile tissue of penis	c. releases secretion just before ejaculation to cleanse urethra
4. glans penis	d. tube common to reproductive and urinary system in men
5. penis	e. produces most of the volume of semen; fluid contains sugar, amino acids and a coagulant
6. prostate gland	f. duct to transport sperm from testes to urethra
7. seminal vesicle	g. spongy connective tissue filling with blood during erection that allows for intercourse
8. testis	h. area of the penis rich in sensory nerves
9. urethra	i. produces alkaline secretions that activate sperm
10. vas deferens	j. ducts where sperm mature and are stored

C. Complete the following paragraph describing the development of sperm.

The sequence of events that gives rise to sperm is called ____________. It occurs in the wall of the ____________ ____________. First, undifferentiated diploid cells divide to produce two new identical ____________. These two spermatogonia follow different paths. One remains in the tubule and divides again to form two new ____________, thus maintaining a constant supply of cells with the potential to develop into new ____________. The other spermatogonium enlarges to form a ____________ ____________. So far, all the cells are diploid. The primary spermatocyte undergoes two cell divisions that result in haploid cells called ____________ ____________. These mature into ____________. When these cells fully mature and are motile, they are called ____________. The mature sperm has three distinct parts: the ____________, ____________, and ____________. A membranous sac that covers the sperm head and contains digestive enzymes is called the ___________.

D. Use the following terms to label the drawing of the female reproductive system shown below: cervix, clitoris, endometrium, labium majora, labium minora, ovary, oviduct, uterine wall, uterus, and vagina.

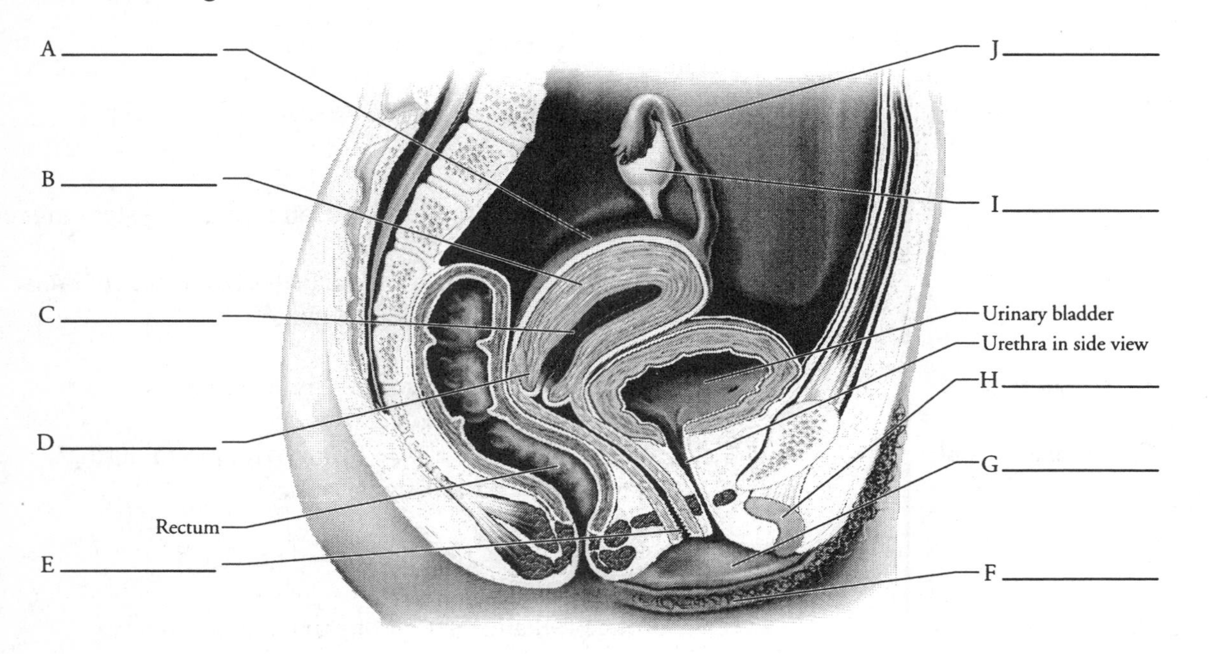

E. Study the text material and then match the organ or structure in Column A with the function in Column B. If you need help, refer to Figure 17-6 and Table 17-3.

Column A	Column B
1. cervix	a. inner skin folds of the external genitalia
2. clitoris	b. houses and nourishes the developing fetus
3. endometrium	c. muscular layer that contracts during delivery
4. labium majora	d. opening of the uterus
5. labium minora	e. produces eggs and the hormones estrogen and progesterone
6. ovary	f. receives penis during intercourse, birth canal
7. oviduct	g. outer skin folds of the external genitalia
8. uterine wall	h. rich in sensory nerves for sexual arousal
9. uterus	i. lining of the uterus, usual site of implantation
10. vagina	j. usual site of fertilization

F. Mark each of the following statements about the ovarian cycle as true or false. If the statement is false, change it so that it is true.

1. ________ Changes in the ovary that produce the egg are called the uterine cycle.

2. ________ The primary follicle contains the primary oocyte. These are formed before the woman is born.

3. ________ As soon as the woman is born, the primary follicles begin the ovarian cycle and release an egg and release an egg about once a month.

4. ________ The primary oocyte divides forming a secondary oocyte and the first polar body.

5. ________ The release of the oocyte from the ovary is called ovulation.

6. ________ The secondary oocyte completes meiosis, a process halving the number chromosomes, when it first leaves the follicle.

7. ________ The cells that formed the outer sphere of the mature follicle remain in the ovary as the fertilized egg.

8. ________ The corpus luteum secretes both estrogen and progesterone.

G. Answer the following questions about the menstrual cycle. Refer to Figure 17-9 in your text if you need to.

1. What happens on the first day of the menstrual cycle? ____________________________

2. As the follicle develops and estrogen levels begin to rise, what happens within the uterus?

   ____

3. What event marks the middle of the uterine and ovarian cycles? ___________________

4. What happens to the endometrial lining if fertilization does not occur? __________

   ____

5. What happens to the endometrial lining if fertilization does occur? _____________

   ____

H. Write the method of birth control in Column A that is described in Column B. Try to complete this review without using your text. If you need help, refer to Table 17-6 in your text.

	Column A	Column B
1.	______________________	Vas deferens is cut and sealed so no sperm are in the semen.
2.	______________________	Small plastic device inserted into the uterus that prevents fertilization and implantation.
3.	______________________	Not having sexual intercourse at all
4.	______________________	Hormone pill taken daily that contains both estrogen and progesterone.
5.	______________________	Oviducts are tied and cut, preventing passage of egg and sperm.
6.	______________________	Close-fitting cover over the penis, usually made of latex
7.	______________________	Covers the cervix and prevents the sperm from entering

Critical Thinking

Read each of the following questions carefully. Circle the most important words in the question before formulating your answer.

1. Diagram or describe the events leading to the release of a single egg at ovulation. Account for each cell that is formed.

2. Your text describes how to perform a breast examination and uses a female model (as do most illustrations). Does breast cancer occur in men? Should men be performing the complementary testicular exam on a regular basis? Begin your search for information at the American Cancer Society web page (http://www.cancer.org). Read about male breast cancer and get the key statistics. Then read about testicular cancer.

3. Several students were discussing various methods of birth control. There were, of course, varying opinions on just about all aspects of the risks and benefits of each method. Use your text, your campus resources and local clinics to gain additional information. Then address these questions/positions raised by the students by discussing the risks and benefits.
   1) One young woman wanted to be "totally natural" and asked, "What's wrong with natural family planning? It keeps chemicals out of my body."
   2) Another student was totally opposed to abortion and felt that all methods of birth control aborted the newly fertilized egg.
   3) A male student said that he always used a condom, not only to prevent pregnancy but also to prevent the spread of STDs. He couldn't think of a better way to prevent both from occurring.

4. Using the Internet or local resources investigate the possibilities of vasectomy reversal. With a small group of peers, discuss the questions one might ask before getting a vasectomy and any actions that might be taken before a vasectomy just in case there may be a desire for another child.

Practice Test

Choose the one best answer to each question that follows. As you work through these items, explain to yourself why each answer you discard is incorrect and why the answer you choose is correct.

1. The cell created by fertilization is called the ____________.
   a. egg
   b. secondary oocyte
   c. sperm
   d. zygote

2. What is the primary male sex hormone?
   a. estrogen
   b. progesterone
   c. testosterone
   d. uterine

3. Where are sperm stored and allowed to mature?
   a. epididymis
   b. seminal vesicles
   c. testis
   d. vas deferens

4. Which organ is common to the reproductive system and urinary system in men?
   a. bladder
   b. vas deferens
   c. rectum
   d. urethra

5. Which gland is responsible for most of the semen released in an ejaculation?
   a. bulbourethral gland
   b. Glans penis
   c. Prostate gland
   d. Seminal vesicles

6. Which gland releases a small amount of fluid just prior to ejaculation to decrease acidity in the urethra caused by urine?
   a. bulbourethral gland
   b. Glans penis
   c. Prostrate gland
   d. Seminal vesicles

7. During spermatogenesis, the secondary spermatocyte develops into ____________.
   a. primary spermatocytes
   b. spermatids
   c. spermatogonia
   d. spermatogozoa

8. The hormone produced in the seminiferous tubules that is an indicator of sperm count is ____________.
   a. follicle stimulating hormone (FSH)
   b. inhibin
   c. luteinizing hormone
   d. testosterone

9. Where is fertilization most likely to occur?
   a. ovary
   b. oviduct
   c. penis
   d. uterus

10. Which of the following statements is true about the uterus?
    a. It becomes the placenta when the fetus is big enough.
    b. It is an endocrine gland secreting progesterone.
    c. It is connected to both the oviducts and the vagina.
    d. It produces the eggs and is the site of fertilization.

11. When do the primary oocytes develop within the primary follicles?
    a. before birth of the woman
    b. between age 10 and puberty
    c. during puberty
    d. from age 14-26

12. What hormones stimulate the growth of mammary glands and the initial steps of the milk-secreting process?
    a. estrogen and progesterone
    b. estrogen and testosterone
    c. progesterone and adrenaline
    d. progesterone and testosterone

13. When does ovulation occur?
    a. just after menstruation
    b. about the middle of the uterine cycle
    c. just prior to menstruation
    d. no one knows for sure

14. What happens to the corpus luteum if pregnancy does not occur?
    a. It degenerates.
    b. It remains and continues to secrete progesterone and estrogen.
    c. It remains and develops a new secondary oocyte.
    d. It remains to bring on menopause.

15. What is the purpose of the endometrium?
    a. hold the sperm until the time of fertilization
    b. produce eggs
    c. provide a suitable site for implantation
    d. provide the muscular strength for delivery

16. Which of the following describes endometriosis?
    a. a common problem following multiple pregnancies
    b. an allergic reaction in the female to the semen of the male
    c. it is the same as menstruation
    d. tissue from the lining of the uterus is found on surfaces outside the uterus

17. Which of the following is most likely to prevent both pregnancy and the spread of STDs?
    a. abstinence
    b. diaphragm
    c. oral contraceptive
    d. sterilization

18. How does the pill containing both estrogen and progesterone prevent pregnancy?
    a. increasing levels of testosterone in the woman
    b. preventing implantation of the fertilized egg
    c. suppressing FSH and LH and preventing ovulation
    d. thickening the mucus of the cervix

19. What is a significant drawback to sterilization in both the male and female?
    a. It is major abdominal surgery.
    b. It is permanent.
    c. It is seldom completely successful and pregnancies often result.
    d. It must be done in both the male and female to prevent pregnancy.

20. What is the primary advantage to using the rhythm method or natural family planning?
    a. It allows for intercourse during 3 weeks of each month.
    b. It does not introduce additional hormones as does the pill.
    c. It is appealing to men because it does not involve a condom.
    d. It is highly effective because of the predictable release of the egg from the ovary.

Short Answer Questions

Read each of the following questions carefully. Review the appropriate section of the text; if necessary; jot down the main points. Then formulate your answer using complete sentences in your explanation.

1. Trace the path of sperm from its development to ejaculation.

2. Compare the effects of luteinizing hormone (LH) and follicle-stimulating hormone (FSH) in men and women.

3. Describe the rise and fall of estrogen and progesterone through a monthly cycle that does not lead to pregnancy.

4. The function of the penis is to deliver sperm to the female. This is accomplished through sexual intercourse. Describe what causes an erection and then explain how medications that treat ED work.

5. Describe how each of the following methods of birth control prevent fertilization: abstinence, estrogen and progesterone pill, diaphragm with spermicidal jelly, and the condom.

ANSWER KEY

REVIEW QUESTIONS

A.
- A. seminal vesicle
- B. prostate gland
- C. bulbourethral gland
- D. epididymus
- E. testis
- F. glans penis
- G. penis
- H. erectile tissue of penis
- I. urethra
- J. vas deferens

B.
1. c
2. j
3. g
4. h
5. b
6. i
7. e
8. a
9. d
10. f

C. The sequence of events that gives rise to sperm is called spermatogenesis. It occurs in the wall of the seminiferous tubule. First undifferentiated diploid cells divide to produce two new identical spermatogonia. These two spermatogonia follow different paths. One remains in the tubule and divides again to form two new spermatogonia, thus maintaining a constant supply of cells with the potential to develop into new sperm. The other spermatogonium enlarges to form a primary spermatocyte. So far, all the cells are diploid. The primary spermatocyte undergoes two cell divisions that result in haploid cells called secondary spermatocytes. These mature into spermatids. When these cells fully mature and are motile, they are called spermatozoa. The mature sperm has three distinct parts: the head, midpiece, and tail. A membranous sac that covers the sperm head and contains digestive enzymes is called the acrosome.

D.
- A. uterus
- B. uterine wall
- C. endometrium
- D. cervix
- E. vagina
- F. labium majora
- G. labium minora
- H. clitoris
- I. ovary
- J. oviduct

E.
1. d
2. h
3. i
4. g
5. a
6. e
7. j

8. c
9. b
10. f

F. 1. F, Changes in the ovary that produce the egg are called the ovarian cycle.
2. T
3. F, When the woman reaches puberty, the primary follicles begin the ovarian cycle and release an egg about once a month.
4. T
5. T
6. F, The secondary oocyte completes meiosis, a process halving the number of chromosomes, when it is penetrated by a sperm.
7. F, The cells that formed the outer sphere of the mature follicle remain in the ovary as the corpus luteum.
8. T

G. 1. Menstrual flow begins.
2. The cells of the endometrial lining begin to divide and thicken in preparation for pregnancy.
3. Ovulation occurs.
4. The blood vessels that were nourishing the endometrial cells collapse, the endometrial cells die, and are sloughed off as menstrual flow.
5. Estrogen and progesterone levels are high enough to prevent endometrial shedding and the embryo implants in the uterine wall.

H. 1. vasectomy
2. intrauterine device
3. abstinence
4. oral contraceptive
5. tubal ligation
6. male condom
7. cervical cap

Critical Thinking Questions

1. A diagram is the easiest way to follow oogenesis, or the development of an egg. During fetal development oogonia divide millions of times by mitosis. In the third month of fetal development, the oogonia enlarge and begin to store nutrients. Each primary oocyte is surrounded by granulose cells. Together the primary oocyte and the granulose cells form the primary follicle. Beginning at puberty, about a dozen primary follicles continue development each month. Within about 10-14 days, the follicle is mature. The primary oocyte then completes the first meiotic division, forming two cells of very different sizes. The larger cell is called the secondary ooctye and the smaller cell is called the first polar body. The first polar body does not develop further. However, the secondary oocyte divides again when fertilization occurs. This cell division results in the second polar body and the mature ovum. The second polar body does not develop. The mature ovum contains one-half the number of chromosomes of the woman who produced it. Since fertilization has occurred, the zygote contains a full complement of chromosomes.
2. Breast cancer is rare in men. It is estimated that 1,450 new cases will be diagnosed in 2004 and that 470 men will die of breast cancer in 2004. The survival rate for men and women with breast cancer is nearly equal. Testicular cancer is one of the most curable forms of cancer with an overall cure rate of over 90%. It occurs in young men usually between the age of 15 and 40. Lance Armstrong is one of the more well-known cases of testicular cancer survival. There will be almost 9,000 new cases of testicular cancer in the United States in 2004 and about 360 men will die of the disease in the same year.

3. 1) The young woman is correct—natural family planning is chemical-free; however, it is also very unpredictable and may not be the best choice for those who absolutely do not want an unexpected pregnancy.
   2) Although the IUD and the morning after pill may be considered to cause an abortion since the egg may already be fertilized, all other methods of birth control prevent fertilization.
   3) This person is correct; a latex condom is a reliable means of preventing both pregnancy and the spread of STDs. Abstinence is the only other method that prevents both.
4. Careful consideration should be given before having a vasectomy and it should be considered permanent sterilization. However, reversal procedures have been effective. As an alternative, the removal of sperm directly from the testis by needle aspiration for use in alternative insemination is a possibility.

Practice Test

1. d	11. a
2. c	12. a
3. a	13. b
4. d	14. a
5. d	15. c
6. a	16. d
7. b	17. a
8. b	18. c
9. b	19. b
10. c	20. b

Short Answer Questions

1. Sperm are produced in the seminiferous tubules found in the testes. They next enter the epididymis where they mature and are stored. When the sperm are mature, they may be stored in the vas deferens for a short time but are carried out of the body through the urethra during ejaculation.
2. In men, LH stimulates the interstitial cells of the testis to produce testosterone. In women, it triggers ovulation and causes the formation of the corpus luteum. In men, FSH enhances the formation of sperm and it women it has a similar effect, stimulating the development of a follicle in the ovary.
3. The follicle begins to develop caused by increasing levels of FSH. As it develops, it produces estrogen which steadily rises until ovulation when it drops. Ovulation causes a surge in LH, which maintains the cells of the corpus luteum. The corpus luteum secretes estrogen and progesterone causing a rise in both. When there is no pregnancy, the corpus luteum degenerates causing estrogen and progesterone levels to decline. This stimulates FSH and the cycle begins again.
4. An erection results when blood fills the spongy tissue of the penis, usually following sexual arousal. Without an erect penis, sexual intercourse is nearly impossible and the sperm cannot travel to the oviduct. Erectile dysfunction is a man's inability to achieve and maintain an erection. Drugs used to treat ED help a man achieve and maintain an erection when sexually aroused. They prolong the effect of nitric oxide that is naturally released during sexual arousal. Nitric oxide increases blood flow to the penis and causes an erection.
5. Abstinence prevents the sperm from entering the woman, thus there is no opportunity for fertilization. The estrogen and progesterone pill prevents ovulation so that, although sperm are in the woman, there is nothing to fertilize. The diaphragm in combination with spermicidal jelly uses a barrier method to prevent most sperm from entering the cervix and a chemical method to kill the sperm that might be able to get past the barrier. A condom catches the sperm following ejaculation and if removed carefully will prevent fertilization.

Chapter 17A

Sexually Transmitted Diseases and AIDS

Objectives

After reading the text and studying the material in this chapter, you should be able to:

- Describe the long-lasting effects of STDs.
- Differentiate between the treatment and/or cure for STDs caused by bacteria and those caused by viruses.
- Explain how STDs are transmitted and what tissue types are most vulnerable.
- Describe the symptoms of chlamydia, gonorrhea, syphilis, genital herpes and genital warts.
- Describe the three stages of syphilis.
- Diagram and label the structure of HIV and explain how it infects a cell and infuses its RNA into the host cell.
- List the most common means of HIV transmission.
- Describe the progression of an HIV infection and its diagnosis as AIDS.
- Explain the relationship between the decline in helper T-cells and the rise in viral load.
- Describe current treatment for AIDS and explain why these treatment methods are effective.
- Identify safe and risky behaviors regarding the transmission of STDs.

Chapter Summary

Sexually transmitted diseases (STDs) are transmitted by sexual contact. They affect women more severely than men, causing sterility, ectopic pregnancy and cervical cancer. STDs can be caused by bacteria or viruses.

Chlamydia, gonorrhea and syphilis are caused by bacteria and can be cured with antibiotics. **Chlamydia** is the most common reportable infectious disease in the United States with an estimated 3 to 10 million new cases annually. Chlamydia is highly contagious and causes no noticeable symptoms in most people. *Chlamydia trachomatis* is the causative bacteria. In men it usually infects the urethra, causing a burning sensation upon urination. In women the symptoms include vaginal discharge, bleeding of the cervix, and pain during intercourse. If the infection scars the vas deferens or the oviducts, it can result in sterility. If the oviduct is only partially blocked and fertilization occurs, then an ectopic pregnancy may result. A chlamydia infection can be transferred to a fetus at birth or can cause the protective membranes around the fetus to rupture killing the fetus. Detection of *Chlamydia* DNA in the urine leads to positive identification and treatment with antibiotics.

Gonorrhea is caused by *Neisseria gonorrhea*. Gonorrhea bacteria are generally transferred by direct contact from one infected mucus membrane to another. Symptoms are not always present, but usually include urethritis and discharge from the urethra. Urination becomes increasingly painful as the infection spreads. Gonorrhea can cause sterility and ectopic pregnancies similar to chlamydia. Gonorrhea can be diagnosed by a urine test to detect the *Nesseria* DNA or by a smear of cells taken from the mucosa. In the past, gonorrhea could be easily cured with antibiotics, but now some strains of the gonococcal bacteria have become drug resistant.

Syphilis is caused by the bacterium *Treponema pallidum*. Infection is only by direct contact with an infected sexual partner. The bacteria can invade any mucous membrane, enter through a crack in the skin or be transferred across the placenta to the growing fetus. If untreated, syphilis progresses through three stages. The first stage is characterized by a chancre that forms at the site of contact. The chancre is not always noticed and may disappear in a few weeks, but the person remains infected and contagious. However, if diagnosed at this time, treatment with antibiotics is effective. If not detected and treated, syphilis can progress to the second stage, which is characterized by a rash covering the entire body that develops bumps that rupture oozing fluid with thousands of bacteria in it making this stage the most contagious. The infected person may experience flulike symptoms, warty growths on the genitals, and hair may fall out. If diagnosed by a blood test at this stage, treatment by antibiotics possible. Again, the symptoms may go away without treatment. In the third stage, lesions called gummas may appear on the skin or internal organs including the aorta. The bacteria that cause syphilis can also infect the nervous system damaging the spinal cord, brain and optic nerve. In the third stage, syphilis becomes more difficult to diagnose and very difficult to treat. Treatment requires massive doses of antibiotics over a prolonged period of time.

The symptoms caused by viral STDs can be treated but not cured. **Genital herpes** is caused by herpes simplex viruses (HSV). Herpes simplex virus is very contagious and can enter the body through mucous membranes or breaks in the skin. The virus causes fever, aching muscles and swollen glands in the groin. Blisters appear and may ulcerate leaving sores. These symptoms may subside and become dormant. Unfortunately, most people experience no symptoms and therefore do not know they are contagious. The herpes infection sometimes can spread to a growing fetus and cause miscarriage or stillbirth or it may be transmitted to the fetus during delivery. Delivery by Cesarean section avoids exposure of the baby to the virus. The presence of HSV-2 can be detected with blood tests and DNA identification. Antiviral drugs can help ease the symptoms and can reduce the frequency of recurrences. Unfortunately, strains resistant to the antiviral drugs are appearing.

Genital warts can lead to cervical, penile or anal cancer. They are caused by several human papilloma viruses (HPVs). Genital warts is the most common of the viral STDs in the United States. Diagnosis of warts is usually on the basis of their appearance and by the appearance of precancerous cells appearing in a Pap test. Treatments for genital warts are intended to kill the cells that contain the virus and include cold and hot cautery, laser, surgery and chemical treatment.

Acquired immune deficiency syndrome (**AIDS**) is a compilation of symptoms caused by the **human immunodeficiency virus** (**HIV**). The immune system is compromised when the helper T-cells are destroyed by the infection allowing opportunistic infections to settle in and cause death. HIV/AIDS is a global epidemic.

HIV consists of RNA, a reverse transcriptase enzyme, a protein coat and an envelope with glycoprotein units embedded in a lipid membrane. The glycoprotein units that bind the virus to the host cell at CD4 receptor sites are found predominantly on helper T-cells and macrophages. Once attached, the HIV fuses with the host cell membrane and releases its contents into the host cell. Then the HIV protein coat degrades and the viral RNA and **reverse transcriptase** are released into the host cell cytoplasm. This causes the viral RNA to be rewritten as DNA. The new viral DNA is spliced into the host DNA. After the viral DNA has been incorporated into the host cell DNA, it is called the **HIV provirus**. The HIV genes can remain dormant or may be activated. If activated they make more HIV RNA which directs the synthesis of the viral proteins. These proteins may be reassembled to form multiple copies of new HIV which can leave the host cell and are infectious.

HIV is found in all bodily secretions but is only transmitted by blood, semen, vaginal secretions and breast milk. Major modes of transmission are unprotected sexual contact allowing the virus to enter through the mucous membranes or tears in the tissues, contact with contaminated blood through intravenous drug use, and from an infected mother to her fetus. It cannot be transmitted by casual contact.

An HIV infection progresses through several stages as helper T-cell numbers decline and the viral load increases. During the initial stages of infection, the virus actively replicates and the circulating level of the HIV rises. Usually within 6 weeks, the body's immune system produces antibodies against the virus. These may be detected and the person is now said to be HIV-positive. Many people have no symptoms when first infected; others have typical flulike symptoms of other viral infections while a few show signs of encephalopathy (brain inflammation). The next stage is often asymptomatic as helper T-cell production increases. However, the virus is in the lymph nodes infecting millions of cells.

As helper T-cells numbers gradually drop, the disease symptoms set in. Wasting syndrome is an unexplained weight loss, lymphadenopathy (swelling of the lymph nodes), and neurological symptoms including weakness, dementia, and paralysis. As T-cell numbers continue to decline, early signs of immune failure including thrush, shingles, and hairy leukoplakia can be seen.

The final stage of AIDS may take up to 10 years to develop. An AIDS diagnosis is made when an HIV^+ person develops one of the following conditions: 1) helper T-cell count below $200/mm^3$ of blood, 2) one of the most common opportunistic infections occur, 3) loss of more than 10% of body weight, or 4) dementia. A fungal infection of the lungs caused by *Pneumocystis carinii* is the leading cause of death in AIDS patients.
Since the cause of death in most AIDS patients is an opportunistic infection, treating or preventing the infection can improve and lengthen the quality of life of the AIDS patient. At the same time it is important to slow the progression of the HIV infection with antiviral drugs. These drugs, often given in combination, include protease and fusion inhibitors. They are not a cure and the virus remains latent in the infected person.

Key Concepts

- STDs are extremely common and can lead to sterility, ectopic pregnancy, and cervical cancer.

- STDs are often without symptoms.
- STDs caused by bacteria can be cured with antibiotics.
- Chlamydia, caused by the *Chlamydia trachomatis* bacteria, is the most common reportable infectious disease in the United States. It is transmitted through contact of mucous membranes and usually infects the urinary tract or reproductive systems.
- Gonorrhea is caused by the bacterium *Neisseria gonorrhea*. It infects the mucous membranes of the genital or urinary tract, throat or anus. Symptoms include urethritis and discharge. Although it can be treated with antibiotics, several strains have become drug resistant.
- Syphilis is caused by the *Treponema pallidum* bacterium. It invades the mucous membrane or enters through a break in the skin. Untreated, syphilis proceeds through three stages of increasing severity.
- STDs caused by viruses can be treated but not cured.
- Genital herpes is caused by the herpes simplex viruses and lead to painful, fluid-filled blisters. Genital herpes is most contagious when the active sores are present. The symptoms can be treated with antiviral drugs, but it cannot be cured.
- Genital warts caused by several human papilloma viruses, can lead to cervical, penile or anal cancer. Genital warts is the most common viral STD in the United States. Treatments are intended to kill the cells containing the virus.
- The human immunodeficiency virus (HIV) causes acquired immune deficiency syndrome (AIDS) resulting in a compromised immune system that allows for opportunistic infection.
- HIV consists of RNA and reverse transcriptase enzymes encased in a protein coat and wrapped in a lipid envelope. The envelope has glycoprotein units that fit into host cell receptor sites.
- HIV enters the cell, rewrites its RNA as DNA, inserts the new DNA into the host chromosome and replicates.
- HIV is transmitted through unprotected sexual contact, intravenous drug use, or from an infected mother to her child before or during birth.
- Targets of the HIV infection include the helper T-cells, macrophages and the brain. The HIV infection progresses through several stages as helper T-cell numbers decline and the immune system is compromised.
- AIDS is the final stage of an HIV infection. The diagnosis is made when several symptoms are present at one time and usually include fungal, protozoan, bacterial and viral infections as well as cancers.
- Treatments for HIV infection are designed to block specific steps in the HIV life cycle thus slowing the infection.

Study Tips

The information found in this chapter is usually of great interest to students because STDs are most common in people under age 25. You may be interested in augmenting the text material with current publications or reliable information from the Internet.

The information in this chapter can be organized onto note cards or can be summarized in a flow chart similar to an organizational chart seen in industry. First identify each disease as caused by a bacterium or virus. Then list the disease, the causative organism, symptoms and progress of the disease. Finally include how diagnosis is made and the usual treatment.

HIV infection and progress to AIDS are described in detail. Be sure to understand the reproductive cycle of a virus and its dependence on a host cell as summarized in Figure 17A-8. Using Figure 17A-10, describe the progress of the HIV infection as it relates to a gradual weakening of the immune system. Finally, define AIDS and describe how a diagnosis is made.

Devote one note card or section of the organizational chart to safe and risky behaviors. Review how the diseases are transmitted and then identify ways to prevent spread of the disease. Identify risky behaviors that should be avoided.

Review Questions

A. List three long-term consequences of an STD infection.

1. ______________________________

2. ______________________________

3. ______________________________

B. Complete the following table.

STD	Causative Organism	Bacteria or Virus	Treatment Is there a cure? Yes or No
Chlamydia	*Chlamydia trachomatis*		
Gonorrhea		Bacteria	
Syphilis			Large doses of antibiotics, difficult to cure if beyond stage two
Genital herpes			Antiviral drugs relieve symptoms, no cure
Genital warts	Several human papilloma viruses		
AIDS			

C. Match the term in Column A with the function in Column B.

Column A	Column B
1. CD4	a. cells targeted by the HIV
2. glycoprotein	b. encases the viral contents
3. helper T-cells and macrophages	c. genetic material in HIV
4. protein coat	d. receptor site that fits the HIV surface proteins
5. RNA	e. proteins found on the viral surface

D. Number the following steps so that they are in the correct order of the life cycle of the HIV.

1. ______ Many copies of the HIV RNA are made.
2. ______ Reverse transcriptase rewrites the viral RNA as viral DNA.
3. ______ HIV attaches to a CD4 receptor site.
4. ______ HIV RNA and proteins are reassembled to form new viruses.
5. ______ HIV fuses with host cell membrane and releases contents into host cell.
6. ______ New viruses leave the host cell acquiring an envelope as they do so.
7. ______ HIV protein coat degrades and contents mix with host cell cytoplasm.
8. ______ HIV DNA is inserted into host DNA.

E. List the three ways that HIV is most commonly transmitted.

1. ______________________________
2. ______________________________
3. ______________________________

F. Answer the following questions regarding HIV and AIDS.

1. Why is HIV considered to be a retrovirus? ______________________________

____

2. Why is it so damaging that the HIV attacks helper T-cells? You may need to refer to the chapter on immunity for more information and detail. ______________________________

____

3. How soon after initial infection with HIV can someone be identified as HIV-positive? Why is this important? ____

____

Critical Thinking

Read each of the following questions carefully. Circle the most important words in the question before formulating your answer.

1. Why aren't there cures for all of the sexually transmitted diseases?
2. Design a questionnaire that asks about the causes, transmission, symptoms and cures of common STDs. Give it to a group of students not in this class, possibly those in your dormitory or in another class, and summarize the results.
3. Use information from your health clinic or the Center for Disease Control website to identify the statistics surrounding various STDs. Try to identify the number of people in your state or city who have been positively diagnosed with a specific STD. Is there a trend? Are there new treatments? New approaches to prevention?
4. Use the Internet to learn more about the development of new antiviral drugs. Many of these drugs will include information on AIDS treatment, however, antiviral drugs are used in other contexts as well.
5. Explain what AIDS stands for and what the term syndrome means. If asked whether a patient died of AIDS, what would you say?

Practice Test

Choose the one best answer to each question that follows. As you work through these items, explain to yourself why each answer you discard is incorrect and why the answer you choose is correct. Do not refer to your text or notes as you take this practice test.

1. What age group has the highest incidence of STD infection?
   a. 15-25 years
   b. 25-35 years
   c. 35-50 years
   d. all have about the same incidence of infection

2. What is the most frequently occurring bacterial STD?
   a. AIDS
   b. Chlamydia
   c. gonorrhea
   d. syphilis

3. Chlamydia can be cured by ___________.
   a. antibiotics
   b. antiviral drugs
   c. creams put onto the lesions
   d. it cannot be cured

4. What is the long-term reproductive consequence of either Chlamydia or gonorrhea?
   a. chronic urethritis
   b. inflammation of the cervix
   c. pain during intercourse
   d. scarring of the vas deferens and oviduct

5. Which of the following is true regarding gonorrhea?
   a. It always shows clear symptoms.
   b. It can spread causing inflammation in several reproductive organs.
   c. It can be cured by avoiding sexual intercourse.
   d. It is caused by the virus, *Gonorrheus*.

6. What is a likely result when the oviduct is only partially blocked and sperm can pass but an egg cannot?
   a. ectopic pregnancy
   b. endometriosis
   c. menstrual cramping
   d. sterility

7. What is the organism that causes syphilis?
   a. herpes simplex virus
   b. HIV
   c. *Neisseria gonorrhoeae*
   d. *Treponema pallidum*

8. During which stage of syphilis is it most likely to be cured?
   a. first
   b. second
   c. third
   d. all stages are equally treatable

9. Damage to nerve tissue such as the optic nerve and brain occur during which stage of syphilis?
   a. first
   b. second
   c. third
   d. it never occurs in syphilis

10. What is the cause of genital herpes?
    a. herpes simplex virus type 1
    b. herpes simplex virus type 2
    c. *Herpes zoster*
    d. HIV

11. Which of the following will cure genital herpes?
    a. antibiotic therapy
    b. antiviral drugs
    c. creams put onto the lesions
    d. it cannot be cured

12. Which of the following is the most common viral STD in the United States?
    a. AIDS
    b. Chlamydia
    c. genital herpes
    d. genital warts

13. Which of the following is linked to cervical, penile and anal cancer?
    a. herpes simplex virus
    b. HIV
    c. human papilloma virus
    d. *Treponema pallidum*

14. Caesarean section delivery is helpful in preventing the transmission of ____________ to the newborn.
    a. AIDS
    b. herpes simplex virus
    c. syphilis
    d. all of the above

15. Which of the following statements is true regarding the human immunodeficiency virus?
    a. It is able to live on any surface such as a table top.
    b. It must link to a bacterium to reproduce.
    c. It must reproduce inside another cell.
    d. It requires a second virus to sexually reproduce.

16. Reverse transcriptase is an enzyme in the virus that ____________.
    a. attaches to the host plasma membrane
    b. dissolves the plasma membrane of the host cell
    c. rewrites the host DNA as viral RNA
    d. rewrites the viral RNA as DNA

17. Full-blown AIDS is present when ____________.
    a. all antibiotic treatments have failed
    b. helper T-cell count falls below $200/mm^3$ of blood in an HIV^+ person
    c. someone contracts pneumonia with a dry cough
    d. the HIV first enters the cell

18. CD4 is the receptor site found on helper T-cells and on macrophages that is used for ___________.
    a. control and development of new viruses
    b. development of antibiotics
    c. identification of cells within the immune system
    d. transmission of the genetic material between virus and host

19. In general, what is the relationship between the number of helper T-cells and the viral concentration in the blood?
    a. as the concentration of HIV decreases, the number of helper T-cells increases
    b. as the number of HIV increase, the number of helper T-cells decreases
    c. as the number of HIV increase, so do the numbers of helper T-cells
    d. there is no relationship between the concentration of HIV and the number of helper T-cells

20. Which of the following characteristics of HIV delay the development of effective antiviral medication?
    a. high mutation rate
    b. multiple strains of the virus
    c. reproduction within host cells
    d. all of the above

SHORT ANSWER QUESTIONS

Read each of the following questions carefully. Jot down the main points you want to include in your answer. Then write a well-organized explanation.

1. Identify several reasons for the rapid spread of STDs. Which of these can you control?

2. During World War II, soldiers would contract syphilis while in another country and it would go untreated. For many years, there would be no symptoms and then the veteran might develop an aneurism, difficulty walking, insanity, or blindness. Use your knowledge of the progression of syphilis to explain how that can happen.

3. Describe the life cycle of HIV from the time it enters a host cell until new viruses leave the host cell.

4. Describe how intravenous drug use can lead to the transmission of HIV.

5. What are the characteristics of HIV that make it particularly difficult to treat?

Answer Key

Review Questions

A. 1. Sterility due to blockage of the vas deferens or the oviduct.
   2. Ectopic pregnancy usually in the oviduct
   3. Cancer of the cervix, penis or uterus

B.

STD	Causative Organism	Bacteria or Virus	Treatment Is there a cure? Yes or No
Chlamydia	*Chlamydia trachomatis*	Bacteria	Antibiotics, yes can be cured
Gonorrhea	*Neisseria gonorrhea*	Bacteria	Antibiotics, yes can be cured
Syphilis	*Treponema pallidum*	Bacteria	Large doses of antibiotics, difficult to cure if beyond stage two
Genital herpes	Herpes simplex virus type 2 (HSV-2)	Virus	Antiviral drugs relieve symptoms, no cure
Genital warts	Several human papilloma viruses	Virus	Freezing, burning, laser surgery of infected cells, no cure
AIDS	Human immunodeficiency virus (HIV)	Virus	Antiviral drugs in combination, antibiotics to fight opportunistic infections, no cure

C. 1. d
   2. e
   3. a
   4. b
   5. c

D. 1. 6
   2. 4
   3. 1
   4. 7
   5. 2
   6. 8
   7. 3
   8. 5

E. 1. unprotected sexual activity
   2. intravenous drug use and the sharing of needles and the blood that is on them
   3. infected mother to child before or during birth or through breast milk

F. 1. HIV carries RNA that must be rewritten into DNA. Usually RNA is copied from DNA, so this mechanism is backward or *retro* to what is most common.
   2. Helper T-cells trigger the rest of the immune system. Therefore, by damaging helper T-cells, the body is left vulnerable to all infections.

3. The antibodies produced to fight HIV can be detected by laboratory tests. However, the body takes time to produce the antibodies and to have a level that is detectable. The soonest that the antibodies can be detected is usually 8 weeks, but it may be 6 months or more. During this time the infected person may not have definitive symptoms and cannot have confirmation that he or she is infected and contagious. The virus can be unknowingly spread during this time.

Critical Thinking Questions

1. The STDs discussed in your text are caused by both bacteria and viruses. Most antibiotics are effective only against bacteria and are not effective against viruses. Also, several of the organisms that cause STDs have developed a drug resistance making even antibiotic treatment ineffective.
2. The questionnaire will vary but should include knowledge level questions about STDs and their spread as well as attitudinal questions concerning responsibilities.
3. The Center for Disease Control and Prevention has statistics by state and city that show the incidence of each STD by age group. They also highlight treatment plans, prevention strategies and new research.
4. An Internet search will provide basic information on the current anti-HIV drugs and how they are used. Antiviral drugs are also used to prevent the flu and to treat other diseases caused by the herpes virus such as shingles.
5. AIDS stands for Acquired Immune Deficiency Syndrome. A syndrome is a set of symptoms that tend to occur together. AIDS is the collection of symptoms resulting from a compromised immune system. AIDS patients die of other infections or cancer.

Practice Test

1. a
2. b
3. a
4 d
5. b
6. a
7. d
8. a
9. c
10. b
11. d
12. d
13. c
14. b
15. c
16. d
17. b
18. b
19. c
20. d

Short Answer Questions

1. Sexually transmitted diseases have no symptoms or have symptoms that are common of other infections. Sometimes these symptoms diminish and the causative organism becomes latent. Unprotected sex with multiple partners is more common making it more difficult to identify people who might have become infected if an STD is identified. Finally, the bacteria that cause some of the STDs are becoming drug resistant making them difficult to cure and remove from the population.
2. If untreated, syphilis progresses through three stages. The first stage is characterized by a chancre that is often difficult to see and which may disappear. The second stage is characterized by a reddish brown rash and flulike symptoms. If untreated, the symptoms may go away but syphilis will progress to the third stage when bacteria can infect the nervous system, damaging the brain and spinal cord resulting in difficulty walking, blindness, paralysis or insanity.
3. HIV attaches to a CD4 receptor site on the host cell. The virus fuses with the cell membrane and releases its contents into the cell. The viral coat degrades releasing the reverse transcriptase into the host cell cytoplasm. The viral DNA is inserted into the host cell's DNA. Then many copies of single-stranded HIV RNA are made. They direct the synthesis of the viral proteins. The HIV proteins are reassembled to form new viruses that acquire an envelope as they leave the host cell ready to attach to other CD4 receptor sites.

4. When sharing a single syringe of a drug, intravenous drug users will draw a small amount of blood into the needle to prevent even one drop of the drug from spilling as it is passed from person to person. Even when the drug is not shared and only the needle is shared, the blood on the needle from an injection into the vein will carry blood borne viruses.
5. The mutation rate of HIV is very high. The proteins on the envelope change with various mutations, making it very difficult to develop an antibody response that can neutralize the virus. Even without mutation, there are several strains of HIV and the development of an antibody that could recognize all of the strains would be difficult to develop. It is very difficult to treat anything that is inside a host cell without damaging the patient.

Chapter 18

Development and Aging

Objectives

After reading the text and studying the material in this chapter, you should be able to:

- Differentiate between prenatal and postnatal periods of development.
- List the three periods of prenatal development and the main events of each.
- Describe the process of fertilization including how the sperm enters the egg and the prevention of polyspermy. Explain how identical, conjoined and fraternal twins are produced.
- Describe the formation of the morula and blastocyst.
- Explain the process of implantation and define an ectopic pregnancy.
- List the extraembryonic membranes and describe their role in development.
- Describe the formation of the placenta.
- Describe the development of the neural tube and problems that can arise in the development of the central nervous system.
- Explain what controls the development of sex organs and when it happens in development.
- Describe fetal circulation and how it differs from postnatal circulation; include the special structures and explain what organs are being bypassed and why that can happen in the fetus.
- List the stages of labor and describe the birthing process. Explain control over milk production and letdown.
- Describe postnatal growth and development and the factors that influence aging and high quality of life through old age.

Chapter Summary

Human development begins with **fertilization**, the union of a sperm and an egg. The period of development before birth is the **prenatal period** and the period after birth is the **postnatal period**.

Prenatal development is divided into three periods. The **pre-embryonic period** begins with fertilization and continues through the second week. Fertilization occurs in the oviduct within 24 hours after ovulation. Only a few of the sperm released near the cervix during intercourse actually make it into the oviduct where they encounter the corona radiata, the layer of cells surrounding the secondary oocyte. Enzymes from the acrosome disrupt the attachments between cells of the corona radiata and the zona pellucida allowing the sperm to reach the oocyte. A change in the electrical charge of the membrane of the oocyte and a hardening of the zona pellucida prevent **polyspermy**, fertilization by more than sperm. The secondary oocyte now completes its second meiotic division and is considered an ovum. The fertilized ovum is a **zygote**, containing equal amounts of genetic material from the mother and father. **Cleavage** is a series of mitotic cell divisions beginning as the zygote moves down the oviduct. By day 4, the pre-embryo is a solid ball of 12 or more cells called a **morula**. Each cell of the morula is capable of forming a complete individual. Thus, if any dividing cells separate from the morula, **identical twins** are formed. If this separation is incomplete, then the twins are **conjoined**. **Fraternal twins** are formed when two secondary oocytes are released from the ovaries and fertilized by different sperm. They are non-identical. By day 6, the pre-embryo is a hollow ball of cells called a **blastocyst** made of two layers: the **inner cell mass** that becomes the embryo proper and some of the embryonic membranes and the **trophoblast**, the layers of cells that give rise to the extraembryonic membrane.

About 6 days after fertilization, the blastocyst becomes imbedded into the **endometrium**, a process termed **implantation**. Usually implantation occurs high on the back wall of the uterus. An **ectopic pregnancy** results if the blastocyst implants outside the uterus. When conditions are right, the blastocyst completes implantation by the end of the second week of the pre-embryonic period. The blastocyst produces **human chorionic gonadotropin (HCG)**, a hormone that maintains the corpus luteum and stimulates continued production of progesterone, essential for maintenance of the endometrium.

Extraembryonic membranes lie outside the embryo and protect and nourish it. The **amnion** encloses the embryo in a cavity filled with amniotic fluid. The **yolk sac** is the site of early blood cell formation and contains **primordial germ cells** that migrate to the gonads where they will become sperm or oocytes. The **allantois** is a small membrane whose blood vessels become part of the **umbilical cord**. The **chorion** is the outermost membrane. It develops from the trophoblast and becomes the embryo's contribution to the **placenta**.

The placenta functions to allow oxygen and nutrients to diffuse from maternal blood into embryonic blood and wastes such as carbon dioxide and urea from embryonic blood into maternal blood. The placenta also produces HCG, estrogen and progesterone necessary for continued pregnancy and responsible for **morning sickness**. **Chorionic villi**, fingerlike processes that grow into the endometrium, provide exchange surfaces for diffusion of materials between the mother and embryo. Although many substances cross the placenta, maternal and fetal blood do not mix. If the placenta forms in the lower half of the uterus covering the cervix, a condition called **placenta previa**, then delivery by Cesarean section is necessary.

The embryonic period extends from the third to eighth week. By the end of the embryonic period all organs have formed and the embryo has a distinctly human appearance. **Morphogenesis** is the development of the body form following the third week of development. The inner cell mass becomes a platelike structure called the **embryonic disk.** Cells within the embryonic disk differentiate and migrate to form the three primary germ layers known as the ectoderm, mesoderm and endoderm. All tissues and organs develop from these three germ layers. These cell movements are called **gastrulation** and during this period embryo is called a **gastrula**.

The **ectoderm** forms the nervous system and the outer layer of the skin and its derivatives. The **mesoderm** gives rise to muscle, bone, connective tissue, and organs such as the heart, kidneys, ovaries and testes. It also forms the **notochord**, which defines the long axis of the embryo and gives it some rigidity. The **endoderm** forms some organs and glands, and the lining of the urinary, respiratory and digestive tracts.

The formation of the central nervous system from ectoderm is called **neurulation**, and the embryo is called a **neurula**. The developing notochord induces the overlaying ectoderm to thicken and form the neural plate. The neural plate folds to form a groove that later closes to form the **neural tube**, which will form the central nervous system. Mesoderm cells near the neural tube develop into **somites**, which will form the skeletal muscles of the neck and trunk, connective tissues and vertebrae. Failure of the neural tube to develop and close properly leaving part of the spinal cord exposed results in **spina bifida**. A neural tube defect that results in incomplete development of the brain is called **anencephaly**.

The gender of an embryo is determined at fertilization by the chromosomes carried in the sperm and egg. The development of sex organs begins about 6 weeks into development when the SRY region of the Y chromosome initiates development of the testes. The testes begin to produce testosterone, which directs the development of the male reproductive organs. In the absence of testosterone, female sex organs develop.

The fetal period extends from the ninth week after fertilization until birth and is characterized by rapid growth. The lungs, kidneys and liver do not perform their postnatal functions in the fetus; therefore, blood is shunted past these organs through temporary vessels or openings. Most of the blood bypasses the fetal liver and enters the inferior vena cava by the **ductus venosus**. Most of the blood entering the right atrium passes to the left atrium through the **foramen ovale** bypassing the lungs. Likewise, blood flowing out of the right ventricle into the pulmonary trunk is shunted away from the lungs and into the aorta through the **ductus arteriosus**. At birth when all organs take on their postnatal functions, fetal circulation converts to the postnatal pattern.

Birth usually occurs about 38 weeks after fertilization and marks the transition from prenatal to postnatal development. The fetus is expelled from the uterus by a process called **labor** that occurs in three stages. **Dilation stage** begins with the onset of contractions and continues until the cervix is fully dilated. The **expulsion stage** begins with full dilation and ends with delivery of the baby. The final stage of labor is the **placental stage**, which begins with delivery of the newborn and ends when the placenta is expelled from the mother's body. The signal for delivery comes from **corticotropin-releasing hormone (CRH)** formed in the placenta. Most babies are born head first, but some are born buttocks first in a process called a **breech birth**. Babies born after 38 weeks of development are considered **full-term infants**, while those born after 37 or fewer weeks of development are called **premature infants**.

Environmental agents, also called **teratogens**, encountered during periods of rapid cell differentiation cause major birth defects. When they are encountered after week 16, the effects are greatly reduced.

Lactation is the production and ejection of milk from the mammary glands. Prolactin from the anterior pituitary promotes milk production and oxytocin stimulates milk ejection. Immediately after birth, the breasts produce **colostrum,** followed about 3 days later with milk. Maternal milk and colostrum contain antibodies and special proteins that boost the immune system of the newborn. Oxytocin, produced in response to the newborn's sucking during nursing, helps the uterus to shrink.

The postnatal period begins with birth and continues through infancy, childhood, puberty, adolescence and adulthood. Growth and development stop at about age 25 as the aging process begins. Scientists believe that **aging** is caused by several interactive processes. Medical advances and a healthy lifestyle can help achieve a high-quality old age with fewer diseases and disabilities. Genes, environment and lifestyle determine the human life span. We have the most control over a healthy lifestyle, which includes proper nutrition, plenty of exercise and sleep, no smoking, and routine medical checkups.

Key Concepts

- Human life has two main periods of development: prenatal period, which is before birth, and the postnatal period, which is after birth.
- The prenatal period is divided into three periods: 1) the pre-embryonic period from fertilization through the second week, 2) the embryonic period from week 3 through week 8, and 3) the fetal period from week 9 until birth.
- The pre-embryonic period is characterized by fertilization, cleavage, and implantation.
    - ✓ Fertilization is the union of the sperm and secondary oocyte.
    - ✓ Cleavage is the series of mitotic cell divisions that form a solid ball called a morula and the hollow blastocyst. The blastocyst contains the inner cell mass that will become the embryo and the trophoblast that will give rise to the extraembryonic membranes, including the amnion, yolk sac, chorion, and allantois. The placenta forms from the chorion and provides for the nourishment of the fetus and the removal of wastes.
    - ✓ Implantation is when the blastocyst attaches to the endometrium.
- During the embryonic period gastrulation occurs, which leads to the formation of the three germ layers from which all organs and tissues will develop.
    - ✓ Ectoderm forms the nervous system and the outer layer of skin and its derivatives. Neurulation occurs as ectodermal tissue forms the neural tube, which later develops into the central nervous system.
    - ✓ Mesoderm gives rise to muscle, bone, connective tissue, and other internal organs. During gastrulation, it develops into the notochord. About 6 weeks into development, the SRY region of the Y chromosome initiates the development of testes. The testes produce testosterone, which directs the development of other male reproductive organs from the mesoderm. In the absence of testosterone, female reproductive organs develop from the same tissues.
    - ✓ Endoderm forms some organs and glands and the lining of the urinary, respiratory, and digestive tracts.
- The fetal period is characterized by rapid growth. Various parts of the body grow at different rates giving rise to a well-proportioned body.
- Some organs in the fetus do not perform their postnatal functions and therefore blood is shunted away from them to increase efficiency. Blood bypasses the liver by a temporary vessel, the ductus venosus, and the lungs by a hole between the atria called the foramen ovale and a shunt from the pulmonary trunk to the aorta called the ductus arteriosus.
- Birth is the transition from prenatal to postnatal development and usually occurs after about 38 weeks of development.
    - ✓ The signal to begin labor comes from the placenta in the form of corticotropin-releasing hormone (CRH).
    - ✓ True labor is divided into three stages: the dilation stage when the cervix opens, the expulsion stage when the newborn is delivered, and the placental stage when the placenta is expelled from the mother's body.

- Environmental and genetic factors can cause birth defects. Environmental agents have their greatest effects on the developing embryo during periods of rapid differentiation.
- Lactation, the production and ejection of milk from the mammary glands, is stimulated by prolactin. The breasts initially produce colostrum until the sucking of the infant stimulates the release of oxytocin from the posterior pituitary gland, which causes milk letdown.
- The postnatal period begins with birth and continues through infancy, childhood, puberty, adolescence and adulthood. Growth and the formation of bone are usually completed by age 25 when the aging process begins.
- Medical advances and a healthy lifestyle including proper nutrition, plenty of exercise and sleep, no smoking, and routine medical checkups help us ward off disease and disability, improving the quality of life as we age.

Study Tips

This chapter describes several developmental processes that occur over time. The figures in the text graphically present the information. As you read the text, highlight that information as it is presented in the figures. Many students benefit by putting text information into their own words. A good study tool is the timeline, a chart that shows what developmental events are occurring, in what order and over what span of time. The pre-embryonic stages of development, from ovulation through implantation, are described in detail in your text. The review section below will structure a timeline for you for this period. The next stage is the embryonic period when the organs, systems and limbs rapidly develop. A timeline is provided for you in text Figure 18-16. Your text focuses on the processes of this period more than the sequence of events. There are fewer developmental events that occur during the fetal period and there may be no need to develop a detailed timeline.

The text focuses on a few critical processes that need to be understood in detail. Spend time understanding the processes of fertilization, implantation, formation of the placenta, neurulation and the development of sex organs. Study the figures in the text, explain them to a study partner and write out an explanation of each. Finally, it is important to understand the differences between fetal circulation and postnatal circulation. Study Figure 18-12 and follow the path of blood in the fetus noting each shunt and why that makes a more efficient flow of blood. The study guide review exercises will help you accomplish this summary.

As you complete your study materials for this chapter, focus on the stages of postnatal development and the aging process. Understand the changes that occur during aging and what lifestyle changes will make aging more enjoyable and less restrictive.

Review Questions

A. Study the text material and then match the term Column A with the definition in Column B.

Column A	Column B
1. blastocyst	a. period of development before birth
2. cleavage	b. more than one sperm enters an egg

3. fertilization c. fertilized ovum

4. human chorionic gonadotropin d. blastocyst becomes imbedded in the endometrium

5. implantation e. developmental period after birth

6. morula f. union of sperm and egg

7. prenatal development g. hollow ball of cells with fluid-filled cavity

8. polyspermy h. produced by blastocyst and used in pregnancy tests

9. postnatal development i. solid ball of 12 or more cells

10. zygote j. mitotic cell divisions of the zygote

B. Complete the following paragraph describing the events just prior to and during fertilization.

At ____________ the ovary releases a secondary oocyte. It is swept into the ____________ by fimbriae. About 500 million sperm are released during ____________ into the vagina and on the cervix. About 200 sperm reach the oviduct. The surface of the ____________, the cap around the tip of the sperm, is altered by secretions from the oviduct or ____________. The first membrane around the secondary oocyte encountered by the sperm is the ____________ ____________. The thick noncellular layer encountered next by the sperm is the ____________ ____________. When the cytoplasm of the secondary oocyte is penetrated by the sperm, it undergoes its second ____________ division and is called an ____________.Then the nucleus of the sperm and ovum fuse to form a ____________.

C. Complete the following table describing the events of the pre-embryonic period.

Time after fertilization	Stage/Process	What happens
First 24 hours		Sperm and secondary oocyte unite to form a zygote
	Cleavage	Mitotic cell divisions occur in the zygote
Day 4		Solid ball of 12 or more cells

	Blastocyst	
Day 6 – 14		Imbedding of the blastocyst in the endometrium

D. Use the following terms to label the drawing of the membranes surrounding the pre-embryonic embryo shown below: allantois, amnion, chorion, chorionic villi, fetal portion of placenta, maternal portion of placenta, umbilical cord and yolk sac.

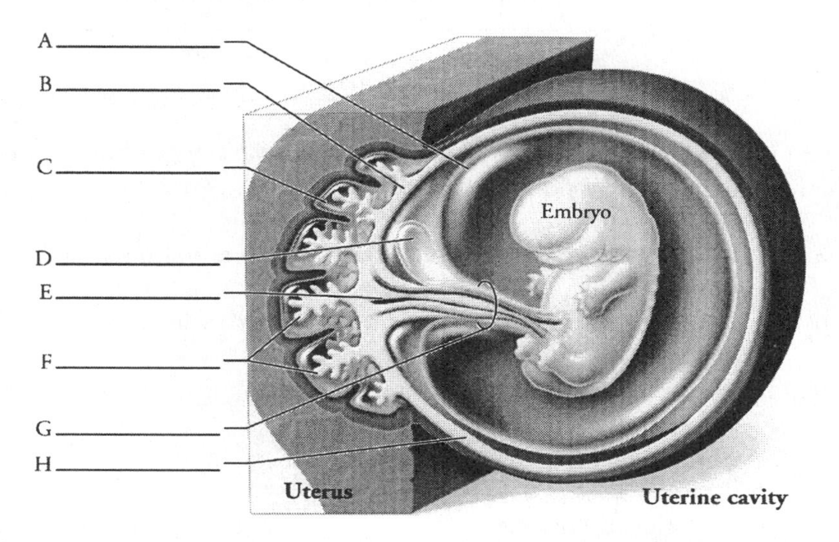

E. Complete the following summary of gastrulation and the development of organs and tissues from the primary germ layers during the embryonic period.

1. The inner cell mass from the blastocyst → ____________ ____________ → three primary germ layers.

2. Ectoderm → ____

3. Mesoderm → ____

4. Endoderm → ____

F. Write a short description of neurulation, the development of the central nervous system. Include these terms: brain, ectoderm, neural folds, neural plate, neural tube, and spinal cord. Then answer the questions that follow.

1. The process of neurulation begins when ____

____

____

____

2. The mesoderm gives rise to somites. What do they form? ____________________________

____

3. Describe two conditions that arise when the spinal cord does not form properly. ________

____

____

____

G. Study the text material and Figure 18-12 and then trace the flow of blood from the placenta through the fetus including the fetal circulatory bypasses.

Placenta → __________ vein → some blood enters the liver but most of the blood bypasses the

fetal liver by the ____________ ____________ → inferior vena cava → right ____________

→ some blood passes to the right ventricle and on to the lungs, but most of the blood moves

through the ____________ ____________ into the left ____________ → left ____________

→ aorta → large arteries of legs → ____________ arteries → placenta. The blood that

entered the right ventricle and passed into the pulmonary trunk passed through the

____________ ____________ on its way to the ____________.

H. Circle the correct word in the sentences below that correctly completes them.

1. The first stage of delivery is the dilation stage referring to the size of the opening of the (uterus, cervix).

2. The second stage of delivery is the expulsion stage. It begins with full dilation and ends with (delivery of the baby, expulsion of the afterbirth).

3. The third stage of labor is the placental stage. This is when uterine contractions expel the (baby, afterbirth) from the uterus.

4. A baby born after only 35 weeks would be considered (full term, premature).

5. The hormone that initiates labor is (human chorionic gonadotropin, corticotropin-releasing hormone).

6. Milk production in the mother is stimulated by (oxytocin, prolactin).
7. The initial substance produced by the breasts immediately after delivery is (colostrum, milk).
8. The hormone that stimulates both milk letdown and contraction of the uterus is (ocytoxin, prolactin).
9. Some say that aging begins the minute one is born; however, growth stops and physical changes indicative of aging begin at about age (25, 50).
10. Scientists believe that aging is caused by a combination of factors. The factors that we can control are (genetic, behavioral).

Critical Thinking

Read each of the following questions carefully. Circle the most important words in the question before formulating your answer.

1. The placenta is the connection between the mother and baby inside. You have learned that many substances cross the placenta and that a few, for example red blood cells, do not. Investigate what materials cross the placenta and then make a list of best practices for those contemplating pregnancy and for the mother during pregnancy. Why should someone consider behaviors before pregnancy?

2. Fetal circulation is different from that of the newborn because the fetus does not use all of its organs while in the womb. The placenta allows for an exchange of gases, nutrients and wastes. The foramen ovale is the hole between the two atria of the heart. What are the consequences if it does not close shortly after birth? Can this be corrected and how often does it occur? Use the Internet or your campus resources to gain more information.

3. Your text described the stages of delivery and several ways to manage the pain associated with it. There is more variety in the methods of delivery and the environment provided for the mother and family than most people expect. Contact your local hospital and discuss the options available for delivery.

4. Form a small group of students and discuss the pros and cons of breast feeding.

5. More and more of the population is living longer and longer. It is well documented that the respiratory and musculoskeletal systems decrease in efficiency and capacity as one grows older. Visit a local senior center or the AARP web site to learn how seniors are staying active. What are some positive lifestyle choices you might make as you age that will help ensure many healthy and disease-free years?

PRACTICE TEST

Choose the one best answer to each question that follows. As you work through these items, explain to yourself why each answer you discard is incorrect and why the answer you choose is correct.

1. Where does fertilization usually occur?
   a. ovary
   b. oviduct
   c. uterus
   d. vagina

2. Which protective layer around the ovum prevents polyspermy?
   a. acrosome
   b. corona radiata
   c. polyona protectiva
   d. zona pellucida

3. How long after fertilization is the first cell division likely to occur?
   a. immediately, within an hour
   b. one day
   c. four days
   d. one week

4. The solid ball of undifferentiated cells that enters the uterus is called the ___________?
   a. blastocyst
   b. embryo
   c. gastrula
   d. morula

5. Which of the following would be the *best* choice for an organ donor?
   a. conjoined twins
   b. fraternal twins
   c. identical twins
   d. siblings

6. Which of the following best describes the blastocyst?
   a. fluid-filled morula
   b. hollow ball of cells with an inner cell mass and a trophoblast
   c. mass of undifferentiated cells
   d. solid ball of cells all made of inner mast cells

7. What happens if implantation does not occur?
   a. implantation will occur later, usually in the oviduct
   b. the blastocyst is reabsorbed or expelled
   c. the cervix begins to dilate
   d. the pregnancy continues, this is not a problem

8. What hormone is detected in a pregnancy test?
   a. estrogen
   b. human chorionic gonadotropin
   c. progesterone
   d. testosterone

9. The placenta is formed from the ____________ of the embryo and the ____________ of the mother.
   a. amnion, endometrium
   b. chorion, endometrium
   c. umbilical cord, oviduct
   d. yolk sac, uterus

10. Placenta previa is a serious condition in which ____________.
    a. the placenta forms in the lower half of the uterus
    b. the placenta forms in the oviduct
    c. the umbilical cord forms around the fetus' neck
    d. the yolk sac prematurely bursts

11. What embryonic structure becomes the central nervous system?
    a. derivative of the anterior pituitary
    b. neural tube
    c. notochord
    d. somites

12. Cases of spina bifida were greatly reduced when mothers were given ____________.
    a. carotene
    b. folic acid
    c. selenium
    d. zinc

13. What causes female sex organs to develop in the embryo?
    a. absence of testosterone
    b. decreasing levels of testosterone and increasing estrogen levels
    c. increasing levels of progesterone and estrogen
    d. increasing levels of testosterone

14. Which of the following organs is bypassed by the ductus venosus in the fetus?
    a. heart
    b. liver
    c. lungs
    d. stomach

15. Which of the following will not cross the placenta and enter the fetal circulation?
    a. alcohol
    b. blood cells
    c. caffeine
    d. urea

16. True labor ends with which of the following stages?
    a. dilation
    b. delivery of the newborn
    c. expulsion
    d. placental

17. If prolactin is responsible for milk production and it is present in increasing levels throughout pregnancy, then why isn't milk produced before the baby is born?
    a. colostrums must be produced first
    b. estrogen and progesterone inhibit the action of prolactin
    c. oxytocin causes milk production to occur
    d. testosterone levels are too high until after delivery

18. When is exposure to teratogens most likely to cause a baby to be born with severe birth defects?
    a. pre-embryonic
    b. embryonic
    c. fetal
    d. the effect is the same in all developmental periods

19. Growth and development are usually completed by age ____________.
    a. 2
    b. 13
    c. 21
    d. 25

20. Chances of aging well can be increased by ____________.
    a. medical checkups
    b. proper nutrition
    c. regular exercise
    d. all of the above

SHORT ANSWER QUESTIONS

Read each of the following questions carefully. Review the appropriate section of the text if necessary; jot down the main points. Then formulate your answer using complete sentences in your explanation.

1. Draw the structure of a blastocyst including the inner cell mass and the trophoblast. Describe the destiny of cells in the inner cell mass and the trophoblast as development continues.

2. Describe the three ways that twins might develop.

3. The development of a fertilized egg into a baby takes about 9 months and is a very complex process. Genetic factors or environmental agents may cause birth defects. At what point during development might an environmental agent be most damaging?

4. Describe the process of implantation and the complication of an ectopic pregnancy.

5. Discuss the role of the Y chromosome in the development of sex organs.

Answer Key

Review Questions

A. 1. g
2. j
3. f
4. h
5. d
6. i
7. a
8. b
9. e
10. c

B. At ovulation the ovary releases a secondary oocyte. It is swept into the oviduct by fimbriae. About 500 million sperm are released during ejaculation into the vagina and on the cervix. About 200 sperm reach the oviduct. The surface of the acrosome, the cap around the tip of the sperm, is altered by secretions from the oviduct or uterus. The first membrane around the secondary oocyte encountered by the sperm is the corona radiata. The thick noncellular layer encountered next by the sperm is the zona pellucida. When the cytoplasm of the secondary oocyte is penetrated by the sperm, it undergoes its second meiotic division and is called an ovum.Then the nucleus of the sperm and ovum fuse to form a zygote.

C.

Time after fertilization	Stage/Process	What happens
In the first 24 hours	Fertilization	Sperm and secondary oocyte unite to form a zygote
Day 1	Cleavage	Mitotic cell divisions occur in the zygote
Day 4	Morula	Solid ball of 12 or more cells
Day 6	Blastocyst	Fluid-filled hollow ball of cells forms
Day 6 – 14	Implantation	Imbedding of the blastocyst in the endometrium

D. A. amnion
B. fetal portion of placenta
C. maternal portion of placenta
D. yolk sac
E. allantois
F. chorionic villi
G. umbilical cord
H. chorion

E. 1. The inner cell mass from the blastocyst → embryonic disk → three primary germ layers.
2. Ectoderm → nervous system and outer layer of skin and its derivatives
3. Mesoderm → muscle, bone, connective tissue, heart, kidneys, ovaries, testes, and notochord
4. Endoderm → lining of the urinary, respiratory and digestive tracts; pancreas, liver, thyroid gland and parathyroid glands

F. 1. The process of neurulation begins when the notochord induces changes in the ectoderm that lead to the development of the central nervous system. First some cells of the ectoderm thicken to form the neural plate. The neural plate folds inward to form a groove with the neural folds on the sides. The neural folds meet and join to form the neural tube. The anterior portion of the neural tube develops into the brain and the posterior portion forms the spinal cord.
2. Somites form skeletal muscles of the neck and trunk, connective tissues, and vertebrae.
3. Spina bifida occurs when there is an exposed portion of the spinal cord. This can cause paralysis and is susceptible to infection. Often it can be avoided by adding folic acid to the mother's diet. Anencephaly is a neural tube defect that involves incomplete development of the brain.

G. Placenta → umbilical vein → some blood enters the liver but most of the blood bypasses the fetal liver by the ductus venosus → inferior vena cava → right atrium → some blood passes to the right ventricle and on to the lungs, but most of the blood moves through the foramen ovale into the left atrium → left ventricle → aorta → large arteries of legs → umbilical arteries → placenta. The blood that entered the right ventricle and passed into the pulmonary trunk passed through the ductus arteriosus on its way to the aorta.

H. 1. The first stage of delivery is the dilation stage, referring to the size of the opening of the cervix.
2. The second state of delivery is the expulsion stage. It begins with full dilation and ends with delivery of the baby.
3. The third stage of labor is the placental stage. This is when uterine contractions expel the afterbirth from the uterus.
4. A baby born after only 35 weeks would be considered premature.
5. The hormone that initiates labor is corticotropin-releasing hormone.
6. Milk production in the mother is stimulated by prolactin.
7. The initial substance produced by the breasts immediately after delivery is colostrum.
8. The hormone that stimulates both milk letdown and contraction of the uterus is oxytocin.
9. Some say that aging begins the minute one is born; however, growth stops and physical changes indicative of aging begin at about age 25.
10. Scientists believe that aging is caused by a combination of factors. The factors that we can control are behavioral.

Critical Thinking Questions

1. It is very important for those contemplating a pregnancy to have a physical exam to determine current immunizations and the possible presence of viral STDs that might affect the baby. Since a pregnancy cannot be immediately identified, and since the early developmental processes are the most critical for the health of the baby, it is very important to begin providing the best possible environment for the baby before it is conceived. Both before and during a pregnancy, it is important to avoid substances that can cross the placenta and harm the fetus such as cigarette smoke, alcohol, caffeine, illegal drugs and some workplace toxins.
2. The foramen ovale remains open in about 10-15% of adults and shows no signs or symptoms. However, in some cases there can be clotting of blood that leads to stroke. Medical treatment may include aspirin or a prescription anticoagulant. Surgical procedures can be used to close the opening, but this will not be attempted unless there are symptoms of compounding problems. The most recent advances include closure by catherization, a less invasive procedure.
3. Many hospitals have family-friendly delivery rooms that provide a more relaxed environment. In addition to a traditional delivery, many communities offer a birthing center which usually has access to a hospital should there be complications. Homebirth is a consideration for low risk deliveries.
4. Colostrum and breast milk provide the newborn with nutrients and antibodies. Infants have an undeveloped immune system and are more susceptible to infection and therefore can benefit from the antibodies. Babies who are fed breast milk also tend to have fewer digestive problems. The suckling activity of the baby stimulates the production of oxytocin, which causes uterine contractions and helps the mother's uterus shrink. In addition, breast feeding builds a social bond

between the mother and infant. However, breast feeding is not always convenient. It requires that the mother be present to feed or use a pump to store milk. It is sometimes messy and is difficult for those who work long days or who travel. Some couples prefer not to breast feed so that the father can be involved more in the care of the infant.

5. It is important to remain physically active as one ages not only for the physiological benefits but for the social benefits as well. Many senior centers have volunteer programs as do hospitals, libraries and schools. Seniors who are around young people seem to be especially happy and feel as though they are making a contribution to others. Most senior centers have exercise programs as do centers associated with health clubs and gyms. Seniors participate in a great variety of sports and enjoy training, competing, and meeting friends.

Practice Test

1. b
2. d
3. b
4. d
5. c
6. b
7. b
8. b
9. b
10. a
11. b
12. b
13. a
14. b
15. b
16. d
17. b
18. b
19. d
20. d

Short Answer Questions

1. Your drawing should be similar to that shown in Figure 18-2 or 18-5. The blastocyst is a hollow ball of cells with a fluid-filled cavity. It is made up of the inner cell mass and the trophoblast. The inner cell mass is a cluster of cells at the point of attachment to the uterus that will become the embryo and some of the extraembryonic membranes. These cells will form the embryonic disk and the primary germ layers of the gastrula. The trophoblast is a thin layer of cells enclosing the fluid that will give rise to the extraembryonic membranes that are the embryo's contribution to the placenta.
2. Identical twins develop when cells of the morula separate and two embryos are formed. These twins are genetically the same. Conjoined twins result in the same way, when cells of the morula separate but the separation is incomplete and some organs are shared. These twins are also genetically identical. Fraternal twins occur when two secondary oocytes are released from the ovaries and fertilized by different sperm. These twins are not genetically identical and are more like siblings.
3. Environmental agents that disrupt development are called teratogens. They have their greatest effects on the developing embryo during periods of rapid differentiation when organs are forming. Thus exposure to disruptive agents during the embryonic period is most likely to cause birth defects.
4. Implantation is the process by which the blastocyst becomes imbedded in the endometrium of the mother. As the blastocyst continues to develop, extraembryonic membranes form. The chorion develops chorionic villi that invade the endometrium. This leads to the development of the placenta by the end of the third month. Implantation usually occurs high on the back wall of the uterus. However, sometimes, the blastocyst implants outside the uterus, which is called an ectopic pregnancy. This usually occurs in the oviduct, but has been known to occur on the abdominal wall.
5. The sperm carry either an X or a Y chromosome. If a secondary oocyte is fertilized by a sperm carrying a Y chromosome, then the embryo will be a male; if it is fertilized by a sperm carrying an X chromosome, then it will be a female. Until 6 weeks into development, embryos are sexually indifferent. When the SRY region of the Y chromosome is activated, it initiates development of the testes in XY embryos. The testes soon produce testosterone, which stimulates the development of male reproductive organs. In the absence of the Y chromosome, ovaries develop.

Chapter 19

Chromosomes and Cell Division

Objectives

After reading the text and studying the material in this chapter, you should be able to:

- Describe the human life cycle using the terms mitosis, meiosis and fertilization.
- Draw, label and explain the structure of a chromosome.
- Explain the cell cycle including interphase, cell division and cytokinesis.
- List the events of interphase, prophase, metaphase, anaphase and telophase as completed in mitosis.
- List the events that occur in meiosis I and meiosis II.
- Diagram and explain how haploid cells result from meiosis.
- Explain crossing over and how it results in genetic variation.
- Describe when chromosomes can be shuffled to ensure genetic variability through independent assortment.
- Compare the events and products of meiosis and mitosis. Explain the importance of each process.
- Describe how nondisjunction results in an abnormal number of chromosomes in the resulting cells.
- Explain how each of the following conditions arise and describe the characteristics of someone with the condition: Down syndrome, Klinefelter syndrome, Turner syndrome, Jacob syndrome, and triple X syndrome.

Chapter Summary

A **chromosome** is a combination of a DNA molecule and specialized proteins called histones. In eukaryotic cells, chromosomes are found in the cell nucleus and contain the information that directs cell processes. The histones provide support and some control of gene expression. A **gene** is a specific segment of DNA that directs the synthesis of a protein. Genes are arranged along a chromosome in a specific sequence. The point on a chromosome where a particular gene is found is called its **locus**, similar to an address. In humans, all cells except the sperm and egg have 46 chromosomes, one set of 23 from each parent. Each cell contains two chromosomes with genes for the same traits called **homologous pairs** of chromosomes. A cell with two sets of chromosomes is described as being **diploid** (2n). The alternative forms of a gene are called **alleles**. One pair of chromosomes of the 23 pairs is the **sex chromosomes**. They determine whether a person is male (XY) or female (XX). The other 22 pairs of chromosomes are called the **autosomes**; they determine the expression of most of a person's inherited characteristics.

The **cell cycle** is the predictable set of events from the creation of a new cell until that cell divides again. The cell cycle consists of interphase and cell division. The process of cell division for body cells is **mitosis** when one nucleus divides into two daughter nuclei with the same number and kinds of chromosomes.

Interphase is a period of growth and preparation for cell division. The cell spends most of its time in one of the three parts of interphase. During G_1, chromosomes consist of a strand of DNA and proteins. DNA is replicated during the S phase and the two copies of the chromosome, called **chromatids**, remain attached at the **centromere**. G_2 is the time after DNA is synthesized and before mitosis begins.

Cell division consists of two processes: the division of the cytoplasm called **cytokinesis** and the division of the nucleus called mitosis. Mitosis occurs in four phases: prophase, metaphase, anaphase and telophase. Mitosis begins with **prophase**. During prophase the chromosomes begin to condense as the DNA wraps around histones forming a compact, visible structure. When the DNA is in this condensed state it cannot be replicated and all gene activity is stopped. The nucleolus disappears and the nuclear membrane begins to break down. In the cytoplasm, the centrioles move to opposite poles of the cell and the mitotic spindle forms.

Metaphase is the next stage of mitosis. The chromosomes form a line at the center of the cell along the **equatorial plate**. At this point, the two sister chromatids are attached and the chromosomes can be seen with a light microscope. They are sometimes stained, photographed, and arranged by size at this stage to form a **karyotype**, which can be used for diagnostic purposes. During **anaphase** the chromatids of each chromosome begin to separate, splitting at the centromere. Now the two sister chromatids are considered independent chromosomes. During **telophase**, the nuclear envelope forms around each group of chromosomes. The mitotic spindle disassembles and nucleoli reappear. **Cytokinesis**, the division of the cytoplasm, is completed during telophase. A band of microfilaments in the area that was the equatorial plate contracts and forms a furrow, eventually pinching the cell in two.

Gametes differ from somatic cells in that they are **haploid** (n). They are formed by **meiosis**, a type of cell division that is actually two divisions resulting in four haploid daughter cells. Haploid gametes join to form a diploid zygote, thus keeping the chromosome number constant through each generation. Meiosis also increases genetic variability by generating new genetic combinations in offspring.

Meiosis involves two cell divisions, meiosis I and meiosis II, each with a prophase, metaphase, anaphase and telophase. Meiosis I is preceded by interphase during which each chromosome was copied and now consists of two attached chromatids. **Meiosis I** is a **reduction division** because it produces two haploid cells. Each daughter cell contains one of each chromosome. During prophase I the chromosomes condense and homologous chromosomes pair. During metaphase I the homologous chromosomes line up at the equatorial plate, a process called **synapsis**. Spindle fibers attach to the chromosomes. During anaphase I the homologous pairs of chromosomes separate and move to opposite ends of the cell. During telophase I the nuclear membrane forms around the chromosomes and cytokinesis occurs to form two haploid cells. There is a brief period of rest called interkinesis before meiosis II begins.

During meiosis II, the chromosomes condense in prophase II, and line up on the equatorial plate in metaphase II. However, in anaphase II, the centromere holding the two sister chromatids comes apart and the two sister chromatids are pulled to opposite poles of the cell by spindle fibers. During telophase II, the nuclear membrane forms around each haploid set of chromosomes and cytokinesis occurs. The important feature of meiosis II is that it separates the two sister chromatids of each chromosome.

Crossing over and independent assortment ensure new combinations of genetic material during meiosis. **Crossing over** occurs when corresponding pieces of chromatids of maternal and paternal homologues are exchanged during synapsis. Equivalent genetic material is exchanged because the two homologous chromosomes are lined up side by side. However, the genetic material is not identical because the genes on the homologous chromosome from the mother and the genes on the homologous chromosome from the father are different.

Independent assortment provides for a new combination of chromosomes. Because one homologue is from the mother and one from the father and they are pulled to opposite poles of the cell during anaphase I, the two daughter cells resulting from meiosis II are not identical.

On occasion, the two sister chromatids or two chromosomes adhere too tightly and may not move to opposite poles of the cell, a phenomenon called **nondisjunction**. Nondisjunction results in too many, or too few, chromosomes in a cell, a condition called aneuploidy. **Trisomy** is a condition in which cells contain three copies of one chromosome. **Monosomy** is the condition in which somatic cells contain only one of a chromosome pair. Most of the time, an abnormal number of chromosomes will result in spontaneous abortion.

However, occasionally, an infant will be born with three copies of chromosome 21, a condition known as **Down syndrome**. These children have multiple physical and mental abnormalities. The risk of having a baby with Down syndrome increases with age. Recall that a woman's eggs are formed before she is born. Meiosis precedes to prophase I where the chromosomes are lined up while she is still in her mother's womb. Then they wait until ovulation to finish meiosis. It might be that the longer the chromosomes remain paired in prophase I the higher the chances will be that nondisjunction will occur as meiosis proceeds following ovulation. Another reason may be that an older mother has been exposed to more environmental factors that may affect normal chromosomal separation. One final consideration is that older mothers are more likely to carry an abnormal fetus to full term and, therefore, older women are more likely to give birth to children with genetic abnormalities.

Nondisjunction can occur with the sex chromosomes (X and Y) as well as the autosomal chromosomes. So that a female might carry an egg with two X chromosomes and another with no X chromosomes and a male might carry a sperm with both the X and Y chromosomes and another sperm with no X or Y chromosomes. Each abnormal gamete can combine with a normal gamete and will form a viable zygote in all cases except one (when a Y sperm fertilizes an egg with no X chromosome). People with **Turner syndrome** have only one X chromosome. They look like girls but are infertile. A similar situation occurs when a male inherits an extra X chromosome and is XXY. This is called **Klinefelter syndrome**. Secondary sex characteristics do not develop normally and XXY males are sterile. Nondisjunction of the sex chromosomes can also result in an XXX female (triple X syndrome), or an XYY male (Jacob syndrome).

KEY CONCEPTS

- Men and women produce gametes by a process called meiosis. These gametes carry one-half the number of chromosomes as other cells and are called haploid. The egg and sperm join during fertilization to form a diploid zygote, which grows by mitotic cell division to an adult who is then capable of producing eggs or sperm completing the cycle.
- Chromosomes consist of genetic material called DNA and proteins called histones. A gene is a specific sequence of DNA that controls the expression of a trait. It is located at a specific locus on the chromosome. The alternative forms of a gene are called alleles; that is, one allele is received from each parent for a given trait.
- Human somatic cells contain 46 chromosomes that consist of 2 sets of 23 pairs of homologous, or similar, chromosomes. One set is carried in the gamete received from each parent. Of the 23 pairs of chromosomes, 22 pairs are autosomal and 1 pair determines sex.
- The cell cycle consists of interphase and cell division.
- Interphase is the period of the cell cycle between cell divisions when the DNA is duplicated in preparation for cell division. Two copies of a chromosome, called sister chromatids, remain joined by the centromere following interphase.
- Cell division consists of nuclear division either by mitosis or meiosis and division of the cytoplasm called cytokinesis.
- Mitosis is cell division where one cell divides into two cells with the same number of chromosomes. It consists of four phases:
  - ✓ Prophase: the chromosomes condense, nuclear membrane breaks down, spindle fibers form.
  - ✓ Metaphase: chromosomes align at the equatorial plate.
  - ✓ Anaphase: sister chromatids separate forming two sets of chromosomes that move toward opposite poles of the cell.
  - ✓ Telophase: nuclear membrane forms around each set of chromosomes and cytokinesis occurs.
- Meiosis is cell division where one cell divides two times to form four cells each with one-half the number of chromosomes of the original cell. Meiosis keeps the chromosome number constant after fertilization and increases genetic variability in the population.
  - ✓ During meiosis I homologous chromosomes, each with sister chromatids, separate to form two haploid cells. Each haploid cell has one-half the number of chromosomes, but each of those chromosomes still has two sister chromatids.
  - ✓ During meiosis II the cells divide again separating the two sister chromatids. Each new cell is formed when the sister chromatids come apart at the centromere forming the same number independent chromosomes.
- Genetic recombination occurs due to crossing over during metaphase of meiosis I and independent assortment as chromosomes are sorted into new cells.
- Nondisjunction is the failure of homologous chromosomes to separate during meiosis I or sister chromatids to separate during meiosis II. This results in gametes with an abnormal number of chromosomes.
- Down syndrome results when someone inherits three 21st chromosomes (trisomy 21).
- Nondisjunction can occur in the chromosomes that determine sex resulting in an egg with two X chromosomes or an egg with no X chromosomes and a sperm with both an X and a Y chromosome or no chromosomes. Nondisjunction may also result in Turner syndrome (0X), Klinefelter syndrome (XXY), triple X syndrome (XXX), or Jacob syndrome (XYY).

Study Tips

This chapter begins with a description of the chromosome. Read the description and then draw and label a chromosome. The meat of the chapter is on the two mechanisms of cell division that allow humans to complete their life cycle without continually doubling the number of chromosomes of each future generation – mitosis and meiosis. It is absolutely essential that you understand the importance of each type of cell division. Begin by reading your text and studying the figures that show mitosis and meiosis. Then gather some materials so that you can make a model of what happens at the chromosomal level during cell division. Models are used to describe events that are too fast, too slow, too small or too large to observe. You will need scissors, a glue stick and two pieces of paper of different colors. If you don't have different colored paper, then use a marker to make the chromosomes different colors. Cut four strips of paper about 1 cm x 10 cm from each piece of paper. The strips as they are now will be chromosomes. You will bend the strips so they form a right angle and then use a glue stick to join them at the bend to represent two sister chromatids. One set of chromosomes will be inherited from the father and the other from the mother. Read the text material and study the figures, then set up your model and move the chromosomes as shown in the figure. When you are ready, practice explaining the processes to a study partner.

Two additional important concepts are discussed in this chapter. One is genetic variation. It is important that you understand how genetic variation is ensured during cell division. As you work with your model, notice whether the same chromosomes always end up in the same cells after cell division. The other means of genetic variation is crossing over. Read that portion of your text and try to demonstrate it with your model chromosomes. Finally nondisjunction is described and the resultant syndromes of people with an abnormal number of chromosomes. You will develop a summary table that will help you study that material.

Review Questions

A. Complete the following paragraph describing the human life cycle.

Since we are describing a cycle, we can begin anywhere. We will begin with the adult man and woman who each produce gametes by the process of ____________. By this process, the number of chromosomes in the egg and sperm are ____________ the number of chromosomes in the body cells. A cell with one-half the number of chromosomes of the adult is called n, haploid. The egg and sperm combine by fertilization to produce a zygote that is diploid (2n). This single cell becomes a multicelled organism by the process of mitosis. When the infant becomes an adult, he or she will make sperm or eggs by the process of meiosis.

B. Match the term in Column A with the function in Column B.

Column A	Column B
1. allele	a. body cells except eggs and sperm
2. centromere	b. gene location on a chromosome
3. chromosome	c. specific segment of a chromosome
4. diploid	d. replicated chromosome attached at the centromere
5. gametes	e. linear arrangement of DNA and histones
6. gene	f. alternate form of a gene
7. haploid	g. having two sets of chromosome per cell (2n)
8. locus	h. egg and sperm
9. sister chromatids	i. having one set of chromosomes (n)
10. somatic cell	j. attachment between two sister chromatids

C. Answer the following questions regarding the cell cycle.

1. What are the two parts of the cell cycle? ____________________

____

2. What are the two parts of cell division? ____________________

____

3. What type of nuclear division leads to cells with one-half the number of chromosomes as the original cell? ____________________

____

4. What type of nuclear division leads to cells with the same number of chromosomes as the original cell? ____________________

____

D. First identify each phase of mitosis shown on the left. Then describe the events of that phase of cell division.

1. Interphase: DNA replication to form two sister chromatids, duplication of centriole.

2. ______________________________

____

____

3. ______________________________

____

____

4. ______________________________

____

____

5. ______________________________

____

____

6. Cytokinesis: cytoplasm divides results in two separate daughter cells.

E. Put in Column A the correct letter of the explanation given in Column B for each of the lettered phases of meiosis shown below.

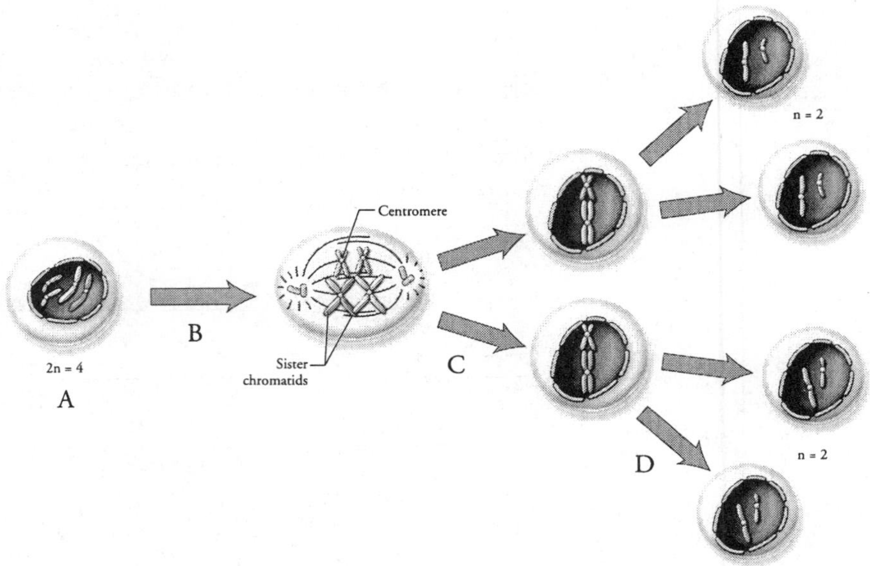

Column A	Column B
A. ________	a. During meiosis I, homologous chromosomes separate into two haploid cells.
B. ________	b. This diploid parent cell contains two homologous pairs of chromosomes.
C. ________	c. During meiosis II, sister chromatids separate and four haploid daughter cells are formed that will develop into gametes.
D. ________	d. DNA replication during interphase I forms two copies of each chromosome held together by a centromere.

F. Indicate whether the statement below is true of meiosis or mitosis or both processes by putting a check in the appropriate column.

	Mitosis	Meiosis	
1.	______	______	Involves one cell division
2.	______	______	Results in the production of gametes
3.	______	______	Daughter cells are genetically dissimilar
4.	______	______	Results in an increase in number of genetically similar cells
5.	______	______	Involves two cell divisions

6. ______ ______ Produces four haploid cells

7. ______ ______ During anaphase I, homologous chromosomes separate

8. ______ ______ Parts of chromosomes may be exchanged during crossing over

9. ______ ______ Results in diploid cells

10. ______ ______ Occurs as a child grows during the first few weeks of life

G. Draw a picture below and show how two chromosomes can exchange genetic material during crossing over.

H. Complete the following table to summarize the information from your text on chromosomal abnormalities.

Syndrome	**Chromosome Characteristics**	**Frequency of live births**	**Description**
	Trisomy 21	1/800-1000	
Turner		1/2000 female births	
	XXY		After puberty, small testes and penis, low levels of testosterone, lacking secondary sex characteristics, slight breast development, sterile.
Triple X		1/1000	
	XYY	1/1000 male births	

Critical Thinking

Read each of the following questions carefully. Circle the most important words in the question before formulating your answer.

1. Explain to a friend why the siblings who are not twins look similar but not identical. Use the model chromosomes you made to demonstrate independent assortment of chromosomes.

2. Often statistics regarding abnormal births such as Down syndrome or Triple X syndrome are stated in a frequency "per 1000 live births." Why is it important to include the statement "per live births"?

3. Explain the process of nondisjunction and then show how it can happen using the model chromosomes you made.

4. A couple had a child born with Klinefelter syndrome. They wanted to place blame with one parent. Show how Klinefelter syndrome could result from an abnormality in either the egg or sperm.

5. Describe a human karyotype and then use the Internet to locate some photographs of karyotypes.

Practice Test

Choose the one best answer to each question that follows. As you work through these items, explain to yourself why each answer you discard is incorrect and why the answer you choose is correct.

1. What is a specific segment of DNA that directs protein synthesis?
   a. chromatid
   b. chromosome
   c. gene
   d. locus

2. Which of the following is true regarding interphase?
   a. Chromatids separate during this phase.
   b. It is the longest phase of cell division.
   c. Male and female gametes combine to form a zygote
   d. The cell cytoplasm divides.

3. During which phase of the cell cycle does growth occur?
   a. anaphase
   b. cytokinesis
   c. interphase
   d. telophase

4. What structure dissolves so that the sister chromatids can move to opposite ends of the cell during cell division?
   a. allele
   b. centromere
   c. chromosome
   d. sister chromatids

5. Which type of cell division is responsible for the repair of your skin following a sunburn?
   a. meiosis
   b. mitosis
   c. both meiosis and mitosis

6. During which phase of mitosis would you expect to see chromosomes consisting of two chromatids within the nucleus?
   a. anaphase
   b. nondividing cell
   c. prophase
   d. telophase

7. In a cell with 4 chromosomes, how many sister chromatids are visible during telophase?
   a. 0
   b. 4
   c. 8
   d. 16

8. During which phase of mitosis do spindle fibers pull the chromosomes to opposite poles of the cell?
   a. anaphase
   b. metaphase
   c. prophase
   d. telophase

9. How many chromosomes, arranged in how many pairs, would be shown in a normal human karyotype? (number of chromosomes, number of pairs)
   a. 4, 2
   b. 23, 23
   c. 46, 23
   d. 92, 46

10. Which of the following is not associated with mitosis?
    a. growth of the fetus in the womb
    b. growth in height
    c. production of egg and sperm
    d. repair of a cut

11. Cytokinesis is associated with which of the following?
    a. meiosis
    b. mitosis
    c. both a and b
    d. neither a nor b

12. Which of the following is a result of meiosis?
    a. 2n egg and sperm
    b. double the number of chromosomes in the daughter cells
    c. half the number of chromosomes in the daughter cells
    d. zygote

13. How many nuclear divisions occur during meiosis?
    a. 0
    b. 1
    c. 2
    d. an unknown large number

14. Which of the following processes results in only one of the haploid cells being viable?
    a. meiosis
    b. mitosis
    c. oogenesis
    d. spermatogenesis

15. Meiosis is called reduction division. When does the reduction in chromosome number occur?
    a. anaphase I
    b. anaphase II
    c. both metaphase I and metaphase II
    d. interphase

16. Crossing over results in ____________.
    a. cancerous growths
    b. gametes
    c. new genetic combinations
    d. new species

17. What is independent assortment?
    a. exchange of genetic material between sister chromatids
    b. rearrangement of alleles along the chromosome
    c. rearrangement of chromosomes during meiosis
    d. survival of the sperm as they travel to the egg

18. What is nondisjunction?
    a. failure of chromosomes or sister chromatids to separate during mitosis
    b. failure of chromosomes or sister chromatids to separate during meiosis
    c. failure of chromosomes to separate during mitosis
    d. failure of the cells to separate to form twins

19. Assuming fertilization by a normal sperm, nondisjunction in the female could lead to which of the following?
    a. X0
    b. XXX
    c. XXY
    d. all of the above

20. Which of the following is not a correct match?
    a. XXX – Turner syndrome
    b. XY – Normal male
    c. XXY – Klinefelter syndrome
    d. XYY – Jacob syndrome

Short Answer Questions

Read each of the following questions carefully. Jot down the main points you want to include in your answer. Then write a well-organized explanation.

1. Explain why the process of meiosis is necessary for the formation of gametes to maintain the number of chromosomes constant from one generation to the next.

2. Suppose that you could conduct a laboratory analysis that indicated the amount of DNA in a cell. During what part of the cell cycle would you expect the amount of DNA to increase and why?

3. Describe what a normal karyotype would look like and the difference between the karyotype of a male and a female.

4. What steps might be taken to determine if a fetus has an abnormal chromosome number?

Answer Key

Review Questions

A. Since we are describing a cycle, we can begin anywhere. We will begin with the adult man and woman who each produce gametes by the process of meiosis. By this process, the number of chromosomes in the egg and sperm are 1/2 the number of chromosomes in the body cells. A cell with one-half the number of chromosomes of the adult is called n, haploid. The egg and sperm combine by fertilization to produce a zygote that is diploid (2n). This single cell becomes a multicelled organism by the process of mitosis. When the infant becomes an adult, he or she will make sperm or eggs by the process of meiosis.

B.
1. f
2. j
3. e
4. g
5. h
6. c
7. i
8. b
9. d
10. a

C.
1. interphase and cell division
2. nuclear division and cytoplasmic division
3. meiosis
4. mitosis

D.
1. Interphase: DNA replication to form two sister chromatids, duplication of centriole.
2. Prophase: chromosomes condense, nuclear envelope breaks down, mitotic spindles form, centrioles move to opposite ends of cell
3. Metaphase: chromosomes consisting of two sister chromatids line up at equatorial plate
4. Anaphase: sister chromatids separate at the centromere becoming independent chromosomes and move toward opposite poles of the cell
5. Telophase: nuclear envelope forms, spindle disassembles, chromosomes become threadlike
6. Cytokinesis: cytoplasm divides results in two separate daughter cells.

E.
A. b
B. d
C. a
D. c

F.
1. mitosis
2. meiosis
3. meiosis
4. mitosis
5. meiosis
6. meiosis
7. meiosis
8. meiosis
9. mitosis
10. mitosis

G. Student drawing should show something similar to Figure 19-15.

H.

Syndrome	Chromosome Characteristics	Frequency of live births	Description
Down	Trisomy 21	1/800-1000	Mental retardation, short stature, characteristic facial features, many have defects due to small openings in the heart and digestive tract.
Turner	X0	1/2000 female births	After puberty, shorter than peers, wide chest, underdeveloped breasts, sterile.
Klinefelter	XXY	1/500-1000 male births	After puberty, small testes and penis, low levels of testosterone, lacking secondary sex characteristics, slight breast development, sterile.
Triple X	XXX	1/1000	Menstrual disturbances, early menopause, fertile
Jacob	XYY	1/1000 male births	Taller than normal, slightly lower intelligence

Critical Thinking Questions

1. When homologous pairs of chromosomes line up at the equatorial plate during metaphase I in meiosis, it is random which one of the two homologues is drawn toward a certain pole. For example, chromosome 1 that was originally from the mother (egg) may go to pole A, while chromosome 2 originally from the father (sperm) is drawn toward the same pole. This rearranges the chromosomes into the sperm or egg presents a new set of chromosomes available for each fertilization.
2. The body naturally aborts abnormal embryos very early in development. Therefore, the true frequency of occurrence is unknown. All that can be documented are the number of newborns with the syndrome.
3. Nondisjunction occurs when homologous chromosomes fail to separate during meiosis I or sister chromatids fail to separate during meiosis II.
4. If nondisjunction occurs in the egg so that the egg with two X chromosomes is fertilized by a normal Y sperm, then the child will be XXY. If nondisjunction occurs in the formation of the sperm and an XY sperm fertilizes a normal egg, then the child will be XXY. The abnormal gamete could be either an egg or sperm.
5. Rapidly dividing cells are grown in culture and then treated with a chemical that destroys the mitotic spindle halting cell division at metaphase. These cells are stained, fixed and photographed. Then the images of the chromosomes are arranged by length and the location of the centromere. The chromosomes can then be identified and irregularities in the number of chromosomes can be identified. Entering the search words of a specific abnormality followed by karyotype should elicit some human karyotypes.

PRACTICE TEST

1. c
2. b
3. c
4. b
5. b
6. c
7. a
8. a
9. c
10. c
11. c
12. c
13. c
14. c
15. a
16. c
17. c
18. b
19. d
20. a

SHORT ANSWER QUESTIONS

1. During fertilization the nuclei of the sperm and egg unite. The total number of chromosomes found in the zygote is the number in the egg and the number in the sperm. If the egg and sperm did not carry one-half the number of chromosomes characteristic of the adult, then there would be a doubling in the number of chromosomes with each successive generation.
2. DNA replication occurs during interphase. This is when the DNA is duplicated to form sister chromatids. This occurs during the S, synthesis, phase of interphase.
3. A karyotype is a photograph of chromosomes that have been stained after cell division has been stopped during metaphase. This is when the two sister chromatids are yet attached, so most chromosomes would have a centromere and arms of varying lengths. The karyotype would show the chromosomes paired and arranged by size and location of the centromere. A male and female karyotype would show 22 pairs of autosomal chromosomes and one pair of sex chromosomes. In the female the sex chromosomes would be XX and in the male they would be X with a much smaller Y.
4. Fetal cells would need to be collected, probably through amniocentesis. The cells are centrifuged and collected, then stained and photographed. They are then arranged into a karyotype. The chromosomes are counted to determine if and where an abnormality might occur.

Chapter 20

The Principles of Inheritance

Objectives

After reading the text and studying the material in this chapter, you should be able to:

- Draw, label and explain the structure of a chromosome.
- Define, give an example, and use the terms: homozygous, heterozygous, dominant, recessive, genotype and phenotype.
- Explain the laws of segregation and independent assortment and tell why they are important to genetic variability.
- Predict the phenotype and genotype ratios of offspring of a genetic cross using a Punnett square.
- Interpret a pedigree and identify a carrier.
- Explain the relationship between the dominant and recessive alleles in cases of complete dominance, codominance and incomplete dominance. Use a Punnett square to show the inheritance patterns in each case.
- Define and give an example of pleiotropy, multiple alleles and polygenic inheritance.
- Explain what it means for genes to be linked. Demonstrate the inheritance pattern of X-linked genes.
- Describe how a gene might be sex-influenced.
- Explain how chromosome deletions and duplications affect the production of proteins.
- Identify the inheritance patterns and describe common genetic disorders in humans.
- Describe the procedures used in amniocentesis and chorionic villi sampling and explain the pros and cons of each procedure.

Chapter Summary

Chromosomal genetics follows the inheritance of simple traits from one generation to another. Genetic information is carried on chromosomes that are carried in the egg and sperm in equal numbers. The 23 chromosomes received from one parent pair with the 23 chromosomes from the other parent to form **homologous pairs of chromosomes**. Segments of DNA, called **genes**, code for a specific protein. The actions of one or more proteins results in a **trait**, or inherited characteristic. Although the same gene is located on both chromosomes of a homologous pair, each copy of the pair is often slightly different and are called **alleles**. One allele for a given trait is inherited from each parent. Individuals with the same two alleles of a given gene are said to be **homozygous**. Those with different alleles of the same gene are heterozygous. When the effects of an allele can be detected regardless of the alternative allele, then that allele is described as **dominant**. If the effects of an allele are masked in the **heterozygous** condition, then the allele is described as **recessive**. **Genotype** refers to the alleles that are present, the genetic composition of an individual, whereas, **phenotype** refers to the observable physical traits of an individual.

The **law of segregation** says that during gamete formation the two alleles separate as the homologous chromosomes move toward opposite ends of the cell during meiosis. In addition, each chromosome is inherited independent of the other chromosomes following the **law of independent assortment**.

Mendel studied how single genes are inherited from parent to offspring first using **monohybrid crosses**. A **Punnett square** is a useful tool for determining the probable outcome of genetic crosses. It is a matrix where the rows represent the possible gametes of one parent, the columns the possible gametes of another parent, and the boxes the possible combinations of gametes. A **pedigree** is a chart showing the genetic connections among individuals in a family. It is especially useful in following recessive alleles that are not visible in the heterozygote. Someone who displays the dominant phenotype but is heterozygous for a trait is a **carrier** of the recessive allele.

A dominant allele often produces a protein that the recessive allele does not. This is the case in **albinism** where the ability to produce the brown pigment melanin is lacking. The ability to produce melanin depends upon the enzyme tyrosinase, which is produced by the dominant allele. Since albinism is a recessive condition, there are no alleles to produce the enzyme. This is an example of an inheritance pattern where one allele demonstrates **complete dominance** over the other in the heterozygous condition.

Codominance is the case when the effects of both alleles are apparent to their fullest extent in a heterozygote. This is the case in the blood type AB where the protein products of both the A and B alleles are expressed on the surface of the red blood cell.

Incomplete dominance is the expression of the trait in a heterozygous individual that is in between the way the trait is expressed in the homozygous dominant or homozygous recessive person. The sickle-cell allele shows incomplete dominance. The dominant and recessive alleles are both present in the heterozygote, but one allele is not clearly dominant, nor do they blend. Instead, under most circumstances, the dominant allele takes over and the person is healthy, but under certain circumstances of low oxygen content, the blood will clump and sickle. The heterozygote is described as having sickle-cell trait (Hb^AHb^S). Sickle-cell anemia is an example of one gene having many effects, called **pleiotropy**. The effect of the sickle hemoglobin affects many areas of the body.

Many genes have more than two alleles. When three or more forms of a given gene exist, they are referred to as **multiple alleles**. Although these multiple forms of the allele exist across many people in the population, only two alleles can exist for one trait in any one person. The ABO blood group has three alleles, I^A, I^B and I^O. However, any one person can only have two of these alleles.

The expression of most traits shows great variability because they are controlled by many genes as is the case for height, hair color, and eye color. This variation, independent of environmental influences, results from **polygenic inheritance** or the involvement of two or more genes in the determination of the trait.

Genes on the same chromosome tend to be inherited together because an entire chromosome moves into a gamete as a unit. Genes that are usually inherited together are described as being **linked**. However, crossing over can change the linked pattern of genes.

The two sex chromosomes, X and Y, are very different in size and in the amount of genetic information they carry. The Y chromosome is much smaller than the X chromosome and therefore carries fewer genes. Thus most genes on the X chromosome have no corresponding alleles on the Y chromosome and are known as **X-linked genes**. Since the Y chromosome is so small, virtually all of the alleles on the X chromosome in a male will be expressed, whether dominant or recessive. On the other hand, the female (XX) demonstrates the same patterns as complete dominance. Common disorders caused by X-linked recessive alleles are red-green color blindness, two forms of hemophilia, and Duchenne muscular dystrophy.

Sex-influenced genes are autosomal genes whose expression is influenced by sex hormones. Male pattern baldness is more common in men than in women because its expression depends on both the presence of the allele for baldness and the presence of testosterone. The allele for baldness acts as a dominant allele in males because of their high level of testosterone and as a recessive allele in females because females have a much lower testosterone level. A male will develop male pattern baldness whether he is homozygous or heterozygous for the trait. However, only women who are homozygous for the trait will develop male pattern baldness.

Chromosome breakage is usually caused by chemicals, radiation or viruses and results in changes in the structure and function of the chromosome. The loss of a piece of chromosome is called a **deletion**. The most common deletion is when the tip of a chromosome breaks off. An added piece of chromosome is called a **duplication**. The effects of a duplication depend on its size and position. Genetic disorders also occur when certain sequences of DNA nucleotides are duplicated multiple times as in fragile X syndrome.

Certain genetic disorders can be detected before birth by **amniocentesis** or **chorionic villi sampling (CVS)**. Prenatal genetic testing is recommended if a defective gene runs in the family or when the mother is older than 35 due to increased risks of nondisjunction. In amniocentesis 10-20 ml of amniotic fluid are withdrawn which contain epithelial cells of the fetus. These cells are cultured and then examined for abnormalities in the number of chromosomes and the presence of certain alleles that are likely to cause specific diseases. Biochemical tests are also done on the fluid to look for certain chemicals that are indicative of problems. This test is usually performed 14-18 weeks into the pregnancy. CVS involves taking a small piece of chorionic villi, fingerlike projections of the chorion. The tissue sample is then examined for abnormalities. This procedure can be done 6-8 weeks into the pregnancy, but carries a slightly higher risk of miscarriage than amniocentesis.

Many predictive genetic tests are now available or are being developed. Some of these tests identify people who are at risk for a specific disease before symptoms appear. Other predictive gene tests look for alleles that might predispose one to a disorder.

Key Concepts

- Humans have 46 chromosomes in their body cells arranged in 23 homologous pairs with one member of each pair inherited from each parent.
- A segment of a chromosome that directs the synthesis of a protein is a gene. The gene-determined protein can influence whether a certain trait will develop. The different forms of genes, called alleles, are inherited from different parents.

- Homozygous individuals have two copies of the same allele for a gene while heterozygous individuals have different alleles for the same gene.
- When the effects of dominant allele can be detected, regardless of whether an alternative allele is also present. The effects of a recessive allele are masked in the heterozygous condition.
- Genotype refers to the alleles that are present, whether homozygous or heterozygous, and phenotype refers to the observable physical traits of an individual.
- Because of the mechanisms of meiosis, alleles segregate and assort independently during gamete formation.
- Monohybrid crosses involve only a single trait.
- A pedigree shows the genetic connections among individuals in a family. It is especially useful in the identification of a carrier, a heterozygote showing the dominant trait but possessing a recessive trait.
- Albinism is an example of when the dominant allele results in the production of a critical protein, such as tyrosinase, that is not produced by the recessive allele.
- Complete dominance is when the heterozygote shows the dominant trait. Codominance is when the effects of both alleles are separately apparent in the heterozygote. Incomplete dominance occurs when the expression of the trait in the heterozygote is in between the expression of the trait in the homozygous recessive and the homozygous dominant individual.
- Pleiotropy is when one gene has many effects.
- Although multiple alleles, three or more forms of a given gene, may control a trait, an individual can only have two alleles, one from each parent. Some traits showing great variation are polygenic, controlled by two or more genes, as is the case in skin and eye color.
- Genes on the same chromosome are usually linked and therefore inherited together.
- The Y chromosome is much smaller than the X chromosome and carries very few genes, and allows for the expression of the alleles carried on the X chromosome. The genes with no homologue on the Y chromosome are called X-linked genes.
- Sex-influenced genes are autosomal genes whose expression is influenced by hormones.
- Occasionally a deletion or duplication can change the structure of a chromosome.
- Amniocentesis and chorionic villi sampling (CVS) can detect genetic disorders before birth. Other predictive gene tests are available after birth as well.

Study Tips

This chapter contains definitions that are critical to your understanding of inheritance patterns. Make flashcards with the bolded terms from the chapter, the definition and an example and review them daily.

Several inheritance patterns are discussed in the chapter including complete dominance, codominance and incomplete dominance. Practice using a Punnett square to solve problems of these types. Do many examples and describe the results using the terms heterozygous, homozygous, dominant and recessive. Be able to calculate the ratios and percentages of each potential zygote. Become familiar with the symbols used in the pedigree so that you can interpret it. As you review the text and practice genetics problems, note common examples for each inheritance pattern.

Understand the difference between sex-influenced traits and X-linked traits. Be able to solve genetic problems of X-linked traits and know common traits that are X-linked.

Finally, be able to explain how and when amniocentesis and chorionic villi sampling is used. Discuss the pros and cons of these methods and of genetic testing.

Review Questions

A. Label the drawing of a chromosome pair shown below with the terms alleles, gene, homologous pair of chromosomes and single chromosome. Then match the definition in Column A to the labels on the drawing.

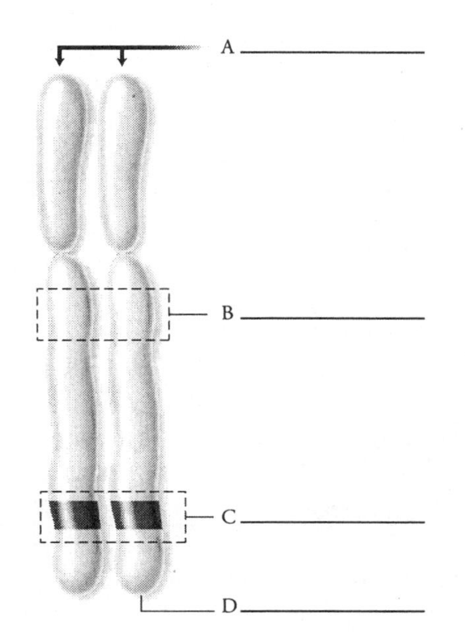

Column A

1. A segment of DNA located in a specific site on a specific chromosome that controls the production of proteins.
2. Structure made of DNA and histones that carries genetic information.
3. Chromosomes that carry the same genes. One chromosome is inherited from the mother and one from the father.
4. An alternative form of a gene that is inherited from either the mother or father.

B. Match the term in Column A with the definition in Column B. Then provide an example for each term in Column A.

Column A	Example	Column B
[There are several good examples for each of these terms.]		
1. dominant	________	a. allele whose effects are masked in the heterozygote
2. genotype	________	b. one gene having many effects
3. heterozygous	________	c. the genetic makeup consisting of both alleles
4. homozygous	________	d. genes on the X chromosome with no Y counterpart
5. polygenic inheritance	________	e. genes found on autosomal chromosomes influenced by sex hormones
6. pleiotropy	________	f. having two different alleles for one gene

7. phenotype ________ g. two or more genes determine a trait

8. recessive ________ h. allele fully expressed in the heterozygote

9. sex-influenced trait ________ i. two identical alleles for one trait

10. X-linked gene ________ j. observable characteristics

C. Complete the Punnett square below and then answer the questions that follow. The trait is the presence or absence of a widow's peak. The presence of a widow's peak is dominant (W) and the absence of it, or a straight hairline, is recessive (w). Show a cross between a homozygous dominant (WW) and a heterozygous (Ww) individual.

1. What are the ratios of the genotypes that might result from this cross? ________________

2. What are the possible phenotypes that might result from this cross? ________________

D. Complete the Punnett square below and then answer the questions that follow. Attached earlobes are a recessive phenotype. Free (unattached) earlobes is a dominant trait. Show a cross between two heterozygous individuals.

1. What are the ratios of the genotypes that might result from this cross? ________________

2. What are the possible phenotypes that might result from this cross? What are the ratios of those phenotypes? ____

E. The allele I^A is codominant with the allele I^B and both I^A and I^B are dominant to I^O. What are the possible phenotypes and genotypes that result from a cross between someone with type B blood and someone with type O blood. Be sure to consider all possible genotypes. Use the results shown in your Punnett squares to answer these questions.

1. What are the ratios of the genotypes that might result from this cross? ________________

2. What are the possible phenotypes that might result from this cross? What are the ratios of those phenotypes? ____

F. A woman who is known to be a carrier of colorblindness and a man who has normal color vision mate. Show the possibilities for offspring in the Punnett square below and then answer the questions that follow.

1. What are the chances that they will have a daughter with colorblindness? ______________

2. What are the chances that they will have a son with colorblindness? ________________

G. Study the pedigree below that shows the inheritance of hemophilia through a family. Then answer the questions that follow. The shading indicates a person with hemophilia. The woman in the original couple was not a carrier.

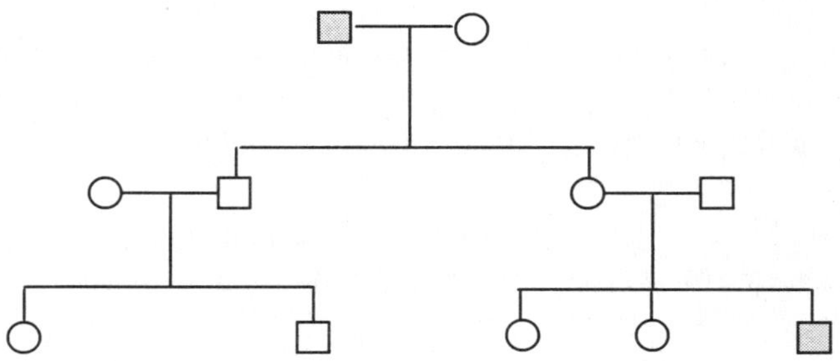

1. Circle the girls who are carriers.

2. Why aren't any of the daughters hemophiliacs? ________________________________

____

3. What would change if the woman in the original couple was a carrier? ________________

____

H. Answer the following questions concerning prenatal genetic testing.

1. What is amniocentesis and how is it performed? ________________________________

____

____

2. What are the risks associated with amniocentesis? ______________________________

____

3. What is CVS? ____

____

4. What are the advantages and disadvantages of CVS? ____________________________

____

5. Does all genetic testing have to be done before birth? ___________________________

____

Critical Thinking

Read each of the following questions carefully. Circle the most important words in the question before formulating your answer.

1. When one allele is completely dominant over another, the phenotype is the same whether the individual is homozygous dominant or heterozygous. What cross would identify the dominant phenotype as either homozygous dominant or heterozygous?

2. Liz is type B blood and Dan is type A blood. Their baby has type AB blood. They could not understand how their baby could have a blood type that neither parent possessed. Explain this to them.

3. As Ashleigh looked about her third grade classroom, she noticed how different people looked. There must have been 10 different shades of skin color with a great variety of hair colors as well. Explain why there are only a few blood types, why someone either has dimples or not, but why there can be such variety in hair, skin and eye color.

4. Male pattern baldness is quite common. Now and then a woman will show the same pattern of baldness, usually as she ages into her sixties. Why are there more bald men than women when baldness is not sex-linked and why would baldness show up later in life for women?

5. There are now many predictive genetic tests. They can identify people who are at risk for getting a disease before the symptoms appear. In some cases, this knowledge can help doctors prevent the disease from occurring, in other cases, it just gives the individual and family more time to plan for the inevitable. Such testing could save millions in medical costs. However, in some instances although the gene may be present for a certain disease, it will never develop. Discuss the pros and cons of genetic testing, or genetic screening as it may be used in industry to influence employability or in the selection of a mate. Specifically deal with the question whether genetic testing should become mandatory.

Practice Test

Choose the one best answer to each question that follows. As you work through these items, explain to yourself why each answer you discard is incorrect and why the answer you choose is correct.

1. What is the segment of DNA inherited from your mother for a certain characteristic called?
   a. allele
   b. chromosome
   c. dominant
   d. trait

2. Which of the following show a heterozygous genotype?
   a. BB
   b. Bb
   c. bb
   d. light blue

3. Which of the following is a phenotype?
   a. BB
   b. Bb
   c. bb
   d. light blue

4. How can you determine the phenotype of someone's eye color?
   a. by looking at them
   b. by test-breeding
   c. by testing for linkages
   d. None of the above will work.

5. In a cross between a homozygous recessive and a heterozygote, one half of the offspring will be ____________.
   a. Homozygous recessive
   b. Homozygous dominant
   c. None of the above
   d. Cannot be determined with this information.

6. All chromosomes that are not X or Y are called ____________.
   a. autosomal
   b. dominant
   c. recessive
   d. sex-linked

7. Tay-Sachs disease results when someone is homozygous recessive for the autosomal trait (tt). If two parents who did not have Tay-Sachs gave birth to a baby with Tay-Sachs, what are their genotypes?
   a. TT x TT
   b. TT x Tt
   c. Tt x Tt
   d. tt x tt

8. If Rh incompatibility arises when the mother and baby are of different Rh types, might this couple have problems with their second child? Assume that their first child was Rh^+ and that the mother is Rh^- and the father is Rh+. Rh^- is a recessive trait.
   a. Yes, problems for sure if the father is heterozygous.
   b. 25% chance if both parents are heterozygous.
   c. Maybe, depending upon the father's genotype.
   d. Absolutely not.

9. In humans, the ability to roll the tongue is a dominant trait. The inability to roll the tongue is a recessive trait. If two individuals homozygous recessive for this trait have a child, what is the chance that the child will be able to roll his tongue?
   a. 0%
   b. 25%
   c. 50%
   d. 75%
   e. 100%

10. What type of blood will the offspring of these parents have?

    Mother: AB   Father: O

    a. AB
    b. A or B
    c. O
    d. Any one of the above.

11. Unattached earlobes is dominant to attached earlobes. Two parents both with unattached earlobes had a child with attached earlobes. What are the chances that their next child will have attached earlobes?
    a. 0%
    b. 25%
    c. 50%
    d. Cannot be determined from this information

12. The AB blood type is an example of ___________.
    a. blending
    b. codominance
    c. complete dominance
    d. incomplete dominance

13. While waiting in line for the school pictures, Alicia noticed that there was a lot of variation in height among her classmates. From this observation what do you know about the inheritance pattern of height?
    a. It is polygenic
    b. It is sex-linked.
    c. Several of the children are taking growth hormones.
    d. Tallness is dominant to shortness.

14. Which of the following is most likely to disrupt the normal linking pattern of genes?
    a. allergic response in the cell membrane
    b. crossing over
    c. desensitization of the alleles
    d. unlinking

15. What is the inheritance pattern shown in the pedigree below?

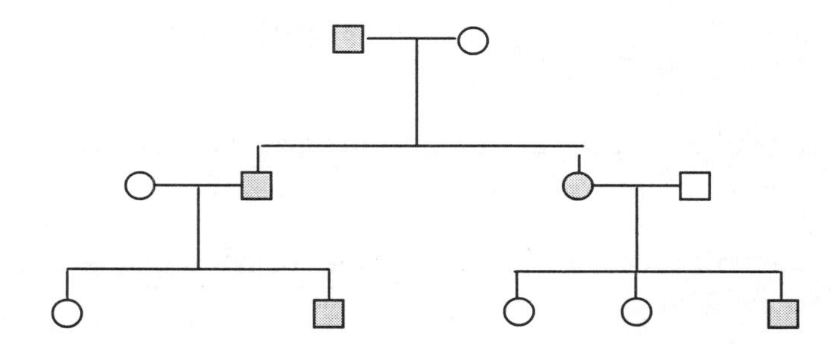

    a. autosomal dominant
    b. autosomal recessive
    c. sex-influenced
    d. sex-linked

16. Both the mother and father of a son with colorblindness have normal vision. From whom did the son inherit the gene for colorblindness?
    a. father
    b. grandfather
    c. mother
    d. unknown

17. Which of the following will be least noticed?
    a. chromosomal addition
    b. chromosomal deletion
    c. chromosomal substitution
    d. all are fatal

18. Which of the following can be done at the earliest stage of development?
    a. amniocentesis
    b. blood testing
    c. chorionic villi sampling
    d. genetic testing

19. Which of the following removes a tissue sample from the placenta?
    a. amniocentesis
    b. blood testing
    c. chorionic villi sampling
    d. genetic testing

20. Which of the following tests would be used to identify the genetic characteristics of an adult?
    a. amniocentesis
    b. chorionic villi sampling
    c. collection of fetal epithelial cells
    d. genetic testing

Short Answer Questions

Read each of the following questions carefully. Jot down the main points you want to include in your answer. Then write a well-organized explanation.

1. What is the difference between a gene and a trait and what is the difference between a gene and an allele?

2. Freckles are a dominant trait. Two people with freckles marry and have a son with no freckles. How can this happen?

3. A couple planning to have children walked into the Sickle Cell Clinic. Their neighbor had a child with sickle-cell anemia and they knew it was a burden. They wanted to know whether they could have a child with sickle-cell anemia. Prior to blood testing, you interviewed them. Both said that no one in their family had ever had sickle-cell and that woman noted that her brothers had both been tested and neither was a carrier. Based on this information, what would you tell them? Could they be a carrier and not know it?

4. Gerald and Stella have four children. They have two boys and two girls. One of the boys and one of the girls are colorblind. What is the genotype of the parents?

5. Many families have genes for hemophilia. Long ago, women who were hemophiliacs died in childbirth. Why didn't this remove the gene from the population?

6. Huntington disease is a recessive condition that does not appear until someone is middle aged and has already reproduced. What is a pedigree and what would be its value to a family with Huntington disease?

Answer Key

Review Questions

A. A. homologous pair of chromosomes
   B. gene
   C. alleles
   D. single chromosome

   1. b
   2. d
   3. a
   4. c

B. There are several good examples for each of these terms.
   1. h, freckles
   2. c, aa, Aa
   3. f, Bb
   4. I, dd, PP
   5. g, skin color
   6. b, sickle-cell anemia
   7. j, blue eyes
   8. a, albinism
   9. e, baldness
   10. d, colorblindness

C.

	W	W
W	WW	WW
w	Ww	Ww

1. WW:2 Ww:2
2. All of the offspring will have a widow's peak.

D.

	F	f
F	FF	Ff
f	Ff	ff

1. 1 FF : 2 Ff : 1 ff
2. 3 Free : 1 attached

E. 1. If the type B parent is homozygous, then all offspring will be heterozygous I^BI^O. If the type B parent is heterozygous then 2 I^BI^O : 2 I^OI^O
   2. If the type B parent is homozygous, then all offspring will be type B. If the type B parent is heterozygous, then there is a 50:50 chance that the offspring will be type B and a 50:50 chance they will be type O.

F. A woman who is known to be a carrier of colorblindness and a man who has normal color vision mate. Show the possibilities for offspring in the Punnett square below and then answer the questions that follow.

	X^C	X
X	XX^C	XX
Y	X^CY	XY

1. There is no chance that a daughter will be colorblind. However, one-half of the daughters will be carriers of the colorblindness gene.
2. There is a 50% chance that they will have a son with colorblindness.

G. 1.

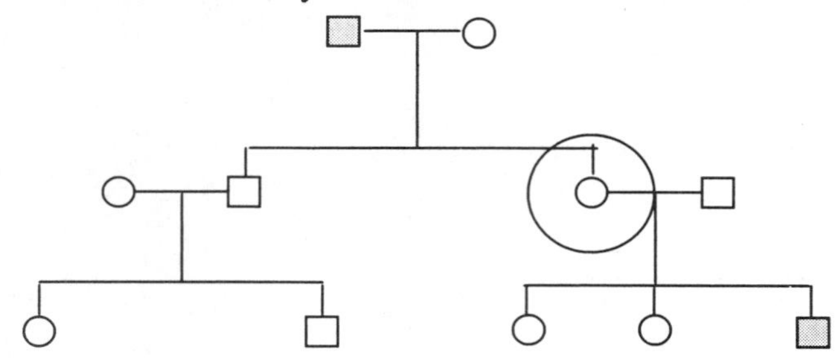

2. Because a daughter would have to inherit a gene from both parents to be hemophiliac and the mother was not a carrier.
3. The first generation daughter would certainly be a carrier and could possibly be hemophiliac.

H. 1. Amnionic fluid is withdrawn about 14-18 weeks into the pregnancy. The cells are grown and then centrifuged and collected. The fetal cells are then examined for abnormalities in the number of chromosomes. Sex can be determined at this time. Biochemical tests are done on the fluid.
2. There is some risk that the needle will puncture the fetus or umbilical cord. Sometimes removal of amnionic fluid causes miscarriage.
3. CVS involves taking a small tissue sample of the chorionic villi and analyzing it for chromosomal abnormalities. It is done in the 6th to 8th week of pregnancy.
4. CVS can be done earlier in the pregnancy than amniocentesis, but there is still a risk of inducing miscarriage.
5. No, there are many genetic tests that can be done after birth that will identify genetic abnormalities that exist at that time and that will predict whether a genetic abnormality will develop later in life.

Critical Thinking Questions

1. If a homozygous recessive individual is crossed with both the homozygous dominant and the heterozygous individual, the genotype can be determined. A homozygous dominant when crossed with a homozygous recessive (TT x tt) yields 100% heterozygous offspring who will show the dominant trait. A homozygous recessive when crossed with a heterozygous individual (Tt x tt) yields 50% heterozygotes that will show the dominant phenotype and 50% homozygous recessive that will show the recessive phenotype. This is called a test cross.
2. To be type AB, the baby needs only to receive the I^B allele from Liz and the I^A allele from Dan. These two alleles combine to form type AB. This could happen if Dan and Liz were homozygous or heterozygous.

3. Many traits such as dimples are determined by single genes with only two alleles. The dominant phenotype is apparent in the homozygous dominant or heterozygous individual and the recessive phenotype is apparent in the homozygous recessive individual. However, other traits are controlled by multiple alleles, sometimes as many as 10 to 15. The various combinations of these multiple alleles yield great variation in the traits they control.
4. Male pattern baldness is a sex-influenced trait. It depends upon the presence of both the allele for baldness and the presence of testosterone. The allele for baldness acts as a dominant allele in men because their high level of testosterone and as a recessive allele in women because of the much lower testosterone level. When a woman develops baldness, it is usually after she reaches menopause and the supply of estrogen declines allowing more testosterone to influence the occurrence of baldness.
5. A primary complication to genetic screening is that not all genes are expressed. Someone may have a detrimental gene that would forecast a terrible disease, but it may never manifest itself because of other masking genes. Genetic screening would falsely discriminate against this person. Employers might be tempted to consider the insurance costs of employing someone with a certain gene set rather than the skills they bring to the job. In addition, people would naturally resist the accumulation and publication of such personal data that may reflect poorly on themselves and their families.

Practice Test

1. a
2. b
3. c
4. a
5. a
6. a
7. c
8. c
9. a
10. b
11. b
12. b
13. a
14. b
15. d
16. c
17. c
18. c
19. c
20. d

Short Answer Questions

1. A gene is a segment of DNA that codes for a specific protein. The expression of this protein is the trait or characteristic as it develops in the person. For example, the trait for eye color is determined by several genes that can produce a great variety of eye colors. Genes occur on both chromosomes of a homologous pair, but they are not always identical. One form of the gene as inherited from the mother can be different from the slightly different form appearing on the chromosome inherited from the father. These two forms of the same gene are called alleles.
2. Both of the individuals with freckles are heterozygotes, Ff. They show the phenotype of freckles but carry the recessive gene. There is a 25% chance in any pregnancy that the child will be homozygous recessive, ff, and show no freckles.
3. Sickle-cell trait is the heterozygous condition and under normal circumstances (air pressure), there are no symptoms; therefore, someone could be a carrier and not know it. Since there have been no cases in the immediate family of sickle-cell anemia, it is unlikely that both of these individuals will be carriers. However, the best advice is to have a simple blood test that will tell exactly whether they carry the gene for sickled cells.
4. Since at least one of the girls is colorblind, both parents must carry that allele on their X chromosome. Therefore the father must be colorblind and have the genotype X^CY. Since only one of the girls is colorblind, the mother must possess at least one X chromosome for normal vision since the girl will receive an X chromosome from the father with the gene for colorblindness. Therefore, the mother must be a carrier, or heterozygote, X^CX.

5. The women had already passed their gene for hemophilia on to their newborn, whether a male or female, before dying.
6. A pedigree is a chart showing the genetic connections among individuals in a family. Family or medical records are used to fill in the pattern of expression of a trait. In the case of a disease that results in the homozygous recessive individual, the pedigree can help identify carriers and provide them with this information prior to conceiving.

Chapter 21

DNA and Biotechnology

Objectives

After reading the text and studying the material in this chapter, you should be able to:

- Describe and draw the structure of a DNA molecule.
- Explain why complementary base pairing ensures reliable replication of the genetic code from DNA to DNA and transcription from DNA to mRNA.
- Describe the purpose and mechanism of DNA replication.
- Compare DNA and RNA in structure and function.
- List the three forms of RNA and their functions in protein synthesis.
- Describe the details of protein synthesis including transcription, translation and the genetic code.
- Differentiate between chromosomal and point mutations. Describe the various types of point mutations and their relative impact on protein synthesis.
- Explain the mechanisms used to control gene activity.
- Define genetic engineering and give examples of how it has been used in various fields.
- Describe the steps involved in making recombinant DNA.
- Explain gene therapy and some potential uses for it.
- Describe a DNA fingerprint.
- Describe the purpose and value of the Human Genome Project.

Chapter Summary

DNA (deoxyribonucleic acid) is a double-stranded molecule twisted in a helix. It is made of repeating units called **nucleotides**, which are composed of one sugar (deoxyribose), one phosphate and one nitrogenous base. If we view DNA as a twisted ladder, the sides of the ladder are alternating sugar and phosphate molecules and the rungs are formed by pairs of bases. Because of the shape of the bases and the pattern of hydrogen bonds, there is always **complementary base pairing** thus, adenine (A) only pairs with thymine (T) and cytosine (C) pairs only with guanine (G). Each pair is held together by hydrogen bonds. Genetic information is encoded in the exact sequence of bases.

Each time a cell divides, the DNA of the original cell must be exactly copied as the daughter cells are formed. During **replication** each original strand of DNA serves as a template for a new strand. Because of the specificity of the pairing of the bases, one strand of DNA can determine the sequence of bases in the other strand. DNA replication begins when an enzyme breaks the hydrogen bonds that hold together the paired bases of the parent molecule. This lets the two strands of the double helix separate and unwind. Bases of nucleotides floating in the nucleus then attach to the bases exposed by this "unzipping." The new strand of DNA is constructed so that A always pairs with T and C always pairs with G (complementary base pairing). **DNA polymerases** link the new nucleotides together along the sugar-phosphate bonds to form a new single strand of the double helix. Now there are two complete strands of DNA, each identical to the parent molecule and each with one strand of the original, parent DNA and with one newly formed strand. This is called **semiconservative replication** because one strand of the original DNA remains in each new DNA molecule.

A specific region of the DNA molecule, a gene, directs the production of a specific protein, which is an inherited trait. For DNA to direct the production of proteins, it uses another nucleic acid, **RNA (ribonucleic acid)**. RNA differs from DNA in that it is single stranded, contains the base uracil (U) and not thymine, and it contains the sugar ribose not deoxyribose. Protein synthesis occurs in two steps. Since DNA occurs in the nucleus and proteins are made at the ribosomes in the cytoplasm, the DNA directions need to get into the cytoplasm, a process called transcription. To accomplish this, the DNA code is first used as the template to generate a smaller **messenger RNA (mRNA)** molecule, a process called **transcription**. The mRNA transcript carries the DNA code into the cytoplasm where it directs the sequencing of the amino acids to form the correct protein, a process called translation. Translation uses another RNA molecule, **transfer RNA (tRNA)** that identifies and transports amino acids to the site of protein assembly. A third type of RNA, **ribosomal RNA (rRNA)**, combines with proteins to form the ribosome, the structure on which the protein is made. Thus the final protein is merely the expression of the original DNA sequence.

Transcription, defined as the copying of the DNA base sequence into a smaller mRNA, is the first step in protein synthesis. First the DNA molecule unzips along the hydrogen bonds between the two paired bases, then free-floating RNA nucleotides pair with their complements (C-G and U-A) along the exposed DNA strand. Next **RNA polymerase** aligns and links together the RNA nucleotides. The specific sequence of DNA bases called the **promoter** starts transcription and another sequence of bases serves as a stop signal. When the mRNA transcript is completed, it is released from the DNA and the DNA closes again. Final processing of the RNA includes removal of introns, internal segments of DNA copied onto the RNA that are not used to make the final protein.

Translation is protein synthesis. The newly formed mRNA carries the DNA genetic message from the nucleus to the cytoplasm where the sequence of nucleotide bases is translated into an amino acid sequence forming a protein. The system of symbols, or code, used to convert the DNA base sequence into an amino acid sequence is called the genetic code. Here each three base sequence, called a **codon**, translates into one amino acid. The mathematics involved in taking four RNA bases and using them three at a time creates an excess of codons, and a few codons are reserved for translation stop signals.

Transfer RNA is the bridge between the language of nucleotides and proteins. Each tRNA identifies a specific amino acid and attaches to it. On the other end of the tRNA are three bases that are complementary to the codon of the mRNA. These bases are called the **anticodon** and allow the tRNA to match with the mRNA that carried the directions from the DNA in the nucleus to the cytoplasm.

Ribosomes are the manufacturing site for proteins. They consist of two rRNA molecules and a protein. The role of the ribosome is to bring the tRNA carrying its specific amino acid close enough to the mRNA to align the codon with the anticodon. This brings the amino acids close enough to each other that enzymes in the ribosomes can link them forming an amino acid chain. This process of adding amino acids to the growing protein chain is called **elongation**. When the amino acid is bound to the protein chain, it dissociates from the tRNA. **Termination** occurs when a stop codon moves into the ribosome. Since there are no tRNA anticodons to pair with the stop codon, protein synthesis is halted.

Changes in the DNA are called **mutations**. Mutations can only be passed to offspring when they occur in a gamete. Mutations occurring in somatic cells can produce problems, but cannot be inherited. **Chromosomal mutations** occur when whole sections of chromosomes become rearranged, duplicated or deleted. **Point mutations** involve changes in one or a few nucleotides in the DNA sequence. Substitutions in the base sequence can result in no change in the protein structure, can change the structure of the protein but not the function, can alter the structure and function of the protein or can make the protein nonfunctional. The effect of a base substitution depends upon the base and where it is located along the DNA strand. Mutations caused by insertion of extra bases or the deletion of bases may be more serious because they can throw off the three-base sequence needed for translation.

Gene activity can be turned on or off. Only certain genes are active in a certain type of cell; most genes are turned off. Coiling and uncoiling of chromosomes regulate gene activity at the chromosome level. When the DNA is tightly coiled, the genes are not expressed. When a certain region of a gene is needed, it unwinds allowing transcription to take place. Gene activity is also regulated by master genes that turn on a battery of other genes. These are especially important during development. Chemical signals from outside the cell can also regulate gene activity. The hormone-receptor complex is one such example. Activator proteins can activate a gene by facilitating the RNA polymerase so that transcription begins.

Biotechnology involves making a living cell perform a task considered useful by humans. **Genetic engineering** is a subset of biotechnology involving the manipulation of DNA for human purposes. The idea behind genetic engineering is to put a piece of a gene that produces something of interest, such as an enzyme or protein, into another piece of DNA forming **recombinant DNA**, or DNA made from two sources. First the gene of interest is identified and its chromosomal location determined. Then a **restriction enzyme** makes a similar cut in both the source and the host DNA. The piece of DNA removed from the source is transferred to the host by a **vector**. Usually the host cell is a rapidly reproducing cell that can be cloned in laboratory conditions, making many copies of the recombinant DNA of interest. The cloned cells can contain a variety of DNA fragments and the clone containing the gene of interest must be identified and isolated. After the gene of interest has been identified, it is helpful to amplify, or duplicate, the gene. Gene amplification can be accomplished during bacterial cloning or a **polymerase chain reaction (PCR)**.

Genetic engineering is a means to produce large quantities of desired proteins. It also allows a gene for a desirable trait to be taken from one species and transferred to the cells of another species, making a transgenic organism. Genetic engineering has had many applications in plant and animal agriculture, medicine and the environment. **Gene therapy** replaces faulty genes with functional genes and may be used to cure inherited diseases. Scientists are refining various methods that can be used to identify and replace faulty genes. Some alter the DNA structure; some snip out erroneous genes and other methods adjust the gene activity of a specific gene. The patterns of DNA fragments that have been cut by the restriction enzymes are called a **DNA**

fingerprint. These patterns are unique to an individual and can be used for identification.

The purpose of the Human Genome Project is to determine the sequence of all the DNA in a human cell, also known as the human **genome**. This sequencing phase of this project been completed and it is now hoped that this information may be helpful in treating genetic diseases.

Key Concepts

- DNA is a double helix consisting of two strings of nucleotides in a shape similar to a twisted ladder. The sides of the ladder are alternating phosphate and deoxyribose molecules. The rungs of the ladder are nitrogenous bases which pair according to the complementary base pairing principle. Adenine always pairs with thymine and cytosine always pairs with guanine. The bases are held together by weak hydrogen bonds.
- DNA must be replicated during cell division. The original strand of DNA serves as a template for a new strand during replication. This is called semiconservative replication because each double strand of DNA in the daughter cells consists of one strand of DNA from the original cell and one newly constructed DNA strand.
- RNA is a nucleic acid used during protein synthesis. RNA is single stranded, contains the sugar ribose, replaces the base thymine with uracil, and functions primarily in the cytoplasm. There are three types of RNA: messenger RNA (mRNA), transfer RNA (tRNA) and ribosomal RNA (rRNA).
- A gene contains the information needed to synthesize a protein molecule which is the expression of that gene. Protein synthesis occurs in several steps.
  - ✓ During transcription, the DNA message is copied onto a strand of mRNA and taken from the nucleus to the cytoplasm.
  - ✓ The DNA message is expressed in a three-base code called the genetic code.
  - ✓ Transfer RNA interprets the genetic code. On one end of the tRNA is a three-base sequence that aligns with the mRNA, on the other end, the tRNA binds to an amino acid.
  - ✓ Ribosomes, composed of rRNA and ribosomal proteins, are the sites of protein synthesis and bring together mRNA and tRNA allowing for protein synthesis.
  - ✓ During translation, the base sequences of the tRNA match to the base sequences of the mRNA at the ribosome. The amino acids attached to the tRNA form a protein growing chain. Translation stops when the stop codon in encountered and the amino acid chain separates from the ribosome.
- Mutations are changes in the base sequence of DNA. They can occur when whole sections of chromosomes become rearranged, duplicated, or deleted. They can also occur when only one or a few nucleotides are substituted in the DNA sequence. If nucleotides are added or deleted, the consequences can be more serious than if they are substituted.
- Gene activity is regulated by the coiling and uncoiling of the DNA molecule, master genes that turn on other genes, enhancers that increase the amount of RNA produced, and regulatory proteins such as hormone receptors.
- Genetic engineering is the manipulation of DNA for human purposes. It can be used to produce large quantities of protein or to put a desirable genetic trait into a certain species.
  - ✓ Recombinant DNA is made from more than one source and contains a gene of particular interest. Recombinant DNA is inserted into the host cell by a vector. Cloning then generates many copies of the gene of interest.

  - ✓ Genetic engineering has many applications in plant and animal agriculture, medicine, and environmental science.
  - ✓ Gene therapy introduces functional genes where faulty genes exist.
  - ✓ Each individual has a unique set of DNA restriction fragments called their DNA fingerprint.
- The Human Genome Project has determined the complete DNA sequence of the human genome.

Study Tips

Detailed molecular mechanisms are at the heart of this chapter. Much of the information in this chapter may be new and, therefore, just the terminology may be a challenge. Begin with an understanding of the structure of DNA. Then apply this knowledge to DNA replication. Next, differentiate between DNA and RNA. Study the processes involved in protein synthesis the role of the different types of RNA. Now is a good time to make a paper model and use it to demonstrate the replication and the steps of protein synthesis. Review the controls of gene activity and demonstrate how they would impact protein synthesis using your model. Explain the processes of replication, transcription, and translation to someone not in this class, describing the similarities and differences between the three processes and the way they work sequentially to express genes.

Next focus on genetic engineering and the field of biotechnology. This is where the molecular basis of genetics is applied. Develop a clear understanding of recombinant DNA and how it is made. Use the figures in the text to clarify your thoughts and explain the process to a study partner. Be able to explain how the polymerase chain reaction is used to mass produce copies of recombinant DNA. Develop a list of the uses of genetic engineering and gene therapy. Reflect on the value of the Human Genome Project.

Review Questions

A. Draw a DNA molecule and label it using the following terms: deoxyribose, phosphate, adenine, cytosine, guanine, thymine, and hydrogen bonds. Circle an entire nucleotide.

B. Indicate whether the statement below is true of DNA, RNS or both by putting a check in the appropriate column.

	DNA	RNA	
1.	______	______	Is a nucleic acid
2.	______	______	Contains the sugar deoxyribose
3.	______	______	Functions primarily in the cytoplasm
4.	______	______	Is single stranded
5.	______	______	Has four types of bases
6.	______	______	Contains the base thymine
7.	______	______	Is a double-stranded molecule
8.	______	______	Contains ribose
9.	______	______	Nucleotides are composed of a sugar, phosphate and base
10.	______	______	Functions primarily in the nucleus

C. Match the term in Column A with the function in Column B.

Column A	Column B
1. anticodon	a. synthesis of new DNA from existing DNA
2. codon	b. mRNA is made from a DNA template
3. intron	c. binds to a specific amino acid and takes it to mRNA
4. messenger RNA	d. unexpressed region of DNA later removed from the mRNA copy
5. promoter	e. process of converting RNA language into a protein
6. replication	f. synthesized from DNA template and then carries the code for amino acid sequencing
7. ribosomal RNA	g. three nucleotide sequence on tRNA
8. transfer RNA	h. sequence of DNA that signals the start of transcription

9. transcription

i. the most abundant form of RNA, structural component of protein synthesis machinery

10. translation

j. three nucleotide sequence on mRNA

C. Use the following terms to label the drawing below illustrating transcription: amino acid chain, large ribosomal subunit, mRNA, ribosome, and small ribosomal subunit.

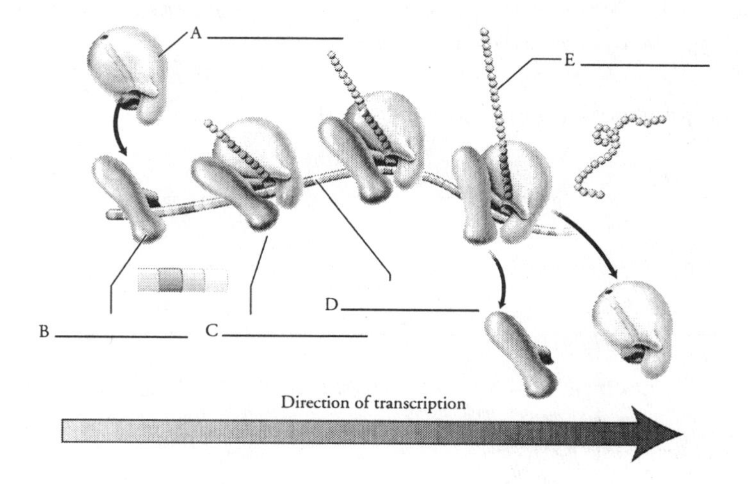

D. Draw and label tRNA molecule. Show the anticodon on the tRNA matching an appropriate sequence of bases on the codon.

E. Order the following steps of protein synthesis in putting the correct number (1-7) before the description.

________ A tRNA with appropriate anticodon pairs with the next codon on mRNA. Enzymes link the amino acids.

________ The stop codon moves into the ribosome.

________ Small ribosomal subunit joins to mRNA at start codon.

________ The first tRNA leaves the ribosome. The ribosome moves along the mRNA, exposing the next codon.

_______ A tRNA with complementary anticodon pairs with start codon. Ribosomal subunits join to form a functional ribosome.

_______ One tRNA after another donates its amino acid to the growing protein, then leaves the ribosome as others move into place along the mRNA. Enzymes bind the amino acids.

_______ Release factors cause the release of the newly formed protein and the separation of the ribosomal subunits and the mRNA.

F. List three ways that gene activity can be controlled.

1. ____________________

2. ____________________

3. ____________________

G. Write a description for each of the events involved in genetic engineering shown below.

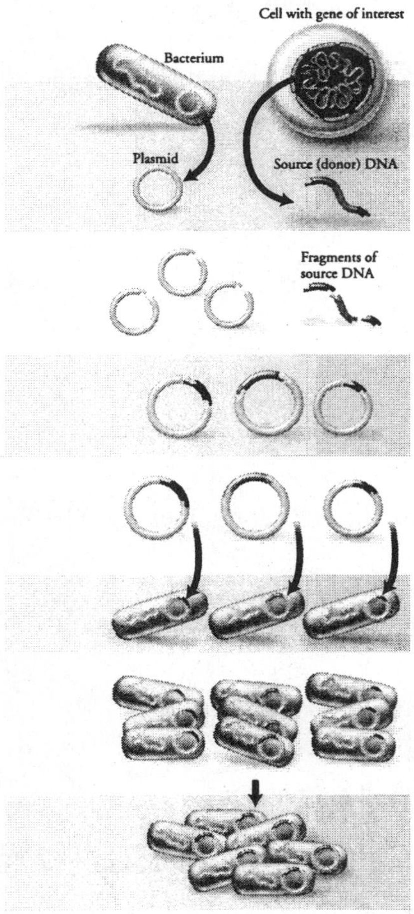

Cell with gene of interest

Bacterium

Plasmid

Source (donor) DNA

Fragments of source DNA

1. ____________________

____

2. ____________________

____

3. ____________________

____

4. ____________________

5. ____________________

____

6. ____________________

7. ____________________

____

G. Answer the following questions regarding genetic engineering and gene therapy.

1. What is one product that has been produced in large quantities as a result of genetic engineering? ______________________________

2. What is a transgenic organism? ______________________________

3. Are any transgenic organisms in existence now? What is an example? ______________

____

4. What is the purpose of gene therapy? ______________________________

____

5. Is gene therapy widely used at the current time? Does it show potential? ______________

____

Critical Thinking

Read each of the following questions carefully. Circle the most important words in the question before formulating your answer.

1. Demonstrate the process of protein synthesis to a study partner. Begin with two strips of paper representing the double-stranded DNA with complementary base pairs. Then demonstrate transcription and make a piece of mRNA with about 21 bases on it. Use smaller strips of paper to represent the tRNA with a three base code and an amino acid.

2. There are terms used in this chapter that are commonly used in other contexts, for example transcription and translation. Explain what is being transcribed and what is being translated in the process of protein synthesis.

3. Demonstrate to a study partner why a gene substitution is not nearly as serious a mutation as most gene additions or deletions. What happens if the addition or deletion is of three bases?

4. Some inherited diseases are very costly to society. If gene therapy was available, should screening for those diseases be mandatory? Should everyone who has the disease be required to undergo gene therapy for the good of society?

5. Gene legislation is an interesting concept. What does that mean to you? Does the United States have gene legislation? Does every country?

PRACTICE TEST

Choose the one best answer to each question that follows. As you work through these items, explain to yourself why each answer you discard is incorrect and why the answer you choose is correct.

1. What are the repeating units of nucleic acids?
   a. bases
   b. nucleotides
   c. phosphate molecules
   d. sugar molecules

2. If you performed a laboratory analysis of DNA, you would find that the amount of adenine is ____________ the amount of thymine.
   a. much greater than
   b. much less than
   c. shows no relationship to
   d. the same as

3. The synthesis of new DNA from existing DNA occurs in the ____________.
   a. cytoplasm
   b. mitochondria
   c. nucleus
   d. ribosome

4. Which of the following is not a valid comparison between DNA and RNA?
   a. adenine - thymine
   b. deoxyribose - ribose
   c. double stranded - single stranded
   d. found in nucleus - found in cytoplasm

5. Gene expression occurs when the ____________.
   a. cell divides
   b. DNA is replicated
   c. protein is made
   d. ribosome engages both the mRNA and tRNA

6. Which of the following best describes transcription?
   a. DNA → DNA
   b. DNA → mRNA
   c. DNA → protein
   d. mRNA → tRNA

7. What does the t in tRNA represent?
   a. transcription
   b. transfer
   c. translation
   d. transpolymerase

8. What determines the sequence of amino acids in a protein?
   a. the sequences of amino acids in another protein
   b. the sequence of amino acids in the DNA
   c. the sequence of nucleotides in the mRNA
   d. the sequence of nucleotides in the tRNA

9. The triplet code of CAT in DNA is represented as ______ in mRNA and ______ in tRNA.
   a. CAT, CAT
   b. GAA, CAT
   c. GTA, CAU
   d. GUA, CAU

10. Why does the genetic code have to be a triplet code?
    a. because there are more base pairs than amino acids
    b. so that each amino acid is coded by one base
    c. so there are enough codons for the 20 amino acids
    d. there is no reason

11. Which of the following identify and move amino acids to the site of protein production?
    a. iRNA
    b. mRNA
    c. rRNA
    d. tRNA

12. Which of the following is mismatched?
    a. DNA - code
    b. mRNA - codon
    c. rRNA - recode
    d. tRNA - anticodon

13. Which of the following is true regarding translation?
    a. DNA is being replicated in the nucleus
    b. mRNA is exposing its anticodon to the ribosome
    c. mRNA is exposing its codon to the tRNA
    d. tRNA remains attached to the amino acid after the peptide chain is formed

14. When does protein synthesis stop?
    a. When the intron has been replaced.
    b. When the stop anticodon is located on the mRNA.
    c. When the stop codon enters the ribosome.
    d. When there is enough protein.

15. Which of the following is most likely to be fatal?
    a. chromosomal mutation
    b. DNA mutation
    c. point mutation
    d. all mutations are fatal

16. Which of the following can regulate gene activity?
    a. coiling and uncoiling of DNA
    b. master genes
    c. proteins
    d. all of the above

17. Genetic engineering ____________.
    a. is a test method only
    b. is a way to make more humans
    c. is commonly used to correct inherited defects
    d. is used to develop improved plant and animal products

18. What is recombinant DNA?
    a. DNA found in transgenic plants
    b. DNA made from a pig and inserted into a human
    c. DNA made from different sources
    d. DNA that results when the sperm enters the egg

19. Which of the following is false regarding gene therapy?
    a. Gene therapy could be used to treat the genetic diseases caused by a defect in a single gene.
    b. Gene therapy is widely used in Europe with great success.
    c. Gene therapy may be effective in the treatment of cancer.
    d. Gene therapy replaces faulty genes with functional genes.

20. What is the purpose of the Human Genome Project?
    a. to archive everyone's DNA fingerprint
    b. to develop an "ideal" genetic code
    c. to determine the sequence of human DNA
    d. to mass produce important sequences of DNA

Short Answer Questions

Read each of the following questions carefully. Jot down the main points you want to include in your answer. Then write a well-organized explanation.

1. Draw the process of DNA replication. Use a DNA strand with at least 20 base pairs.

2. How does the double-stranded nature of DNA relate to its function as the carrier of our genetic identity from generation to generation and how does the smaller, single-stranded nature of RNA enhance the role that it plays in protein synthesis?

3. What does the term gene expression mean? When is the gene expressed?

4. Why does the genetic code used to convert the base sequence of DNA and then RNA into an amino acid sequence have to be a triplet code?

5. What is recombinant DNA and how is it made?

Answer Key

Review Questions

A. The student's drawing should look similar to Figure 21-1. The base pairs should be A-T and C-G. A nucleotide includes the sugar, the phosphate and the base.

B.
1. both
2. DNA
3. RNA
4. RNA
5. both
6. DNA
7. DNA
8. RNA
9. both
10. DNA

C.
1. g
2. j
3. d
4. f
5. h
6. a
7. i
8. c
9. b
10. e

C.
A. large ribosomal subunit
B. small ribosomal subunit
C. ribosome
D. mRNA
E. amino acid chain

D. Student drawing should be similar to Figure 21-5 and should indicate proper base pairing between the codon and the anticodon, A-U and C-G.

E. 3, 6, 1, 4, 2, 5, 7

F.
1. Coiling and uncoiling of the chromosomes regulate gene activity. Tightly coiled DNA cannot be transcribed.
2. Master genes can turn on or off several other genes.
3. Chemical signals such as hormones regulate gene activity.

G.
1. Genes of interest are identified in the donor cell and the vector.
2. Genes of interest are sliced from the donor cell and from the vector using a restriction enzyme.
3. Recombinant DNA is formed.
4. DNA ligase binds the DNA covalently.
5. Recombinant DNA is transferred to the host cell.
6. Cloning of the host cells makes many copies of the gene of interest.
7. Identify and isolate the recombinants and harvest the products.

G.
1. Several hormones, for example insulin, growth hormone
2. An organism that contains genes from a different species
3. Yes, there are many. Transgenic livestock that are more resistant to disease
4. Gene therapy inserts functional genes to overcome deficits caused by faulty genes.
5. Gene therapy is in the experimental stages now. It has been tried in humans, but with varying success. There is tremendous potential in the areas of cancer and genetic diseases.

Critical Thinking Questions

1. The sequence of events should be similar to those shown in Figure 21-3 and 21-7. It is essential that uracil be replaced in the RNA strands and that all base pairs are complementary. Using the actual codon and anticodon sequences will reinforce the concept of the genetic code.
2. When something is transcribed it is copied over. In the case of protein synthesis, the DNA code is being copied into an RNA code so that it can leave the nucleus on a much smaller molecule, the mRNA. When this code reaches the ribosomes, it needs to be translated or changed into another code, the code of protein synthesis. So, at the ribosome, the DNA code is translated into a sequence of amino acids, the language of proteins. This occurs as the tRNA interprets the message of mRNA and brings the appropriate amino acids into place.
3. The genetic code is translated in groups of three bases at a time. Since each amino acid is encoded by multiple codons, a substitution may not change the amino acid sequence coded for by a gene. If a substitution causes the codon to code for a different amino acid than the normal codon, then the substitution may have no, little, or great effect on protein function. An addition or deletion that is not a multiple of three will throw off all the sequencing of all codes that follow. However, a three-base deletion will remove one amino acid without altering the reading frame for all following codons, and loss of one amino acid may not change the structure/function of the protein at all.
4. The problems associated with many genetic diseases result because a faulty gene fails to produce its normal protein product. Gene therapy would replace the faulty gene with a functioning gene. This is certainly a good idea, but shouldn't it be up to the individual whether to submit to gene therapy? There will be many people who will not want to undergo genetic screening for diseases that may or may not manifest themselves because of possible discrimination with regard to health insurance and employment. There are a few examples of gene therapy that have been very successful, and several that have been fatal.
5. Gene legislation is the body of law that regulates things such as patents on newly identified gene sequences, the development of new organisms, such as viruses, genetic screening, the development of human super genes, the development of organisms that could be used as biological warfare. The United States has legislation regulating some aspects of gene research and development. However, most countries do not.

Practice Test

1. b
2. d
3. c
4. a
5. c
6. b
7. b
8. c
9. d
10. c
11. d
12. c
13. c
14. c
15. a
16. d
17. d
18. c
19. b
20. c

Short Answer Questions

1. The drawing should show DNA as a double helix with side structures of deoxyribose and phosphate and bases paired A-T and C-G. The final product should be two identical molecules of double-helical DNA.
2. The DNA strand is "closed" and the nucleotide bases are paired. The molecular structure protects the DNA base sequence. The mRNA must be a physically smaller structure so that it can leave the nucleus and pass into the cytoplasm. Also it must be an "open" structure so that the base sequence can be accessed by the tRNA during protein synthesis.

3. A gene is a sequence of DNA that codes for a protein. Because it is in the nucleus of a cell, it cannot be seen. The language of expression for the gene is the protein that is, or is not, produced under its direction. A gene is expressed when the protein is made for example when insulin is produced or when the curl in someone's hair is visible.
4. The genetic code must be able to code for all 20 different amino acids. There are only four possible bases. If the bases were used one at a time, that is if one base coded for one amino acid, that would only account for four amino acids. If the bases were used two at a time, that is, if two bases coded for each amino acid, then that would account for 16 amino acids ($4^2 = 16$). However, if three bases are used to code for each amino acid, there are 64 possible combinations ($4^3 = 64$) enough codes for each amino acid and quite a few are left over to provide duplicate codes or to serve as stop codes.
5. Recombinant DNA is DNA created from more than one source. A gene of interest is identified and sectioned out of a source DNA strand. It is then incorporated into a vector genome. Now the DNA is composed of both source DNA and vector DNA and is called recombinant DNA. The vector then enters a rapidly dividing host cell such as a bacterium which produces many copies of the gene or of the gene product. The gene or product is then isolated, or harvested.

Chapter 21A

Cancer

Objectives

After reading the text and studying the material in this chapter, you should be able to:

- List the characteristics of precancerous cells.
- Describe how the normal mechanisms of cell self-destruction (apoptosis) are altered in cancerous cells.
- Explain how cancerous cells are able to spread throughout the body.
- Describe the body's immune responses to abnormal cells.
- Describe common causes of cancer including viruses, chemicals and radiation.
- List lifestyle habits that reduce the risk of cancer.
- Describe methods used to screen for and diagnose cancer.
- Differentiate between surgery, radiation, chemotherapy, immunotherapy, and gene therapy as methods to treat cancer.

Chapter Summary

Cancer is uncontrolled cell division. There are over 200 forms of cancer usually named for their cell or tissue origin.

An abnormal growth of cells can form a mass of tissue called a **tumor**, or neoplasm. Tumors can be either benign or malignant. A **benign tumor** is an abnormal mass of tissue surrounded by a capsule of connective tissue that usually remains at the site where it forms. Often they can be removed completely by surgery.

Malignant tumors that invade surrounding tissue and spread to multiple locations are called cancerous. The spread of cancer cells from one part of the body to another is called **metastasis**. Cancer cells result from an accumulation of genetic damage, and even early precancerous cells look different form normal cells. Dysplasia is the term used to describe the ragged edges, large and atypically shaped nuclei, and increased amounts of DNA found in precancerous cells. Cancer cells may have extra copies of some chromosomes, are missing parts of others and have extra parts to still other chromosomes. The cells grow into an unorganized mass called a carcinoma *in situ*, meaning in place. If not removed, the tumor will secrete chemicals that cause blood vessels to invade and nourish the tumor. Besides providing nutrients and removing wastes, the blood vessels are a means to travel throughout the body, or **metastasize**. As the cancer spreads, it causes death by interfering with the ability of body cells to function normally.

Cancer cells escape the normal control mechanisms of healthy cells. Cells contain a **tumor-suppressor gene**, *p53*, that codes for a protein that detects damaged DNA, stops the cell cycle, and works with other gene products to assesses the severity of the damage. If the damage can be repaired, then gene activity is initiated for repair, but if not, still other genes cause cell death. However, if damage repair is unsuccessful or incomplete, the genetic abnormalities accumulate and can lead to cancer.

Cancer cells lose restraints on cell division. **Proto-oncogenes** stimulate cell division, while **tumor-suppressor genes** slow or stop cell division. A mutation in a proto-oncogene changes it to an out of control **oncogene** that speeds the rate of cell division. For example the normal *ras* protoncogene stimulates cell division in response to a growth factor, but the *ras* oncogene stimulates cell division in the absence of a stimulating growth factor. Mutations in the tumor-suppressor gene, *p53*, can lead to uncontrolled cell division. Mutations in other tumor-suppressor genes result in damaged DNA not being repaired. Mutations must occur in at least two genes before cancer can occur.

When most cells contact neighboring cells, they stop dividing. However, cancer cells do not show contact inhibition and continue to divide forming a tumor.

Cancer cells do not self-destruct when their DNA is damaged. Normally, when the cells that regulate cell division become faulty, they initiate a series of changes leading to cell death, **apoptosis**. The defective DNA in cancer cells does not trigger genetic self-destruction. A faulty tumor-suppressor gene, *p53*, fails to initiate the events leading to cellular self-destruction so the cells proliferate uncontrollably. Tumors with damaged *p53* grow aggressively and spread easily and quickly.

Cancer cells divide indefinitely whereas normal cells have a mechanism that limits the number of times a cell can divide. At the ends of the DNA molecules are telomeres. These are produced during embryonic development only by the enzyme telomerase. Each time DNA is replicated a piece of the telomere is lost. When the telomere is totally shaved off, then cell division stops for that cell. However, it is suspected that cancer cells begin to produce telomerase, which repairs the ends of the DNA and allows continued cell division.

Cancer cells release special growth factors that cause capillaries to invade the tumor. Normal cells produce a protein that prevents blood vessels from spreading into tissues. Mutations in *p53* can block the production of that protein and thus blood vessels invade the tumor providing nutrition, waste removal and a mechanism of metastasizing.

When normal cells become separated from other cells, they stop dividing and self-destruct. Cancer cells secrete enzymes that break down the cellular adhesion molecules that hold them to other cells. Then, so that cell death does not occur, their oncogenes send a false message to the nucleus indicating that the cell is properly attached.

Natural killer cells and cytotoxic T-cells usually kill cancer cells that develop in our bodies every day. Some types of cancer cells actively inhibit the defense cells and others grow and multiply too quickly for defense cells to destroy.

The genetic changes that turn normal cells into cancerous cells are usually brought about by viruses or mutations caused by certain chemicals or radiation. A viral oncogene can be inserted into a regulatory gene of the host cell DNA or it may interfere with the function of the immune system. A **carcinogen** is an environmental agent that fosters the development of cancer. Tobacco smoke contains chemical carcinogens as do many organic compounds. Excessive alcohol consumption is another cancer risk factor. These carcinogens may cause mutations in the tumor-suppressor gene or stimulate cell division. Radiation, even from the sun, causes mutations in DNA that can lead to cancer.

Certain lifestyle habits greatly reduce the risk of developing cancer. Avoid exposure to all tobacco smoke. Reduce the amount of saturated fat and smoked foods in your diet. Eat a diet rich in fruits and vegetables because they contain antioxidants, chemicals that can prevent the formation of molecules that damage DNA. And eat a diet high in fiber because fiber will bind to carcinogens and speed their passage out of the body.

Early detection is critical to cancer survival because successful treatment is much more likely before the cancer has spread. Seek medical help if you notice any of the cancer warning signs. Participate in regular cancer screen tests.

Surgery, radiation and chemotherapy are conventional ways to treat cancer. **Surgery** is performed when a tumor is accessible and can be removed without damaging vital tissue. **Radiation therapy** is used to kill localized cancer cells. Radiation damages the DNA in cells that are rapidly dividing triggering programmed cell death.

Chemotherapy is carried by the circulatory system and therefore is used to kill rapidly dividing cells throughout the body. It can block DNA synthesis, damage DNA or prevent cell division. However, if the *p53* gene is mutant then the cells with damaged DNA do not self-destruct and treatment fails. **Immunotherapy** boosts the natural immune responses against cancer cells. Cytotoxic T cells of the immune system continually search for abnormal cells, including cancer cells, and destroy them. The goal of immunotherapy is to boost the patient's immune system with factors normally secreted by lymphocytes. New research is working on a vaccine for cancer and drugs that block blood vessel formation. Researchers are also working on **gene therapy** that would insert normal tumor-suppressor genes into the cancerous cells, or insert a piece of DNA that prevents an oncogene from being effective, or insert a gene that will make the cancer cell more sensitive to a drug that will kill them.

Key Concepts

- Cancer is uncontrolled cell division caused by mutations in proto-oncogenes and tumor-suppressor genes and by loss of contact inhibition.
- Cancer cells do not self-destruct when damaged DNA is detected due to a faulty *p53* gene.
- Cancer cells divide indefinitely possibly because of telomere repair.
- Cancer cells attract a blood supply that provides nutrients, removes wastes, and provides transport for metastasized cells.
- Cancer cells break the cellular adhesion molecules and become separated from other cells.
- Natural killer cells and cytotoxic T-cells destroy cancer cells, but sometimes are ineffective or cannot keep up with the numbers.
- Certain viruses disrupt the genetic control of cell division.
- Carcinogens, environmental agents that foster the development of cancer do so by increasing the rate of cell division or causing mutations in the DNA.
- Radiation can cause mutations in DNA resulting in cancer.
- Lifestyle habits can greatly decrease the risk of developing cancer.
- Early detection of cancer enhances survival of cancer patients.
- Cancer can be treated by surgical removal, radiation therapy to kill localized cells, chemotherapy, immunotherapy, inhibition of blood vessel formation and gene therapy.

Study Tips

This chapter provides an overview of the mechanisms employed by cancerous cells to escape the normal cellular regulatory mechanisms. Understanding of this information is based upon a knowledge of genes and the insertion of viral DNA into host cells. Review this material for clarification. Make a flow chart showing the mechanisms used by mutant cells.

The remainder of the chapter describes common causes of cancer, lifestyle habits to reduce risk, methods used for diagnosis and current treatment options. Careful reading and highlighting of the text material may be sufficient for mastery. Some students may benefit from making an outline of this information that includes lists under each major heading. The material in this chapter is interesting and current; you might supplement it with additional reading on the latest advances in cancer detection and treatment.

Review Questions

A. Answer the following questions regarding the steps that a healthy cell uses when damaged DNA is detected.

1. What is the first step in damage control? ____

   ____

2. What happens if the repair is successful? ____

   ____

3. What happens if the DNA damage is too extensive for repair? ______________________

   ____

4. What happens if the DNA repair is unsuccessful or incomplete and normal cell death does

   not occur? ____

   ____

5. What two types of genes regulate normal cell division and what do they do? ___________

   ____

6. What is the difference between a proto-oncogene and an oncogene? ________________

   ____

B. Complete the following table comparing normal cells to cancerous cells.

Control Mechanism	**Normal Cells**	**Cancerous Cells**
Cell division		Mutations in tumor-suppressor genes and proto-oncogenes, making them oncogenes, lead to uncontrolled cell division.
Contact inhibition	Cells stop dividing.	
Cellular self-destruction (apoptosis)		Faulty *p53* gene allows continued cell division.
Limited number of cell divisions		Genes for telomerase turned on, telomeres are repaired and cell division continues.
Blood supply	Blood vessels do not grow into developed tissues.	
Cellular adhesion	Cells held together by CAMs to form tissues.	
Immune response		Immune response cannot keep up with cell divisions of cancerous cells.

C. List the most common causes of genetic change that lead to cancer and give an example of each.

Cause   Example(s)

1. ____________________   ____

2. ____________________   ____

3. ____________________   ____

D. List the seven warning signs of cancer. Try to complete the list using mnemonic CAUTION.

1. ____________________

2. ____________________

3. ____________________

4. ____________________

5. ____________________

6. ____________________

7. ____________________

E. Match the organ in Column A with the cancer-screening test in Column B.

Column A	Column B
1. breast	a. PAP test
2. cervix	b. Pelvic exam
3. colon and rectum	c. Tissue sample
4. endometrium	d. Self-exam monthly and exam by physician yearly
5. uterus	e. Sigmoidoscopy and occult blood test

F. Put the name of the cancer treatment to the left of the statement describing it.

1. ____________ Boosts the immune responses against cancer cells with factors such as interferons, interleukin-2, and tumor necrosis factor.

2. ____________ Removes tumors from the body.

3. ____________ Inserting pieces of DNA into cancer cells to prevent oncogene action, to insert normal tumor-suppressor genes, or to make tumor cells more sensitive to drugs.

4. ____________ Used to kill a localized group of cells by damaging the DNA and stimulating normal cell death, often follows surgery.

5. ____________ Cuts off the lifelines for tumors and starves them out.

6. ____________ Chemicals intended to damage DNA and incite cell destruction are carried throughout the body.

Critical Thinking

Read each of the following questions carefully. Circle the most important words in the question before formulating your answer.

1. Ed, a 20-year-old accounting major, was diagnosed with non-Hodgkin's lymphoma. The tumor on his thymus gland grew rapidly, he was admitted to the hospital, given radiation therapy and then chemotherapy. The physician suggested that he collect and save his sperm in a sperm bank. Why would the physician suggest this? Would you have followed the physician's advice?

2. Your grandmother has just learned that viruses can cause cancer. Yet she knows that she gets the cold and flu viruses each year and does not have cancer. Explain to her how viruses can cause cancer and clarify why all viruses do not cause cancer.

3. The text lists several lifestyle habits that may reduce the risk of cancer. Review Table 21A-4 and reflect on your own lifestyle. What changes can you make now that will reduce your risk of cancer?

4. Choose any of the cancers listed in Figure 21A-1, possibly one that has affected members of your family, and do further research on the symptoms, measures of prevention and current treatment. Use the Internet for the most current information, but be sure to select reputable sites such as those sponsored by the American Cancer Society, hospitals, research centers, the National Institutes of Health.

5. It is well documented that tobacco smoke contains carcinogens. Gather a group of classmates and discuss whether employers or insurance companies should be able to demand that people quit smoking.

Practice Test

Choose the one best answer to each question that follows. As you work through these items, explain to yourself why each answer you discard is incorrect and why the answer you choose is correct.

1. Cancerous cells and normal cells differ in the amount of ____________.
   a. cytoplasm
   b. genetic damage
   c. organic molecules
   d. proteases

2. Which of the following are characteristic of precancerous cells?
   a. increased amount of DNA
   b. irregular cell shape
   c. large nuclei
   d. all of the above

3. What is the most common cancer among men?
   a. colon
   b. leukemia
   c. prostate
   d. rectal

4. What is cancer of the bone marrow stem cells that produce white blood cells called?
   a. carcinoma
   b. leukemia
   c. lymphoma
   d. sarcoma

5. What is the first thing that happens when a healthy cell detects damage to DNA?
   a. It becomes cancerous.
   b. It dies.
   c. It divides quickly.
   d. It stops the cell cycle.

6. The products of exactly which genes assess cell damage and coordinate the repair process?
   a. activator genes
   b. oncogenes
   c. proto-oncogenes
   d. tumor-suppressor genes

7. Which genes stimulate growth in normal cells?
   a. activator genes
   b. oncogenes
   c. proto-oncogenes
   d. tumor-suppressor genes

8. Why is a mutation in the *p53* gene so important?
   a. cell division is stopped
   b. cell division is uncontrolled
   c. cells grow in size and multiple nuclei develop
   d. the *ras* gene is inhibited

9. Proto-oncogenes mutate into ____________.
   a. oncogenes
   b. *p53*
   c. *RB* genes
   d. tumor-suppressor genes

10. How many genes must be damaged before cancer occurs?
    a. every gene in the cell
    b. more than 1000
    c. more than 1
    d. none

11. Which of the following are NOT characteristic of tumor growth?
    a. cell differentiation and specialization
    b. metastasis
    c. nondifferentiated cell mass
    d. uncontrolled growth

12. Which cells can undergo programmed cell death?
    a. benign tumors
    b. cancerous tumors
    c. metastasized cells
    d. normal cells

13. What is the normal role of telomerase?
    a. builds new DNA in cancerous cells
    b. builds telomeres in embryonic cells
    c. builds and repairs proto-oncogenes
    d. repairs broken strands of chromosomes

14. How do most mutant cancerous cells travel through the body?
    a. along the nerve pathways
    b. carried by cells capable of phagocytosis
    c. in the blood vessels
    d. through the tissue membranes

15. Which cells normally kill defective or mutant cells?
    a. antibody-producing cells
    b. chemokiller cells
    c. cytotoxic T and natural killer cells
    d. no normal cells can kill a mutant

16. Which of the following is not a good match?
    a. estrogen → breast cancer
    b. radiation → skin cancer
    c. smoked foods → mouth cancer
    d. tobacco smoke → lung cancer

17. A healthy diet would include reduction in ____________.
    a. canned fruit
    b. deep sea fish
    c. orange vegetables
    d. saturated fat

18. Which of the following is a mismatch between the screening test and the cancer?
    a. dental x-rays detects cancer of the tongue
    b. mammogram detects breast cancer
    c. PAP test detects cervical cancer
    d. sigmoidoscopy detects colon and rectal cancer

19. Which of the following is not a feature common to radiation therapy and chemotherapy?
    a. affects normal cells as well as cancerous cells
    b. localized in its action
    c. target rapidly growing cells
    d. used in conjunction with surgery

20. Which of the following statements is not true regarding gene therapy use for cancer treatment?
    a. It can insert a healthy *p53* gene into a mutant cell.
    b. It is approved by the FDA.
    c. It is being used in clinical trials on humans.
    d. It uses viruses to insert genes into normal and cancerous cells.

Short Answer Questions

Read each of the following questions carefully. Jot down the main points you want to include in your answer. Then write a well-organized explanation.

1. Describe how the appearance of precancerous cells might differ from normal cells.

2. Explain the role of the *p53* gene in normal cells and why a mutation in that gene is problematic.

3. What function might telomeres play in regulating cell division? What happens that allows continual cell divisions in cancerous cells?

4. Explain why radiation and chemotherapy make cancer patients sick

5. Why is surgery ineffective if the cancerous cells have begun to metastasize?

ANSWER KEY

REVIEW QUESTIONS

A. 1. The cells stop dividing, molecular mechanisms are used to assesses the damage and initiate repair.
2. The cell returns to the normal cell cycle.
3. The cell initiates programmed cell death.
4. The genetic damage to the DNA accumulates and can lead to cancer.
5. Proto-oncogenes stimulate cell division and tumor-suppressor genes slow cell division.
6. A proto-oncogene is a normal gene that stimulates cell division. An oncogene is a mutated proto-oncogene that uncontrollably speeds the rate of cell division.

B.

Control Mechanism	Normal Cells	Cancerous Cells
Cell division	Proto-oncogenes and tumor-suppressor genes are in balance controlling cell division.	Mutations in tumor-suppressor genes and proto-oncogenes, making them oncogenes, lead to uncontrolled cell division.
Contact inhibition	Cells stop dividing.	Cells continue to divide forming a tumor.
Cellular self-destruction (apoptosis)	Abnormalities cause initiation of apoptosis.	Faulty *p53* gene allows continued cell division.
Limited number of cell divisions	When telomere is gone, cell stops dividing.	Genes for telomerase turned on, telomeres are repaired and cell division continues.
Blood supply	Blood vessels do not grow into developed tissues.	Growth factors cause capillary growth.
Cellular adhesion	Cells held together by CAMs to form tissues.	Enzymes destroy CAMs and oncogenes send false message preventing initiation of cell self-destruction.
Immune response	Natural killer cells and cytotoxic T-cells kill abnormal cells.	Immune response cannot keep up with cell divisions of cancerous cells.

C. 1. Viruses: HPVs, hepatitis B and C, herpes viruses, Epstein-Barr virus, cytomegalovirus
2. Chemical carcinogens: tobacco smoke, benzene, formaldehyde, hydrocarbons, pesticides
3. Radiation: sun, tanning lights

D. 1. **C**hange in bowel or bladder habit or function.
2. **A** sore that does not heal.
3. **U**nusual bleeding or bloody discharge.
4. **T**hickening or lump in breast or elsewhere.
5. **I**ndigestion or difficulty swallowing.
6. **O**bvious change in wart or mole.
7. **N**agging cough or hoarseness.

E. 1. d
2. a
3. e
4. c
5. b

F. 1. Immunotherapy
   2. Surgery
   3. Gene therapy
   4. Radiation
   5. Inhibition of blood vessel formation
   6. Chemotherapy

Critical Thinking Questions

1. Both radiation and chemotherapy destroy rapidly dividing cells including those that produce sperm, leaving Ed sterile. If the cancer treatments are successful, the patient may want to have a family later. Collecting and saving sperm would provide the option of *in vitro* fertilization and the birth of a child that is genetically his own.
2. Viruses cannot live on their own so they take over the mechanisms of other cells. Like other viruses, cancer-causing viruses insert their genes into the DNA of the host cell only the genes inserted by cancer-causing viruses include oncogenes. With the viral DNA, the host cell responds as though it has an oncogene. Since damage must occur in at least two genes before cancer occurs, this cell is on its way to becoming cancerous.
3. Although answers will vary, many students should stop smoking and pay more attention to their diet, increasing fiber, fruits and vegetables.
4. It is important for students to learn how to gather current and reliable information on relevant health issues. Research and topics will vary, but the reference citations should reflect reliable sources of information.
5. There are always two sides to an issue. Although requiring employees and those who gain insurance to stop smoking would dramatically cut risks and therefore costs, the individual has rights as well. Unfortunately, it is their prerogative to continue to cause damage to their body, even cancer, if they choose.

Practice Test

1. b
2. d
3. c
4. b
5. d
6. d
7. c
8. b
9. a
10. c
11. a
12. d
13. b
14. c
15. c
16. c
17. d
18. a
19. b
20. b

Short Answer Questions

1. Precancerous cells will show an irregular shape, the nuclei may have become enlarged and atypical in shape, and there may be increased amounts of DNA present in the cell. Precancerous cells will have an unusually high rate of cell division.
2. *p53* is an important tumor-suppressor gene whose product normally slows or stops cell division. Therefore, a mutation in *p53* allows the cell to divide uncontrollably. In addition *p53* prevents the replication of damaged DNA and leads to programmed cell death.
3. The number of cell divisions appears to be related to the length of the telomere at the end of the DNA molecule. Each time the cell divides, a piece of the telomere is lost. When the entire telomere is gone, the cell does not divide again. Telomeres are formed by the enzyme telomerase that is usually only active in the embryo. However, cancerous cells reactivate telomerase so that the telomeres at the ends of the DNA molecules are restored and cell division continues.

4. Both methods of treating cancerous cells destroy DNA in dividing cells. The cells then self-destruct. This kills not only cancerous cells that are rapidly dividing, but also cells that normally divide at a rapid rate. Thus patients undergoing cancer treatments such as these are nauseated, have sores in their mouths, become anemic, and lose their hair.
5. Surgery is only effective in removing tumors that are contained. If the cells have begun to metastasize, then they have spread to many other locations and cannot be surgically chased.

Chapter 22

Evolution: Basic Principles and Our Heritage

Objectives

After reading the text and studying the material in this chapter, you should be able to:

- Describe the formation of organic molecules, genetic material, and early cells.
- Describe the experiments of Oparin and Haldane, and Miller and Urey. Explain how they supported theories that organic molecules formed from inorganic molecules.
- Explain the endosymbiont hypothesis as the origin of some organelles in eukaryotic cells.
- Differentiate between microevolution and macroevolution.
- Explain mechanisms of genetic drift including the bottleneck effect, the founder effect, gene flow, and natural selection.
- Describe the process of natural selection including differential reproductive success and fitness.
- Describe how variation enters the gene pool.
- Describe how each of the following provides evidence for evolution: the fossil record, geography distribution, comparative anatomy and embryology, and comparative molecular biology including the use of the molecular clock.
- Trace human evolution from the primates through hominoids and identify distinguishing characteristics of each group.
- Trace evolution of hominids noting the appropriate fossil evidence.

Chapter Summary

Evolution is change within a line of descent over time. The earth is estimated to be 4.5 billion years old. Physical and chemical evidence suggests that life has existed on early earth for 3.8 billion years. The theory of **chemical evolution** suggests that life evolved from chemicals slowly increasing in complexity over about 300 million years. It is hypothesized that in a low-oxygen atmosphere with energy from lightning and intense UV radiation, small organic molecules formed from inorganic molecules. These small organic molecules formed larger molecules, possibly proteins and then later, genetic material originated. The proteins and genetic material aggregated into droplets, precursors to living cells. The earliest cells were prokaryotes about 3.8 billion years ago. Eukaryotic cells evolved about 2 billion years later. The **endosymbiont hypothesis** states that some of the organelles within the eukaryotic cells were once other, free-living prokaryotes. Multicellularity evolved in eukaryotes.

Microevolution involves changes at the genetic level. A **gene pool** includes all of the **alleles** of all of the **genes** of all individuals in a population. A **population** is a group of individuals of the same species living in a particular area. Sexual reproduction and mutation produce variation in populations by rearranging the alleles into new combinations. Every fertilization produces a new individual with a new combination of alleles. New genetic information is introduced through mutation, a change in the DNA of genes. However, mutation rates are extremely low and have very little influence on genetic diversity in a population.

Genetic drift occurs when allele frequencies within a population change randomly because of chance alone. The **bottleneck effect** is a change in the gene pool that occurs when there has been a dramatic reduction in population size and the alleles in the remaining population may not be representative of those in the original population. Some alleles may have been lost from the population entirely reducing the genetic variability among survivors.

Genetic drift also occurs when a few individuals leave their population and establish themselves in a new, somewhat isolated place. By chance alone, the genetic makeup of the colonizing individuals is probably not representative of the population they left. Genetic drift in new, small colonies is called the **founder effect**.

Gene flow occurs when individuals move into and out of a population interbreeding with the resident population. A **species** is a population or group of populations whose members are capable of successful interbreeding. When populations become separated they may become genetically distinct and are no longer able to successfully interbreed. This results in a new species, the process of **speciation**.

Darwin's ideas on **natural selection** stated that individual variation existed within a species and that some of this variation was inherited. Some individuals live longer and have more offspring than others because their inherited characteristics make them better suited to their local environment. This is the process of natural selection. Evolutionary change occurs as the traits of individuals that survive and reproduce become more common in the population. Traits of less-successful individuals become less common. **Fitness** is the average number of reproductively viable offspring left by an individual. Individuals with greater fitness have more of their genes represented in future generations. Through natural selection, populations become better suited to their environment by a process called **adaptation**. Natural selection occurs slowly and can only work on the variation that is available in the population. Some of this variation comes from mutations.

Macroevolution occurs on a large scale and involves major changes in groups of species. Linnaeus developed the **Latin binomial** scheme for naming organisms whereby the genus name is followed by the species name. He also developed a system for grouping species into five broad categories of increasing similarity called **taxa**: kingdom, phylum, class, order, family, genus and species. **Phylogenetic trees** depict hypotheses about evolutionary relationships among species or higher taxa in graphic form.

Evidence for evolution comes from the fossil record, biogeography, the comparison of anatomical and embryological structures and molecular biology. Preserved remnants and impressions of past organisms are **fossils**. **Fossilization** occurs as hard body parts such as bones, teeth and shells become impregnated with minerals from surrounding water and sediment. Fossils of extinct organisms show both similarities to and differences from living species and are used to assess evolutionary relationships.

Biogeography is the study of the geographic distribution of organisms. Geographic distributions often reflect evolutionary history and relationships because related species are more likely to be found in the same geographic area than are unrelated species. If animals have been separated, biogeography can be used to help identify how long ago they separated. New distributions of organisms occur because either the organisms dispersed to new areas or the areas occupied by the organisms move or are subdivided.

Comparative anatomy and embryology reveal common descent. Species with more shared traits are considered more closely related than are species with fewer shared traits. Structures that are similar and that probably arose from a common ancestry are called **homologous structures**. They arise from the same kind of embryonic tissue. Common embryological origins can be considered evidence of common descent. **Convergent evolution** results when two organisms evolved similar structures because of similar ecological roles and selection pressures. Structures that are similar because of convergent evolution are called **analogous structures**.

Comparative molecular biology also reveals evolutionary relationships. Similar sequence in DNA nucleotides or amino acid chains indicate that organisms diverged fairly recently from a common ancestor. The **molecular clock** hypothesis is based on the assumption that point mutations in DNA occur at a constant rate. Thus the more differences in the DNA sequences between two organisms, the more time that has elapsed since the common ancestor.

Paleoanthropology is the study of human origins and evolution. Human evolution begins with primates. Modern primates are divided into two suborders: **prosimians** (premonkeys) and **anthropoids** including monkeys, apes and humans. One group of the anthropoids are **hominoids**, which includes apes and humans. The term **hominid** refers only to members of the human family including the genera *Australopithecus* and *Homo*. Hominids show **bipedalism**, shortening of the jaw and flattening of the face, reduced **sexual dimorphism**, and increasing brain size along with tool use, language and behavioral complexity. The oldest hominid remains are likely those of *Sahelanthropus tchadensis*, dating 7 million years old. The next hominid remains are those of *Ardipithecus ramidus*, dating about 4.4 million years old. The famous remains of *Australopithecus afarensis*, named Lucy, are about 3.8 million years old. About 3 million years ago, several new hominid species including *A. africanus* appeared.

About 2.5 million years ago, *Homo habilis* appeared in the fossil record in Africa. *H. habilis* had a larger brain size, may have had language, and used stone tools. *Homo erectus* appeared in the fossil record about 1.8 million years ago and may have been the first hominid to migrate out of Africa. *H. erectus* had a brain size close to that of modern humans and may have used fire. *H. erectus* disappeared between 200,000 and 300,000 years ago and *Homo sapiens* appeared. Neanderthals evolved from *H. erectus* living in Europe, lived in caves and were adapted to cold climates. Cro-Magnons evolved next and were quite similar to modern humans.

The **multiregional hypothesis** suggests that modern humans evolved independently in Europe, Asia, Africa and Australia from distinctive local populations of *H. erectus*. The **Out of Africa hypothesis** suggests a single origin for all *H. sapiens* that evolved from *H. erectus* in Africa and later migrated to Europe, Asia and Australia.

KEY CONCEPTS

- Life originated about 3.8 billion years ago through a process of chemical evolution.
    - ✓ Conditions were such in the early atmosphere that small organic molecules could form from inorganic molecules and that, over time, probably in the oceans, the small organic molecules formed larger organic molecules.
    - ✓ Next genetic material, probably RNA formed and accumulated with the organic molecules in droplets, the precursors of cells.
    - ✓ Early cells were prokaryotic. More complex cells formed as smaller organisms were incorporated into the cells becoming organelles.
- The gene pool is a collection of all of the alleles of all of the genes of all of the individuals in a population.
- A species is a population or group of populations whose members are capable of interbreeding to produce fertile offspring under natural conditions.
- Sexual reproduction and mutation produce variations in populations.
- Microevolution involves changes at the genetic level, specifically changes in the frequency of certain alleles relative to others in the gene pool. Microevolution occurs by genetic drift, gene flow, natural selection and mutation.
- Charles Darwin suggested that modern species evolved from ancestral species and that evolution occurred by a process called natural selection.
- Macroevolution is large-scale evolution involving changes among groups of species.
- The Latin binomial system for naming uses the genus name followed by the species name.
- Phylogenetic trees describe the history of life by showing evolutionary relationships among species.
- Evidence for evolution comes from the fossil record, geographic distributions, comparative anatomy and embryology, and comparative molecular biology.
- Humans may have evolved from a branch of primates, the anthropoids, and mores specifically the hominoids (gorillas, chimpanzees and humans). Members of the human family including species of the genera *Australopithecus* and *Homo*.
- One hypothesis of hominid evolution is the following: *Australopithecus afarensis* → *Homo habilis* → *Homo erectus* → *Homo sapiens.*
- The multiregional hypothesis and the Out of Africa hypothesis present two views on the origin of modern humans.

STUDY TIPS

Two major bodies of information are contained in this chapter. One presents the events of evolution and the evidence for those events as they progress along a time line. The other information provides the basis for change – the mechanisms that underlie evolutionary change.

As you study, develop a timeline of events including a description of the event and evidence for it. Begin with changes at the molecular level, next describe the formation of single cells, and finally the advent of multicellular organisms. Use Figure 22-4 as a guide. Next, focus on the evolution of humans as primates, referring to Figure 22-19. Finally study hominid evolution and the fossil evidence carefully. Add notes to your timeline indicating similarities and differences among groups.

Begin a set of note cards that describe the processes of microevolution including genetic drift and the bottleneck and found effect, gene flow and natural selection. Form a study group and explain how these processes can lead to speciation. Add to your note cards the evidence for evolution including the fossil record, biogeography, comparative anatomy and embryology, and molecular biology. Use your study group to develop a firm understanding of these processes, the evidence they provide and the limitations on that evidence.

Review Questions

A. Complete the following table drawing analogy between Miller and Urey's test of the Oparin-Haldane hypothesis of chemical evolution.

	Oparin-Haldane Hypothesis	**Miller and Urey Experiment**
Year	1920s	
Energy Source	Lightning and UV radiation	
Early Atmosphere		Gases in the spark chamber were CH_4, NH_3, H_2, H_2O
Water Environment		Water in flask
Heat		Heat source to boil water
Product	Small proteins	

B. Order the following possible steps in the origin of life on earth. Put the correct number (1-7) before the description.

________ Small organic molecules join to form larger organic molecules.

________ Eukaryotic cells arise.

________ Large organic molecules, including genetic material, aggregate within droplets.

________ Multicellular eukaryotic organisms, such as plants and animals, arise.

________ Inorganic molecules form small organic molecules, such as amino acids.

________ Genetic material, probably RNA, originates.

________ Droplets eventually form true prokaryotic cells.

C. Match the term in Column A with the definition in Column B.

Column A	Column B
1. adaptation	a. group of individuals of the same species living in a particular area
2. endosymbiont hypothesis	b. changes at the genetic level
3. fitness	c. two-part naming system of genus and species
4. gene pool	d. explains origin of some organelles in eukaryotic cells
5. Latin binomial	e. population(s) whose members are capable of successful interbreeding
6. macroevolution	f. process of becoming better suited to the environment by natural selection
7. microevolution	g. large-scale evolutionary changes
8. phylogenetic tree	h. all of the alleles of all of the genes of all individuals in a population
9. population	i. show genealogical relationships among species or higher taxa
10. species	j. average number of reproductively viable offspring left by an individual

D. Define each of the following terms and give an example. Use the text descriptions and glossary.

1. Genetic drift ______________________________

____

2. Bottleneck effect ______________________________

____

3. Founder effect ______________________________

____

4. Gene flow ______________________________

____

5. Natural selection ____

____

E. Identify each of the following as an example of evidence for evolution. Use these terms: comparative anatomy, comparative embryology, comparative molecular biology, fossilization, and geographic distribution.

1. ________________ wooly mammoth tusks and jaw found when excavating for a shopping center

2. ________________ similarity between the bones of a human hand and the bones in the flipper of a dolphin

3. ________________ differences between similar animals found on different continents

4. ________________ gill pouches and tail development in the embryos of humans, chickens, and fish

5. ________________ similarities in DNA nucleotide sequences among different primates

F. Complete the following sequences depicting the proposed evolution of (1) hominoids and (2) hominids.

1. Ancestral primate, possibly mammalian insectivore → ________________ → ________________ → hominids

2. *Sahelanthropus tchadensis* → ________________ → *Australopithecus afarensis*, *Australopithecus robustus*, and ________________ → *Homo habilis* → ________________ → *Homo sapiens neanderthalensis* → ________________

G. Correct these two common misconceptions concerning human evolution.

1. Humans descended from chimpanzees similar to modern chimpanzees as seen in the zoo.

____

____

2. Humans evolved in an orderly, stepwise fashion. ______________________________

____

____

3. Human bones and organ systems evolved all at one time. ___________________________

____

____

Critical Thinking

Read each of the following questions carefully. Circle the most important words in the question before formulating your answer.

1. Form a discussion group of fellow classmates. Honestly describe how you would have defined evolution and described the evidence for it prior to this class. Did you or your classmates hold any of the common misconceptions described in the text?

2. Evolution and creation by an intelligent source are both theories. A theory is based on many observations and experiments. Are both of these theories based on evidence? What is the evidence for the theory of evolution? What would others claim is evidence for the theory of creation by an intelligent source?

3. How would modern transportation affect gene flow among humans? What do you expect to be the results of increased gene flow?

4. The molecular clock is an interesting hypothesis. Explain what it is and how it can be used to determine the relationship among organisms.

5. Explain how genetic variation enters the gene pool. Why is genetic variation important?

Practice Test

Choose the one best answer to each question that follows. As you work through these items, explain to yourself why each answer you discard is incorrect and why the answer you choose is correct.

1. What does the term evolution mean?
   a. change of any kind
   b. genetic offspring
   c. gradual change over time
   d. sudden change in response to environmental changes

2. The first step in organic evolution is ____________.
   a. the creation of prokaryotic cells
   b. formation of organic molecules from inorganic molecules
   c. presence of lightning and UV radiation
   d. sedimentation of nucleotides and amino acids

3. Who were the scientists who experimentally demonstrated that it was possible to generate small organic molecules from an environment similar to the early environment on earth?
   a. Darwin and Mendel
   b. Miller and Urey
   c. Oparin and Haldane
   d. Watson and Crick

4. Which of the following molecules is most critical to the development of cells?
   a. amino acids
   b. genetic material
   c. phospholipids
   d. proteins

5. According to the endosymbiont hypothesis, what is the likely origin of mitochondria?
   a. free-living bacteria
   b. infolded cell membranes
   c. modified nuclei
   d. non-photosynthetic chloroplasts

6. Evolution that results from changes at the genetic level is called ____________.
   a. allelic change
   b. macroevolution
   c. microevolution
   d. species evolution

7. A gene pool is the ____________.
   a. accumulation of mutations in small populations
   b. early primordial soup
   c. genetic set that identifies a species
   d. totality of genes in a population

8. What is the only source for brand new genetic information?
   a. crossing over during meiosis
   b. independent assortment of chromosomes
   c. mutation
   d. sexual reproduction

9. Amish populations show high incidence of genetic abnormalities such as polydactyly, Ellis-van Creveld syndrome and dwarfism. What mechanism of population genetics might explain this?
   a. bottleneck effect
   b. founder effect
   c. natural selection
   d. spontaneous mutation

10. When a population becomes so isolated that it can no longer successfully reproduce with the original population it is called a ____________.
    a. founder population
    b. genus
    c. population outgrowth
    d. species

11. Which of the following statements best describe the theory of natural selection?
    a. All organisms are equally suited to their environment.
    b. Organisms better adapted to their environment have greater reproductive success.
    c. Organisms that produce the most offspring are better suited to their environment.
    d. Random selection will determine which organisms survive.

12. Darwin drew upon his vast knowledge of the natural world and his extensive travels as he put forth the theory of natural selection in his book *On the Origin of Species* published in 1859. Darwin did all of this without a clear understanding of ____________.
    a. molecular genetics
    b. organic evolution
    c. relative skull size
    d. reproductive behaviors

13. Which of the following is true regarding phylogenetic trees?
    a. They are a tool used at the micorevolution level.
    b. They are used to assign genus and species names.
    c. They are used to discover new fossils.
    d. They are used to show relationships among species and higher taxa.

14. The observation that the bone structure found in the flippers of dolphins and whales is similar to that found in a human hand is evidence of a common ancestry based on ____________.
    a. adaptation
    b. analogous structures
    c. behavior
    d. homologous structures

15. Comparative molecular biology uses ____________ to identify phylogenetic relationships.
    a. carbon dating
    b. comparative anatomy of bone structure
    c. DNA sequences
    d. embryological information

16. Which of the following characteristics set hominids apart from primates?
    a. anthropological advantage
    b. bipedalism
    c. flexible shoulders with a ball and socket joint
    d. forward-facing eyes for improved depth perception

17. Which of the following is true regarding the descent of humans?
    a. Humans and chimpanzees share a common ancestor.
    b. Humans evolved ahead of chimpanzees.
    c. Humans evolved from chimpanzees.
    d. Humans evolved from lemurs, not chimpanzees.

18. It is currently believed that *Homo sapiens* evolved most directly from ____________.
    a. *Homo erectus*
    b. *Homo habilis*
    c. *Pan troglodytes*
    d. *Sahelanthropus tchadensis*

19. What would you expect to find living on earth a billion years from now?
    a. Only new organisms; all organisms would have evolved.
    b. Some of the same organisms that are here now and some new ones.
    c. The same organisms that are here now.
    d. The same organisms that are here now and a few that were previously extinct.

20. Changes in species that have occurred in the past cannot be directly observed. Therefore evolution ____________.
    a. must be accepted as fact
    b. will forever be discounted
    c. will never be accepted as valid
    d. will remain a theory and not a proven fact

Short Answer Questions

Read each of the following questions carefully. Jot down the main points you want to include in your answer. Then write a well-organized explanation.

1. Describe the conditions that may have led to the first organic molecules. How was this environment approximated in the Miller-Haldane experiments?

2. Describe the formation of the early eukaryotic cells using the endosymbiont hypothesis.

3. Explain how natural selection allows the alleles of the characteristics making an organism "most fit" to the environment to become more common in the gene pool.

4. Describe the formation of a fossil. What conditions are best for making a fossil? What tissues are preserved? Can we assume that the fossil record is representative of the entire population that existed at the time the fossil was created?

5. Begin with *Homo habilis* and trace the likely evolution to *Homo sapiens* noting characteristics of each.

Answer Key

Review Questions

A.

	Oparin-Haldane Hypothesis	Miller and Urey Experiment
Year	1920s	1953
Energy Source	Lightning and UV radiation	Sparks
Early Atmosphere	Contained H_2, N_2, CO_2, CH_4 and CO. Almost no O_2.	Gases in the spark chamber were CH_4, NH_3, H_2, H_2O
Water Environment	Oceans	Water in flask
Heat	Volcanoes, lava, hot sand	Heat source to boil water
Product	Small proteins	Organic molecules

B. 2, 6, 4, 7, 1, 3, 5

C. 1. f
2. d
3. j
4. h
5. c
6. g
7. b
8. i
9. a
10. e

D. 1. Genetic drift occurs when allele frequencies within a population change randomly because of chance alone. Ex. In an island population, a random mutation could become integrated into the population.
2. The genetic drift associated with dramatic, unselective reductions in population size so that by chance alone the genetic makeup of survivors is not representative of the original population. Ex. A fire kills most of the people in an area; those who survived were all located across the river in an isolated part of the island.
3. Genetic drift associated with the colonization of a new place by a few individuals so that by chance alone the genetic makeup of the colonizing individuals is not representative of the population they left. Ex. A small group from a larger population head west to settle in a new part of the country.
4. Gene flow is the movement of alleles between populations as a result of the movement of individuals in and out of the population. It occurs when the two populations successfully interbreed. Ex. Gene flow occurred as new groups of people came to this country and were able to interbreed with very different populations.
5. Natural selection is the differential survival and reproduction of individuals that results from genetically based variation in their anatomy, physiology and behavior. Ex. When a mixture of antibiotic resistant and antibiotic sensitive organisms cause an infection, antibiotic therapy

causes those that are antibiotic sensitive to be killed while those that are antibiotic resistant are able to reproduce. More and more of the population becomes antibiotic resistant.

E. 1. fossilization
   2. comparative anatomy
   3. geographic distribution
   4. comparative embryology
   5. comparative molecular biology

F. 1. Ancestral primate, possibly mammalian insectivore → anthropoids → hominoids → hominids
   2. *Sahelanthropus tchadensis* → *Ardipithecus ramidus* → *Australopithecus afarensis*, *Australopithecus robustus*, and *Australopithecus africanus* → *Homo habilis* → *Homo erectus* → *Homo sapiens neanderthalensis* → *Homo sapiens sapiens*

G. 1. Modern chimpanzees and humans diverged from the evolutionary tree about 6-8 million years ago. Their common ancestor was different from any modern form of ape. Chimpanzees and humans evolved together, not one after the other.
   2. The path to modern humans includes a great variety of precursors and an unclear path. Some species may have displaced another or evolved into another, some may have become extinct while others persisted.
   3. Various human characteristics evolved at different rates and times. For example bipedalism was one of the first characteristics to evolve, then increased brain size, the use of language, etc.

Critical Thinking Questions

1. Most people know that evolution is change over time. Most also believe that the theory of evolution claims that humans evolved from monkeys in a linear fashion changing from one pre-human to another.
2. Scientists hold that experiments such as those performed by Miller and Haldane and clear evidence such as the fossil record, biogeography, comparative anatomy, and similar embryonic structures all support the theory of evolution. The newest evidence comes from an examination of nucleotide and amino acid sequencing. Those who believe in creation view their holy word as evidence. That is not evidence as we in the sciences view evidence.
3. Gene flow occurs when groups of people move in and out of populations, interbreeding while they are together. This causes a mixing of the genes and has the opposite effect as isolation. We might expect that gene flow would cause the human population to become more homogeneous.
4. The molecular clock hypothesis is based on the notion that single nucleotide changes in DNA occur at a steady rate. Organisms that are more closely related on the evolutionary tree would have more nucleotide sequences that are similar than those that diverged much earlier in evolutionary time.
5. Genetic variation occurs during meiosis when the homologous chromosomes separate and crossing over occurs. It also occurs in the random fertilization of sperm and egg each with an individual set of chromosomes. Point mutations are another source of variation. Variability is the material used in the evolutionary process. Without variation all organisms would have equal fitness and there would be no selection for traits that are better suited to an environment. As environments changed, there would be no mechanism to favor complementary change in the organisms.

Practice Test

1. c
2. b
3. b
4. b
5. a
6. c
7. d
8. c
9. b
10. d
11. b
12. a
13. d
14. d

15. c
16. b
17. a
18. a
19. b
20. d

Short Answer Questions

1. The early atmosphere contained all of the elements found in both amino acids and nucleotides. It also was very low in oxygen, which is important because oxygen tends to attack chemical bonds and not to let them build as is needed for synthesis to occur. There was a source of heat from molten activity inside the earth, a medium for the reaction in the oceans, and a source of energy for bond formation in lightning and UV radiation. The Miller-Haldane experiments duplicated that environment closely as described in the review exercise.
2. The endosymbiont hypothesis states that organelles such as mitochondria were once free-living prokaryotic organisms that were engulfed by primitive eukaryotic cells with which they established a symbiotic (mutually beneficial) relationship.
3. Natural selection is the differential survival and reproduction of individuals that results from genetically based variation in their anatomy, physiology and behavior. Organisms that are better suited for a particular environment have more offspring that survive to reproduce. Organisms that are less suited for the environment have less reproductive success. This causes the alleles of the traits that provide for a good fit to the environment to become more common and those that do not provide a good fit to the environment to decline.
4. Fossils are formed when an organism dies and the soft tissues decay while the hard tissues become impregnated with minerals from the surrounding environment. This usually happens in water or mud so that the dead tissue is not exposed to the air. The fossil record is often not complete. Most organisms die and are eaten by scavengers, making the remains incomplete and scattering the bones. Organisms that die near water have a much higher chance of being preserved than those not in water, therefore there are more fossils of aquatic organisms. Finally, larger organisms with hard, bony body parts are more likely to be preserved than soft-tissued organisms.
5. *Homo habilis* had a small brain volume, but may have used tools and had rudimentary speech. *Homo erectus* appeared next in the fossil record and was much larger than other hominids. *Homo erectus* showed less sexual dimorphism, had a brain capacity similar to modern humans, and used tools, weapons and fire. *Homo sapiens* had a large skull and larger teeth. Cro-Magnons, *Homo sapiens sapiens*, had superior tools and weapons, including arrows, knives, and stone-tipped spears helping them to become accomplished hunters.

Chapter 23

Ecology, the Environment, and Us

Objectives

After reading the text and studying the material in this chapter, you should be able to:

- Describe the earth as a closed system with energy as the only input.
- Define and use the terms biosphere, ecosystem, community, population, niche and habitat.
- Explain how primary and secondary succession can lead to a climax community.
- Describe how energy moves through an ecosystem.
- Differentiate between producers and consumers and a food chain and food web.
- Explain how energy and biomass pyramids are used to describe energy flow and the accumulation of pollutants in food.
- Trace the cycling of water, carbon, nitrogen, and phosphorus through living and nonliving systems.
- Describe the impact humans can have on each of the biogeochemical cycles resulting in shortages, pollution, the green house effect and eutrophication.

Chapter Summary

Ecology is the study of the interactions among organisms and between organisms and the environment. Ecologists study these interactions. The **biosphere** is that part of the earth where life exists. The biosphere consists of many **ecosystems**, each consisting of the organisms in a specific geographic area and their physical environment. All of the living species in an ecosystem that can potentially interact form a **community**. A **population** is all the individuals of the same species that can potentially interact. Thus, a community is formed of populations of different species. The **niche** is an organism's role in the ecosystem and the place where it lives is its **habitat**.

Ecosystems change over time and the sequence of changes in the species composition of a community is called **ecological succession**. **Primary succession** is the change in species making up a community where no previous community existed. It progresses from a pioneer community to a **climax community**. The nature of the climax community depends largely upon geography. **Secondary succession** occurs when an existing community becomes cleared and then undergoes a sequence of events leading once again to a climax community. Soil is already present in secondary succession.

Energy enters the living world as photosynthetic organisms trap the sun's energy and transform it into glucose. The amount of light energy that is converted to chemical energy in the bonds of organic molecules during a given period is called gross **primary productivity**. Net primary productivity is what remains after photosynthetic organisms fuel their own metabolic activities.

Photosynthetic organisms are **producers** and form the first **trophic level**. All other organisms are **consumers** and are grouped based on their food source. **Herbivores** are **primary consumers**, **carnivores** are **secondary consumers**. **Omnivores** eat both plants and animals and **decomposers** consume dead organic material for energy and release inorganic material that can be used by producers. **Food chains** are linear patterns that describe the flow of energy through an ecosystem. **Food webs** are the complex interrelationships of many food chains and more realistically describes the trophic relationships in an ecosystem.

Energy that is not captured, eliminated as waste, or used for cellular activities is lost. Only the energy converted to **biomass** is available to the next higher trophic level. An **ecological pyramid** depicts the energy available to each successive trophic level (**pyramid of energy**) or the biomass at each trophic level (**pyramid of biomass**.) Both nutrients and nondegradable substances are passed from one organism to the next. However, nutrients are being recycled while nondegradable substances are accumulating. The tendency of nondegradable chemicals to become more concentrated in organisms as in each successive trophic level is called **biological magnification**. This is especially serious for chlorinated hydrocarbon pesticides, heavy metals, and radioactive isotopes, since each is broken down slowly and often stored in animal fat. Since only about 10% of the energy captured by one trophic level is available to the next, it takes a greater investment of energy and more energy is lost as one eats higher on the food chain. Thus, eating lower on the food chain allows more people to be fed with the same initial energy investment.

Materials move through a series of transfers from living to nonliving systems and back again in recurring pathways called **biogeochemical cycles**.

Water cycles from the atmosphere to land as precipitation, and back to the atmosphere as it evaporates. Large amounts of water are temporarily stored in living things. Water forms part of the oxygen cycle as the oxygen we breathe is produced from water though photosynthesis.

Carbon cycles between the environment and living organisms. Carbon dioxide is removed from the environment as producers use it to synthesize organic molecules during photosynthesis. The carbon in the organic molecules then moves through the food web. Carbon is returned to the atmosphere as carbon dioxide when organisms use these organic compounds in cellular respiration.

Nitrogen cycles through several nitrogenous compounds. Atmospheric nitrogen is the largest reservoir of nitrogen. It can be converted to ammonia by nitrogen-fixing bacteria, which then convert the ammonia to nitrate, the main form of nitrogen absorbed by plants. Plants use nitrate to produce proteins and nucleic acids. Animals eat the plants and use the plants' nitrogen-containing chemicals to produce their own proteins and nucleic acids. When plants and animals die, their nitrogen-containing molecules are converted to ammonia by bacteria. Denitrifying bacteria return nitrogen to the atmosphere.

Phosphorus cycles between rocks and living organisms. The reservoir of phosphorus is in sedimentary rock. where it is found in the form of phosphate. Phosphates are dissolved from rocks by rainwater. Then they are absorbed by producers and are passed to other organisms in food webs. Decomposers return phosphates to the soil. Some phosphates are carried to the oceans and form marine sediments.

Human activities, especially since the industrial revolution, can cause disturbances in the biogeochemical cycles. The world's supply of water is threatened by overuse and pollution by humans. A lack of clean water causes 80% of disease in developing countries. In most regions, the shortage of water is caused by overuse of a limited supply by agriculture and industry. Atmospheric levels of carbon dioxide are increasing largely due to the burning of fossil fuels and deforestation. Carbon dioxide is one of the greenhouse gases that promote global warming by trapping infrared radiation in the atmosphere. Disruptions to the nitrogen and phosphorus cycles can cause eutrophication, the enrichment of water in lakes or ponds by nutrients. The excess nutrients cause explosive growth of photosynthetic organisms. When these organisms die, organic material accumulates at the bottom of the lake where decomposers deplete the water of dissolved oxygen. As the oxygen concentration decreases, the species of fish changes and other oxygen-dependent organisms are choked out.

Key Concepts

- Ecology is the study of the interactions among organisms and between organisms and the environment.
- The biosphere is the part of the earth where life exists in many ecosystems. Within an ecosystem, populations of different species form a community. Each organism occupies its own niche, or role, within the habitat where it lives.
- Ecological succession is the sequence of changes in the species making up a community that occur over time. A climax community is the final stage that results after primary or secondary succession.
- Energy flows through ecosystems from producers to consumers in complex food webs of interconnected food chains. Producers capture the sun's energy through photosynthesis. All other organisms are consumers that use the energy that producers store.
- Energy is lost at every trophic level and therefore consumers higher on the energy pyramid are more costly. Certain chemicals that are not easily cycled are concentrated as they move up the food chain.
- Chemicals move through ecosystems in biogeochemical cycles. The atmosphere is the reservoir for water, carbon and nitrogen. Phosphorus reserves are in sedimentary rocks.
- Humans can disrupt the normal pattern of cycling. In the case of water, humans have caused shortages and pollution. By increasing the level of atmospheric carbon dioxide, humans, are causing global warming. The addition of nitrogen and phosphate from fertilizer can lead to eutrophication.

Study Tips

This chapter focuses on the interconnectedness of living organisms with one another and the living and nonliving components of the environment. Develop an understanding of an ecosystem. Make flashcards with definitions of the important terms. Construct an organizational chart showing the relationships between populations and communities as they make and ecosystem and the trophic levels within the ecosystem as energy is transferred. Use arrows to show the movement or change that is occurring in these dynamic, but one-way, systems.

Draw each biogeochemical cycle including all of the intermediate steps. Understand the biological, geological and chemical components of each cycle. Indicate where humans have the greatest impact. Identify ways that humans can reduce their impact on each cycle.

Review Questions

A. Match the term in Column A with the definition in Column B.

Column A	Column B
1. biosphere	a. the sequence of changes in the community species that begins in an area where no previous community existed and ends in a climax community
2. climax community	b. the sequence of changes in the community species that occur after devastation of a previously existing community leaving behind soil
3. community	c. the environment where an organism lives
4. ecology	d. part of the earth where life is found
5. ecosystem	e. individuals of the same species that live in the same geographic area and can interact
6. habitat	f. all species in an ecosystem that can interact
7. niche	g. all organisms living in a certain area along with their physical environment
8. primary succession	h. relatively stable community that marks the end of succession if no further disturbances occur
9. population	i. organism's role in an ecosystem
10. secondary succession	j. the study of the interactions among organisms and between organisms and their environment

B. Use the space below to draw a realistic food web using at least the following organisms: algae, apple trees, blue heron, fish, frogs, grasses, insects, mice, osprey (predatory bird of wetlands), raccoons, small aquatic plants, snake, song birds and tadpoles.

C. Answer the following questions concerning the water cycle. Refer to Figure 23-13 as a resource.

1. What happens to the water that falls as precipitation? ______________________________

____

____

2. Besides evaporation from the soil and bodies of water, how does water get returned to the atmosphere? ______________________________

____

____

3. How do the water cycle and carbon cycle interface? ______________________________

____

____

4. Where does most of the water that falls as precipitation end up? Where do we get our drinking water? Is there a problem with using too much fresh water for agriculture and industry? ____

____

____

____

D. Use the following terms to complete the carbon cycle shown below: bicarbonate, dead organisms and animal waste, decay, destruction of vegetation, diffusion, fire, photosynthesis, respiration, runoff, and sedimentation.

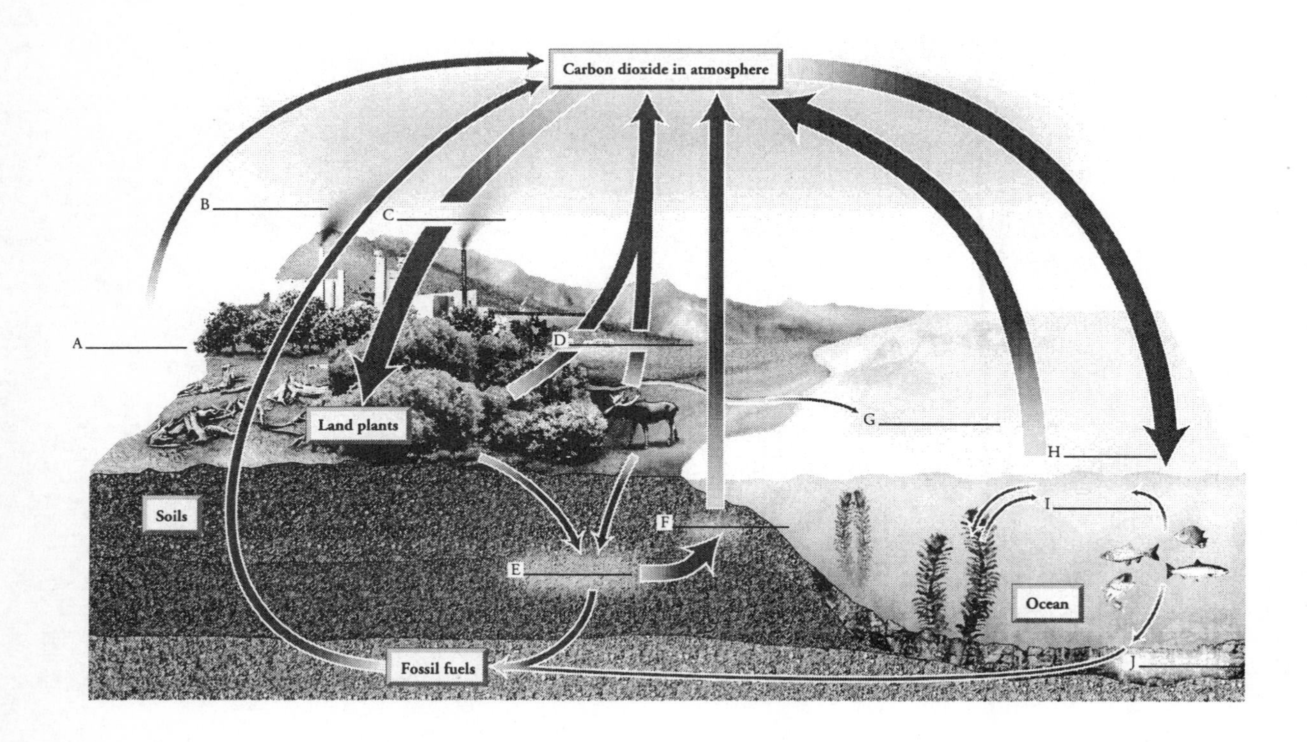

E. Mark each of the following statements regarding the nitrogen cycle as true or false. Correct the false statements so that they are true.

1. _______ Atmospheric nitrogen can be converted to nitrates by nitrogen-fixing bacteria in plants.

2. _______ The final products of nitrifying bacteria found in the soil are nitrites.

3. _______ The nitrogen source that plants use to build proteins and nucleic acids is ammonia.

4. _______ Animals get their nitrogen from the air and return it to the soil when they defecate and die and decay.

5. _______ Denitrifying bacteria found in the soil and water, convert nitrate to nitrogen gas.

F. Complete the following paragraph describing the phosphorus cycle. Refer to Figure 23-16 if needed.

Phosphorus is a critical component of many biological molecules such as ____________, ____________ and phospholipids found in membranes. The phosphorus cycle is different from the other biogeochemical cycles mentioned in the text in that the reservoir is not the atmosphere but is ____________ ____________. Phosphates are ____________ by runoff and rainfall. Phosphates enter the food web by ____________ who then pass it along to consumers. ____________ return the phosphates to the soil or water where they become available to plants and animals. Phosphates that enter the oceans often become ____________ as they remain tied to the marine sediments.

Critical Thinking

Read each of the following questions carefully. Circle the most important words in the question before formulating your answer.

1. Why is the earth considered to be a closed system except for energy? What makes energy different from other resources? What are the implications for human consumption of resources?

2. Research is conducted to quantify the amount of energy used by beef cattle in chewing. Although this may sound silly, why is this use of energy important to the beef industry? Explain your answer in terms of energy transfer.

3. Some people choose to be vegetarians so that they can "eat lower on the food chain." What does this mean and what are the ecological impacts?

4. Use any food chain to demonstrate the carbon cycle. How could you prove that carbon was cycling through this system?

5. How do disruptions in the nitrogen and phosphorus cycles cause eutrophication? What are the consequences of eutrophication?

PRACTICE TEST

Choose the one best answer to each question that follows. As you work through these items, explain to yourself why each answer you discard is incorrect and why the answer you choose is correct.

1. Which of the following is an example of an ecosystem?
   a. group of red-winged blackbirds
   b. pond
   c. snake eating a frog
   d. tadpoles growing into frogs

2. The earth is a closed system except for ___________.
   a. carbon dioxide
   b. energy
   c. nitrogen
   d. water

3. Which of the following describes the niche of a hawk?
   a. field mice
   b. fence row
   c. forest
   d. predator

4. Which of the following is an example of secondary succession?
   a. landscaping at a golf course
   b. reforestation of the cedar forests of the northwest
   c. sedimentation on the bottom of drained lake
   d. scrub grasses growing on a newly exposed dune

5. Which of the following is found only in primary succession?
   a. algae
   b. lichens
   c. pioneer species
   d. soil

6. Which of the following is a producer?
   a. corn
   b. mushroom
   c. snake
   d. toad

7. Humans are ___________.
   a. carnivores
   b. herbivores
   c. omnivores
   d. all of the above

8. Which of the following could be a tertiary consumer?
   a. bear
   b. rabbit
   c. robin
   d. tuna

9. Biomass pyramids always have more organisms on the lower trophic levels.
   a. True
   b. False

10. Which of the following describes a food web?
    a. complex
    b. dependent on one or two organisms
    c. linear
    d. mostly tertiary consumers

11. One reason to eat lower on the food chain is because ____________.
    a. energy is lost at every level
    b. it requires more energy to gain the food
    c. saturated fats provide more energy per gram
    d. toxic wastes are highest in deep see algae

12. When does biological magnification occur?
    a. as biomass increases due to increased metabolism
    b. as pesticides are sprayed on a wheat field
    c. as pollutants accumulate up the food chain
    d. under a microscope

13. DDT is a nondegradable pesticide that weakened the shells of bird eggs. Which of the following would be most affected by DDT?
    a. ducks and geese
    b. eagles and osprey
    c. robins and blue jays
    d. turkeys and pheasants

14. Thinking of the transfer of energy, which of the following would be most expendable?
    a. animals
    b. plants
    c. sunlight
    d. all are equally expendable

15. Within the soil, bacteria function to ____________.
    a. capture light energy
    b. raise the temperature of the water
    c. return inorganic matter from organic matter
    d. synthesize new proteins

16. What do plants contribute to the carbon cycle?
    a. Legumes fix nitrogen in the soil
    b. Plants consume CO_2 from the atmosphere and convert it to glucose.
    c. Plants form bicarbonate in the oceans.
    d. Plants trap sunlight for later use.

17. What is the role of nitrogen-fixing bacteria such as those found in the root nodules of alfalfa?
    a. Nitrogen-fixing bacteria are found in the soil and not in root nodules.
    b. Nitrogen-fixing bacteria capture atmospheric nitrogen and convert it to ammonia.
    c. Nitrogen-fixing bacteria convert ammonia to nitrates.
    d. Nitrogen-fixing bacteria return nitrogen to the atmosphere.

18. Where is the reservoir for phosphorus?
    a. atmosphere
    b. oceans
    c. sedimentary rocks
    d. sewage treatment plants

19. If the current rate of burning fossil fuels continues ____________.
    a. photosynthesis will increase
    b. the temperature of the atmosphere will increase
    c. there will be decreased need for phosphorus
    d. there will be many more areas of primary succession

20. What would be the effect of septic tanks, some old and cracked, along the bank of a small resort lake?
    a. eutrophication
    b. great fishing
    c. greenhouse effect
    d. nothing

Short Answer Questions

Read each of the following questions carefully. Jot down the main points you want to include in your answer. Then write a well-organized explanation.

1. What is the difference between primary and secondary succession?

2. What is the difference between a food chain and a food web? Explain which one is a more realistic representation of energy flow.

3. Focusing on the flow of energy and trophic levels, name an organism that occupies more than one trophic level. Can the number of organisms ever be equal at all tropic levels? Explain your answers.

4. Why would a low-fat diet help reduce the amount of pesticides that are consumed? What are some dietary recommendations that you could make that would avoid biological magnification of pesticides, heavy metals and radioactive isotopes?

5. Explain why promoting a vegetarian diet would allow more people to be fed for the same cost.

ANSWER KEY

REVIEW QUESTIONS

A. 1. d
   2. h
   3. f
   4. j
   5. g
   6. c
   7. i
   8. a
   9. e
   10. b

B. The food web should include an aquatic component and may resemble that shown in Figure 23-7 of the text. It should include all trophic levels.

C. 1. Some water from precipitation evaporates immediately but most collects in ponds, lakes or oceans. Surface water becomes groundwater and collects in aquifers or flows to the oceans. Some water is used by plants and animals.
   2. Plants release water through transpiration and animals release water as water vapor through the skin and lungs.
   3. The water cycle provides water to be used in photosynthesis. The hydrogen from the water is used in making hydrocarbons while the oxygen is released from the plants.
   4. Most of the water that falls as precipitation ends up in the oceans and is not available as fresh water. Drinking water comes from aquifers underground. Aquifers can become depleted if too much water is removed too quickly.

D. A. destruction of vegetation
   B. fire
   C. photosynthesis
   D. respiration
   E. dead organisms and animal waste
   F. decay
   G. runoff
   H. diffusion
   I. bicarbonate
   J. sedimentation

E. 1. F, Atmospheric nitrogen can be converted to ammonia by nitrogen-fixing bacteria in plants.
   2. F, The final products of nitrifying bacteria found in the soil are nitrates.
   3. F, The nitrogen source that plants use to build proteins and nucleic acids is nitrates.
   4. F, Animals get their nitrogen from the plants they eat and return it to the soil when they defecate and die. Bacteria then convert the nitrogen to ammonia.
   5. T

F. Phosphorus is a critical component of many biological molecules such as DNA, ATP and phospholipids found in membranes. The phosphorus cycle is different from the other biogeochemical cycles mentioned in the text in that the reservoir is not the atmosphere but is sedimentary rock. Phosphates are dissolved by runoff and rainfall. Phosphates enter the food web by producers who then pass it along to consumers. Decomposers return the phosphates to the soil or water where they become available to plants and animals. Phosphates that enter the oceans often become unavailable as they remain tied to the marine sediments.

Critical Thinking Questions

1. The earth is considered a closed system because there is normally no transport of resources to or from the earth. All of the elements on earth must be recycled and used repeatedly. The sun provides energy to the earth and that resource does not cycle; the source is continuous. Since the earth is a closed system, theoretically humans could consume all of the resources that are available.
2. The goal of the beef producers is to produce the most meat with the least waste. Of the total energy consumed, some is lost as heat and as feces. Energy is required for the animal to carry out cellular respiration and energy is required to eat. If the animals are penned, then they do not use energy grazing and in fact, chewing is a major source of energy use in cattle. Reducing the amount of energy used in chewing will increase the amount of energy available to turn into biomass for profit.
3. All food chains begins with producers which are able to capture the sun's energy and trap it as chemical energy. Because of the energy lost at each trophic level, it is much less wasteful to eat producers than to eat primary or secondary consumers. More people can be fed using the same amount of energy at lower trophic levels.
4. Carbon enters the food chain as it is captured from the atmosphere and converted to glucose in photosynthesis. The plants convert some of the carbon to plant material that will remain and decay after the plant has died. Some of the plant will be eaten by herbivores. The herbivores will convert some of the carbon in the plant to their own biomass, release some back to the environment as carbon dioxide and excrete some as indigestible waste. This carbon enters the soil where decomposers convert it to carbon dioxide. The plant products become fossil fuels that are burned returning carbon to the atmosphere. Carbon can be traced using the radioactive isotope ^{14}C.
5. Nitrogen and phosphate from fertilizer can wash into streams, rivers, ponds or lakes in runoff. Once in the aquatic system, the nitrogen and phosphate runoff enhances the growth of the algae in the aquatic system. This increases the amount of plant material that not only grows, but dies. The bacteria that break down the dead plant material use oxygen and can cause the dissolved oxygen levels in the water to drop so low that fish cannot survive. The change in species of fish can change the entire character and usefulness of an aquatic system.

Practice Test

1. b
2. b
3. d
4. b
5. c
6. a
7. d
8. d
9. b
10. a
11. a
12. c
13. b
14. a
15. c
16. b
17. b
18. c
19. b
20. a

Short Answer Questions

1. Primary succession is the sequence of events that establish life where none has existed before. It begins with pioneer species that are capable of building soil and ends in a stable climax community. Secondary succession begins when a primary community has been cleared. Soil is left behind which provides a rich nutrient source and a place to anchor plants. Secondary succession ends in a stable climax community providing there is no further destruction.
2. A food chain is a linear sequence showing the energy flow from one organism to another. A food web is an interconnected grid of food chains. It provides many sources of food for each organism. It is more realistic because natural communities are very diverse and organisms must be able to

depend upon more than once source of energy to survive population fluctuations in the supply chain.

3. Several organisms occupy more than one tropic level. For example, humans eat fruits and vegetables as well as beef and pork. Raccoons and bears eat fruits and fish. The number of organisms is usually greater at the lower trophic levels; however, this is not the case for very large organisms. For example, the number of trees in a climax forest may be fewer than the number or herbivores because the biomass of the trees is so large.
4. Chlorinated hydrocarbon pesticides, heavy metals and radioactive isotopes are broken down very slowly in biological systems. In addition, they tend to accumulate in the fatty tissues of organisms. Eating lower on the food chain would help to reduce consumption and accumulation of these materials. Eating low-fat foods, including dairy products, would also reduce the consumption of these materials as they have been stored in the products of other animals. Plants do not accumulate these materials.
5. More people could be fed at the same expense if they ate lower on the food chain. The least amount of lost energy to heat and metabolic wastes occurs at the level of the herbivore.

Chapter 24

Human Population Dynamics

Objectives

After reading the text and studying the material in this chapter, you should be able to:

- Describe factors that affect population size.
- Interpret changes in birth rate, size of the starting population, age structure of the population and the age at which a woman has her first offspring as they relate to increases in population size.
- Define and use the terms immigration, emigration, doubling time, and carrying capacity.
- Explain the factors that effect whether an exponential or logistic growth pattern exists for a population.
- Differentiate between density-independent and density-dependent factors that regulate population size.
- Identify problems directly caused by the increasing human population.
- Describe decisions that might be made that will recognize that resources are limited and must be shared among all living things.

Chapter Summary

The human population consists of all individuals alive at any point in time. **Population dynamics** describes how populations change in size. Population size changes when individuals are added and removed at different rates.

Births increase population size. The **birth rate** is the number of births per a specified number of individuals in the population during a specific time. The size of the starting population affects how quickly new individuals will be added. The **age structure** of a population indicates the number of individuals of each age. Only individuals within a certain age range can reproduce. The relative size of the base of the age structure of a population determines how quickly numbers will be added to the population. The age at which a female has her first offspring has dramatic impact on the rate at which a population grows. This is the most important factor in determining a woman's reproductive potential. Individuals also can be added to a population by **immigration**, the arrival of individuals from other populations.

Deaths decrease population size. The human population continues to grow because the discovery of antibiotics and improved hygiene, sanitation, and nutrition. Population size can also decrease by **emigration**, the exodus of individuals from a population.

Population growth is often expressed as **doubling time**, the number of years it will take for a population to double at that rate of growth.

The human population cannot continue to grow exponentially. The unrestricted growth at a constant rate is called **exponential growth**. When the size of a population growing in this fashion is shown graphically as a plot of the number of individuals against time, we see a **J-shaped growth curve**. Exponential growth occurs in ideal environments, where there are plenty of resources and adequate waste removal. The number of individuals of a given species that a particular environment can support for a prolonged period is called its **carrying capacity**. It is determined by such factors as availability of resources, including food, water, and space; ability to clean away wastes; and predation pressure. Under these more realistic conditions, population growth begins rapidly, but it becomes increasingly slower as the population size approaches the carrying capacity of its environment. Eventually, growth levels off and fluctuates slightly around the carrying capacity. This **logistic growth** pattern produces an **S-shaped growth curve**.

Environmental factors regulate population size. **Density-independent regulating factors** are events that bring about death that are not related to population density such as natural disasters. **Density-dependent regulating factors** are events that have a greater impact on the population as conditions become more crowded such as starvation and disease.

It is difficult to estimate the earth's carrying capacity for humans because it is influenced by both natural constraints and human choices. Estimates of the carry capacity of the earth vary from 5 billion to 20 billion people. We are currently at a world population of 6.2 billion. The growing human population has caused several problems. World hunger, brought on by undernourishment and malnourishment, is one consequence of human population growth. Two diseases affecting children who receive a diet low in calories and protein are marasmus and kwashiorkor. Starvation is associated with irreversible changes in growth and brain development.

One reason for hunger is that food is unequally distributed throughout the world. The **Green Revolution**, the development of high-yield varieties of crops and the use of modern cultivation methods, has boosted crop production. These modern farming practices produce four times more than with traditional methods but use as much as 100 times more energy and mineral resources. Over fishing is leading to depleted fish populations.

Human activities cause pollution. Chlorofluorocarbons (CFCs) are the primary culprit responsible for the destruction of the ozone layer in the stratosphere. This is the layer that shields the earth from excessive UV radiation. The CFCs drift up to the stratosphere where UV radiation causes them to break down to chlorine, fluorine and carbon. The chlorine then reacts with ozone converting it to oxygen, depleting the stratosphere of the protective ozone.

Human activities deplete the earth's resources. Overfarming and overgrazing transform marginal farmlands to deserts, a process called **desertification**. Forests are an essential component of the nutrient cycles and the global ecosystem. **Deforestation** is the removal of trees from an area without replacement. Tropical forests are falling the fastest to clear land for local farming, cattle ranching, mining and to build hydroelectric dams. Commercial logging also results in deforestation. Soil fertility declines after deforestation leaving the local people without a critical resource.

Human activities have reduced **biodiversity**, the number and variety of all living things in a given area. Although tropical rain forests cover only 7% of the land, they are home to between 50% and 80% of the earth's species. habitat destruction to provide living space and economic gain is largely responsible for the loss of biodiversity. Northern rain forests and marine habitats are also being destroyed threatening the loss of species and leading to a reduction in genetic diversity. A reduction in biodiversity may lead to a reduction in the development of new drugs.

Our future depends on the decisions we make today. Humans must realize the resources on earth are limited and must be shared with all living organisms. This will require changes in the way governments and businesses operate and the ways in which individual think.

Key Concepts

- Population dynamics describe how populations change in size. Populations change in size when individuals are added and removed at different rates.
- Population growth is affected by the birth rate, the size of the starting population, the age structure of the population and the age at which a female has her first offspring.
- Deaths decrease population size.
- Population growth rate is also affected by immigration and emigration.
- Population growth rate is expressed as doubling time, the number of years it will take for a population to double at the given rate of growth.
- Unrestricted growth at a constant rate is called exponential growth and is expressed graphically in a J-shaped growth curve.
- The number of individuals of a given species that a particular environment can support for a prolonged period is called its carrying capacity.
- Growth, which begins rapidly but becomes increasingly slower as the population size approaches the carrying capacity of its environment and levels off at the carrying capacity is called logistic growth and produces an S-shaped growth curve.
- Density-independent and density-dependent environmental factors regulate population size.
- The human population may be reaching the earth's carrying capacity as evidenced by pollution and world hunger. Although the Green Revolution has lead to increased crop production, it has consumed resources at a much faster rate.
- Human activities cause pollution and have led to the destruction of the ozone layer.
- Human activities have resulted in desertification of some farmlands and deforestation of the rain forests and forests of the northwest.
- Human activities have reduced biodiversity largely due to habitat destruction.
- Our future depends on the legislative and economic decisions we make today

Study Tips

The material in this chapter builds on your knowledge of ecosystems, biodiversity, and nutrient cycles. It focuses on human population dynamics. Begin your study by identifying the major sections to the chapter, then read the chapter highlighting the most important terms. List the factors that affect increases in population size and those that influence decreases in population size. Next list the problems that have been caused by an increasing human population. For each problem that humans have caused, try to identify a decision that could be made to preserve the earth for the future generations.

REVIEW QUESTIONS

A. Population dynamics describe how populations change in size. The four factors influencing population size are births, deaths, emigration and immigration. Write an equation showing the relationship of those for factors.

B. Circle the word that correctly completes the sentences below correctly stating the relationship between births and population size.

1. Birth rate: A population with a higher birth rate will increase (faster, slower) than a population with a lower birth rate.
2. Starting size of the population: Given the same birth rate, the larger the initial size of the population, the (fewer, more) individuals are produced.
3. Age structure of the population: Given the same birth rate, a population will grow more (rapidly, slowly) as the size of the reproductive class increases.
4. Age of female at her first offspring: The younger a woman is when she begins to reproduce, the (faster, slower) the population grows.

C. Natural populations are regulated by both density-dependent and density-independent factors. Provide three realistic examples of each type of factor that affect the human population.

1. Density-dependent factors: ____
2. Density-independent factors: ____

D. Draw a graph showing exponential growth and a graph showing logistic growth. Then describe the differences in the two growth curves and what causes them. Be sure to label each axis of the graph.

<u>Exponential Growth</u> <u>Logistic Growth</u>

E. Complete the following paragraph describing the effects humans have on the earth's resources.

Overuse and misuse of land tends to reduce soil ____________. As more land is cleared, wind and rain cause ____________ that reduces the topsoil. Overgrazing and overfarming has turned marginal farmlands into deserts, a process called ____________. One way to reduce erosion is to avoid overgrazing and to plant cover crops. The removal of trees from an area is called ____________. Forests in the Pacific Northwest have been cleared for commercial ____________. Avoiding the slash-and-burn techniques and replanting a diverse stand of trees will help prevent the destruction caused by deforestation. Tropical rainforests have been cleared so that native families can feed themselves, for commercial logging and for cattle ranching. Both deforestation and desertification reduce the variety of species called ____________. Habitat preservation and habitat restoration help maintain biodiversity.

F. Mark each of the following statements regarding sustainability as true or false. Correct the false statements so that they are true.

1. _______ Today's society, especially as represented by the United States, is sustainable.
2. _______ Our society primarily uses renewable fuel sources.
3. _______ A sustainable society would avoid excessive resource consumption.
4. _______ To be more sustainable, we should eat more red meat to increase income to the agriculture industry.
5. _______ To become a more sustainable society, we should focus on recycling and reusing products.

Critical Thinking

Read each of the following questions carefully. Circle the most important words in the question before formulating your answer.

1. What is the difference between population size and growth rate? Which is more useful for determining the impact humans have on the environment?
2. The growth curve of the human population remained nearly flat from the beginning of humankind and then rapidly increased. What factors influenced this change in the growth rate of the human population? Refer to Figure 24-1 in your text.

3. Age structure diagrams such as those shown in Figure 24-2 in your text can be used to predict population growth. The age structure for the population of the United States has a narrow base as is characteristic of more-developed countries. The age structure for India has a very broad base as is characteristic of less-developed countries. Form a group of students in your class and discuss the question, "How does the economic condition of a country affect its population growth rate?"

4. How does deforestation relate to human population size and how does it affect the condition of our environment?

5. Form a small group and role model a discussion on the significance of the ozone hole. One group is environmentally aware and can identify the importance of maintaining the ozone layer while the other group is skeptical that a hole in the ozone layer exists or has any impact on their lives.

PRACTICE TEST

Choose the one best answer to each question that follows. As you work through these items, explain to yourself why each answer you discard is incorrect and why the answer you choose is correct.

1. A population is ___________.
   a. birth rate minus death rate
   b. change in the distribution of people within the area
   c. the number of individuals born living to reproductive age
   d. the number of people living in a specific geographic area

2. Population dynamics is used to describe ____________.
   a. birth rates
   b. changes in population size
   c. doubling time
   d. success of rearing children

3. What causes changes in population size?
   a. births
   b. deaths
   c. movement of people in and out of an area
   d. all of the above

4. Which of the following is true of the United States?
   a. agricultural advances have stopped
   b. increase in infant mortality
   c. large prereproductive population base
   d. women are having children later in life

5. If the population size was 8 at time 0, 16 at 5 minutes, 32 at 10 minutes, and 64 at 15 minutes, how long will it take the population to increase to over 1000?
   a. 5 more minutes
   b. about 1.5 hours
   c. total time of 35 minutes
   d. cannot be determined from these data

6. Given the same growth rate, as the starting size of the population increases, the population size ____________.
   a. decreases
   b. increases
   c. remains constant
   d. cannot be determined from this information

7. As population growth rate increases the doubling time ____________.
   a. decreases
   b. increases
   c. remains constant
   d. cannot be determined from this information

8. What is the characteristic pattern of population growth when a new species is introduced to an unused habitat?
   a. J-curve
   b. S-curve
   c. dome-shaped
   d. triangle-shaped

9. What is the maximum number of individuals of a given species that a particular environment can support for a prolonged period of time?
   a. biotic potential
   b. carrying capacity
   c. maximum growth rate
   d. population density

10. Which of the following affects the carrying capacity of an environment?
    a. abiotic factors
    b. biotic factors
    c. technical advances
    d. all of the above

11. Which of the following is a density-dependent factor that regulates population size?
    a. disease
    b. fire
    c. flood
    d. freezing

12. Has the human population reached the carrying capacity of the earth?
    a. Yes
    b. No

13. Which of the following best describes the Green Revolution?
    a. advances in agriculture
    b. increased algae in the ocean as a food source
    c. Industrial Revolution
    d. rebellion of underdeveloped countries

14. Thinning of the ozone layer is caused by ____________.
    a. CFCs
    b. CO_2
    c. O_3
    d. PCBs

15. What causes the greenhouse effect?
    a. CFCs
    b. CO_2
    c. O_3
    d. PCBs

16. As populations increase in less-developed countries, what would you expect to happen to pollution?
    a. dramatic increase
    b. slow increase
    c. slow decrease
    d. There is no way to predict this.

17. Overgrazing and overfarming marginal farmlands have caused ____________.
    a. deforestation
    b. desertification
    c. increases in farm products
    d. increases in net income

18. It is of direct benefit to humans to maintain biodiversity.
    a. True
    b. False

19. Currently the greatest biodiversity occurs in the ____________.
    a. Mexico
    b. northwest cedar stands
    c. temperate forests
    d. tropical rain forests

20. What does sustainability mean?
    a. develop new technology
    b. keep going
    c. maintain at current levels
    d. use wisely

SHORT ANSWER QUESTIONS

Read each of the following questions carefully. Jot down the main points you want to include in your answer. Then write a well-organized explanation.

1. What is doubling time and why is it used in discussions of world populations? Given the following approximate population data, calculate the doubling times. Predict the year when the population will reach 7 billion.

Year	Approximate World Population	Doubling Time (years)
8000 BC	5,000,000	
1800	1,000,000,000	9800
1930	2,000,000,000	130
1960	3,000,000,000	30
1975	4,000,000,000	15
1987	5,000,000,000	12
1998	6,000,000,000	11

2. The carrying capacity of an environment usually limits the population size. How have humans altered the carrying capacity of the earth?

3. What will allow a population to grow even after it has slowed its reproductive rate to replacement only (two parents are replaced by two children)?

4. What is the greenhouse effect and how does it impact our lives? What can be done to curb the greenhouse effect?

5. What are the primary values for maintaining biodiversity?

ANSWER KEY

Review Questions

A. Population size = (births + immigration) - (deaths + emigration)

B. 1. Birth rate: A population with a higher birth rate will increase faster than a population with a lower birth rate.
   2. Starting size of the population: Given the same birth rate, the larger the initial size of the population, the more individuals are produced.
   3. Age structure of the population: Given the same birth rate, a population will grow more rapidly as the size of the reproductive class increases.
   4. Age of female at her first offspring: The younger a woman is when she begins to reproduce, the faster the population grows.

C. 1. Density-dependent factors: starvation, malnutrition, disease, war
   2. Density-independent factors: fires, floods, earthquakes, volcanic eruptions, landslides, avalanches, storms

D. Growth curves should resemble those in Figure 24-4 and 24-5. Vertical axis is population size; horizontal axis is time. When population growth is exponential, it is unlimited. The carrying capacity is the largest number of individuals that can be maintained by a population. It exerts its pressures as the growth curve flattens slightly and then levels off at or near the carrying capacity.

E. Overuse and misuse of land tends to reduce soil fertility. As more land is cleared, wind and rain cause erosion that reduces the topsoil. Overgrazing and overfarming has turned marginal farmlands into deserts, a process called desertification. One way to reduce erosion is to avoid overgrazing and to plant cover crops. The removal of trees from an area is called deforestation. Forests in the Pacific Northwest have been cleared for commercial logging. Avoiding the slash-and-burn techniques and replanting a diverse stand of trees will help prevent the destruction caused by deforestation. Tropical rainforests have been cleared so that native families can feed themselves, for commercial logging and for cattle ranching. Both deforestation and desertification reduce the variety of species called biodiversity. Habitat preservation and habitat restoration help maintain biodiversity.

F. 1. F, Today's society, especially as represented by the United States, is consuming more than it is replacing.
   2. F, Our society primarily uses nonrenewable fossil fuel energy.
   3. T
   4. F, To be more sustainable, we should eat lower on the food chain to reduce costs of production and waste.
   5. T

CRITICAL THINKING QUESTIONS

1. Population size is merely the number of people alive at any given time. Population growth rate is the change in the number of people per a standard number over a specified period of time. For example, the annual growth rate may be 1.2%, slightly over 1 per 100. The growth rate is an indicator of the change in population size in some unit of time. It is a more useful indicator of environmental impact because it is predictive.
2. Disease kept the population in check for most of history. In fact, the bubonic plague in Asia and Europe caused a noticeable decline in the population size. Advances in medicine— especially the discovery of antibiotics, improvements in hygiene, sanitation, and nutrition, and improvements in agriculture including breeding, technology and crop resistance—have been major influences in the growth rate of the human population.
3. Growth rates are slowing in more developed countries as the impact of the industrial revolution are slowing. In addition in more developed countries women are having their first child later in life which also slows population growth. Underdeveloped nations have age structure diagrams that

indicate that they will continue to exponentially increase in population size. This trend of increased population growth will usually increase as death rate decreases but birth rate remains high as nations just begin to industrialize.

4. Deforestation is occurring in the rain forests as land is cleared for cattle ranching, commercial logging, and farming. Although the immediate effects are increased food production, increasing the carrying capacity of that environment, long-term effects leave over grazed and depleted soil behind. Deforestation occurs in the northwest as trees are harvested for industrial use. Often deforestation results in increased air pollution as carbon dioxide is released when debris is burned.
5. The environmentally aware group should note that the ozone in the stratosphere protects the earth from UV radiation. CFCs, volcanic activity and some pesticides have caused destruction of the molecule ozone (O3) leading to a seasonal hole in the ozone layer over Antarctica. The group that is less concerned about the condition of the ozone layer may point out that international regulations are costly to implement and enforce. They may selfishly indicate that there will be little impact in their own lifetime, so why bother to repair a hole that might be a hoax of the environmentalists.

Practice Test

1. d
2. b
3. d
4. d
5. c
6. b
7. a
8. a
9. b
10. d
11. a
12. b
13. a
14. a
15. b
16. a
17. b
18. a
19. d
20. b

Short Answer Questions

1. Doubling time is the number of years it will take for a population to double at that rate of growth. Doubling time is a good expression of population growth because it is predictive. It is an indicator of exponential growth over time. Small changes in doubling time have great impact. Graphing the results would yield the best prediction. Based on doubling times, the population should reach 7 million about 2009. However, your text predicts a population size of 7.03 billion in 2013, a slight decrease in the growth rate.
2. Limits on food, habitat, and waste removal usually slow population growth. Humans have slowed the natural death rate due to communicable disease with improved medical care, which has decreased the death rate. Sanitation has improved greatly and also slowed the death rate. Improved agricultural techniques have provided more food. And industrialization has allowed us to expand our habitat vertically causing densities that would not be tolerable without high-rise apartments.
3. Population size is determined by births, deaths, immigration and emigration. Although the birth rate may be at replacement, people may be moving into the country causing an increase in population size. More likely, however, the death rate has decreased due to improved medical care. Fewer children are dying, fewer women are dying in child birth, and more people are living longer. These factors will allow a population to increase in size with a replacement birth rate.
4. The greenhouse effect is what happens when gases, particularly CO_2, are trapped in the upper atmosphere and allow sunlight to enter, but do not allow the reflected heat to escape. The net effect is that over time, these greenhouse gases will cause global warming. The temperature of the earth is rising and over time, this could cause a melting of polar ice. The resulting rise in sea level could endanger coastal ecosystems and population centers. Increased evaporation could cause changes in climate and precipitation patterns. Reduced consumption of fossil fuels is the single greatest action we can take to reduce greenhouse gases.

5. Maintaining a genetically diverse reservoir of organisms is important because it is the fuel for advances in genetic engineering. A large gene pool will facilitate the development of medicines and agricultural products. We depend on many species for our clothing, food and products. When extinction occurs, those genes are removed forever from the gene pool and both known and unknown benefits will be lost.